AF779501

Diversity and Developmental Science

Dawn P. Witherspoon • Gabriela Livas Stein
Editors

Diversity and Developmental Science

Bridging the Gaps Between Research, Practice, and Policy

Springer

Editors
Dawn P. Witherspoon
The Pennsylvania State University
University Park, PA, USA

Gabriela Livas Stein
University of North Carolina at Greensboro
Greensboro, NC, USA

ISBN 978-3-031-23162-9 ISBN 978-3-031-23163-6 (eBook)
https://doi.org/10.1007/978-3-031-23163-6

© The Editor(s) (if applicable) and The Author(s), under exclusive license to Springer Nature Switzerland AG 2023
This work is subject to copyright. All rights are solely and exclusively licensed by the Publisher, whether the whole or part of the material is concerned, specifically the rights of translation, reprinting, reuse of illustrations, recitation, broadcasting, reproduction on microfilms or in any other physical way, and transmission or information storage and retrieval, electronic adaptation, computer software, or by similar or dissimilar methodology now known or hereafter developed.
The use of general descriptive names, registered names, trademarks, service marks, etc. in this publication does not imply, even in the absence of a specific statement, that such names are exempt from the relevant protective laws and regulations and therefore free for general use.
The publisher, the authors, and the editors are safe to assume that the advice and information in this book are believed to be true and accurate at the date of publication. Neither the publisher nor the authors or the editors give a warranty, expressed or implied, with respect to the material contained herein or for any errors or omissions that may have been made. The publisher remains neutral with regard to jurisdictional claims in published maps and institutional affiliations.

This Springer imprint is published by the registered company Springer Nature Switzerland AG
The registered company address is: Gewerbestrasse 11, 6330 Cham, Switzerland

We dedicate this book to all of the children in the world whose developmental processes have been ignored, pathologized, or mischaracterized by the scientific enterprise. We commit to elevate your experiences, understand your strengths, and identify the structural forces that impact your lives to inform how best to support your growth, thrival, and success.

Acknowledgment

We would like to thank all of the scientists that have paved the way for the development of diversity developmental science – those of you who have fought and made space for our work to continue to do right by our communities. We also would like to thank the Society for Research in Child Development and their ethnic-racial issues (ERI) committee that both editors have chaired. During Dr. Witherspoon's tenure as chair of this committee, SRCD provided financial support for a pre-conference in 2019 that served as the impetus for this book. We would like to thank all of the ERI committee members and SRCD leadership for supporting the ideas that went into this book.

Contents

Chapter 1
Developmental Science in the Twenty-First Century: Eschewing Segregated Science and Integrating Cultural and Racial Processes into Research

Gabriela Livas Stein, Charissa S. L. Cheah, Wonjung Oh, and Dawn P. Witherspoon

A Changing World

Across the globe, countries are experiencing increasing ethnic-racial diversity in their populations due to immigration, migration, and differential fertility and mortality (of native- and foreign-born people) (Vespa et al., 2018). In the United States, the recent 2020 Census demonstrated significant demographic shifts in just the past 10 years with a 9% decline in the White population, and an astounding 276% increase in the Multiracial population (Jones et al., 2021). There was also growth in other racial or ethnic groups including Asian, Latinx,[1] American Indian/Alaskan Native, and Black populations (with 36% growth, 25% growth, 27% growth, and

[1] We acknowledge that for Latinx populations racial categorization may be distinct from ethnic identification (e.g., in the case of Afro Latinx, Indigenous Latinx, or White Latinx) and this distinction shapes the daily experiences of Latinx populations. We use the term Latinx as an umbrella term that is inclusive of race and ethnicity for the purpose of this chapter given the unique racialization of brown and Black Latinx populations and point readers to other chapters in this book that consider the intersectional experience of race and ethnicity for the Latinx population.

G. L. Stein (✉)
University of North Carolina at Greensboro, Greensboro, NC, USA
e-mail: glstein@uncg.edu

C. S. L. Cheah
University of Maryland, Baltimore County, Baltimore, MD, USA

W. Oh
Texas Tech University, Lubbock, TX, USA

D. P. Witherspoon
Pennsylvania State University, University Park, PA, USA

© The Author(s), under exclusive license to Springer Nature Switzerland AG 2023

D. P. Witherspoon, G. L. Stein (eds.), *Diversity and Developmental Science*,
https://doi.org/10.1007/978-3-031-23163-6_1

6% growth, respectively) (Jones et al., 2021). The Census estimates that by 2060, although non-Latinx Whites are projected to remain the single largest race or ethnic group, non-Latinx Whites will constitute only 44% of the US population, whereas the numerical majority of the population is predicted to identify as a member of a "minority" ethnic or racial group (28% Latinx, 15% Black, 9% Asian, 6% Multiracial) (Vespa et al., 2018). These demographic shifts are already evident in the child populations of the United States such that non-Latinx Whites at the time of this publication (2021) make up less than 50% of the under 18 population with Latinx and Black youth being the largest other groups (26% and 14%, respectively) (Child Trends, 2018). Across adults and children in 2020, 58% of the US population identified as non-Latinx White, 19% Latinx 12% non-Latinx Black, 6% non-Latinx Asian, 3% Multiracial, 1% American Indian/Alaskan Native, and 0.6% Native Hawaiian/Pacific Islander populations. In fact, the US population is the most diverse in history with a diversity index of 61% (relative to 55% in 2010), which means that if one person in the United States was selected at random there is a 61% chance that another randomly selected person would be of a different ethnic-racial group (Jensen et al., 2021).

This ethnic-racial diversification of the United States is *partly* a result of immigration patterns with the foreign-born population making up 14% of the US population in 2018 (Budiman, 2020), and continued growth of the foreign-born population is projected such that by 2060 the immigrant population will be highest (at 17%) recorded in the United States since 1850 (at 15%) (Vespa et al., 2018). However, the ethnic-racial diversification of the United States is also due to the differential birth and mortality rates of the foreign-born populations already in the United States and the continued growth of non-White racial and ethnic native-born groups (Vespa et al., 2018). Immigrants to the United States are themselves racially and ethnically diverse with recent shifts in immigration patterns that are also contributing to the rapidly changing ethnic-racial makeup of the United States. For example, in 2009, immigration from Asian countries outpaced Latinx countries and, in 2018, Asians made up the 37% of new immigrant arrivals (relative to 31% from Latin America) (Budiman, 2020). Indeed, in 2018, 278,000 immigrants came from China and India (the top two countries of origin) more than double of the 120,000 that came from Mexico (the third largest country of origin) (Budiman, 2020). Overall, there are differential drivers of the diversification for Asian American and Latinx populations with immigration primarily accounting for the increase in the Asian American population and natural increases of those already in the United States fueling the growth in the Latinx population (Vespa et al., 2018). However, it is important to note that despite these recent trends, the foreign-born population in the United States is still majority Latinx (25% Mexico, 25% other Latin American countries) with the next largest share being Asian (28%) (Budiman et al., 2020).

Although Asian and Latinx populations account for the largest portion of immigrants in the United States (Budiman, 2020), Black immigrants also contribute to the increasing diversity of the United States. In the past 40 years, the Black immigrant population has grown by 500% (Tamir, 2022), and made up 10% of the US Black population in 2019 (Tamir, 2021). In 2019, the majority of Black immigrants were from two regions—the Caribbean (46%) and Africa (42%) (Tamir & Anderson, 2022), but there has also been a shift in immigration patterns such that in recent years, immigration from Africa has grown (Tamir, 2022). Caribbean immigrants are primarily from Jamaica and Haiti, and immigrants from Nigeria, Ethiopia, and Ghana make up the largest portion of immigrants from Africa (Tamir & Anderson, 2022).

These immigration patterns are important to our understanding of the future diversity of the United States. More importantly, these patterns indicate that the ethnic-racial makeup of the children and youth in immigrant households needs to be considered for current research with children and youth in the United States. Indeed, in 2019, one in four US children lived in an immigrant household (Casey, 2021), and the majority of both Asian American and Latinx youth and young adults reside in immigrant households (77% of Asian American and 51% of Latinx youth). However, only a small proportion of children themselves are foreign-born (only 3% of all children). Taken together, the US population under 18 years of age is the most diverse in history, and one quarter of them are growing up in immigrant households. Unfortunately, this diversity is often ignored in developmental science and instead typically relegated to a line or two in the method section describing the sample or is featured prominently in the discussion noted as limitations with very few developmental scientists actually incorporating ethnic-racial diversity into their research designs. Yet our science is incomplete if we do not, as a field, attend to the implications of this diversity for our theories, questions, methods, interpretations, and applications.

While it may seem that this as an American story, and the United States is indeed the home to largest number of immigrants (Budiman, 2020), recent increases in migration and immigration across the world have resulted in other countries having greater proportions of their populations being foreign-born, leading to a similar diversification of their communities. For example, the percentage of foreign-born populations in Canada (22%) and Australia (28%) is significantly larger than in the United States (Connor, 2016), and in Canada, this diversification has resulted in 19% of the Canadian population identifying as "visible minorities" in 2011 (Statistics Canada, 2013). Asian and Middle Eastern immigrants account for the largest immigrant population in Canada (57%) with smaller proportions from Africa (13%) and Latin America/Caribbean (12%). Australian ethnic-racial demographics are not as well reported, but in their 2016 Census, Asian immigrants comprised 40% of the foreign-born population with another large portion coming from North or West Europe (23%) (Australia Bureau of Statistics Census, 2017). Similar demographic shifts can be seen in other majority White countries where their foreign-born populations have doubled in the past 20 years (e.g., Austria, Spain, Greece, the UK), but there have also been large immigrant population increases in other non-Western countries (Pew Research Center, 2019). For example, in the Middle East, countries like Qatar, the United Arab Emirates, and Kuwait have majority foreign-born populations of around 75%. Oman has seen an increase in their foreign-born population from 17% in 1990 to 35% in 2017—largely from South East Asia (Pew Research Center, 2019). As in the United States, we must capture and represent this diversity in the values, customs, beliefs, and daily life experiences of all populations.

Residents across the globe are noticing the changes in the ethnic-racial makeup of their countries as 69% of respondents reported that their countries are becoming more diverse in a 2018 Pew Global Attitudes Survey across 27 countries (Poushter & Fetterolf, 2019). Interestingly, this diversity is felt even in countries with smaller percentages of foreign-born populations. For example, 84% of respondents in South Korea reported perceiving this change, and their foreign-born population in 2017 was 2%, which doubled from <1% in 1990 (Pew Research Center, 2019). Although 45% of respondents of the Pew survey favored these demographic shifts and only 23% opposed it, there was wide variability in the reactions among nations globally (Poushter & Fetterolf, 2019). For example, European respondents indicate generally less favorable views in countries like Greece, Italy, and Germany (62, 45, and 32% opposed, respectively). In contrast, there have been recent changes in these perceptions in countries such as France, the UK, and Spain, where there were small increases in favorable views of immigrants from 2014 to 2018 (Gonzalez-Barrera & Connor, 2019). Other countries in Latin America, Africa, and the Middle East tend to be less polarized with more moderate levels of support for the increases in diversity (Poushter & Fetterolf, 2019).

Canada and the United States share similar levels of support for the increases in diversity in this global survey (62 and 61% favoring; 19 and 17% opposing it, respectively), but this belies some deeper concerns apparent in other surveys—at least in the United States. When explicitly asked about the country's demographic being majority "Blacks, Asians, Hispanics, and other racial minorities in 2050," the majority of those Americans surveyed said it was "neither good nor bad" (42%) or "very/somewhat bad" (23%) with 35% saying it was "very/somewhat good" (Parker et al., 2019). These views differ by respondents' ethnicity-race with only 26% of Whites saying this change is good for the United States relative to 53% of Blacks and 55% of Latinxs (no data provided on Asian American respondents). White respondents perceived this shift as a potential threat to American customs and values (with 46% reporting that this shift would weaken them relative to 24% of non-Whites) and perceived these changes as inevitably leading to increased racial and ethnic conflict (with 53% endorsing this concern compared to 43% of non-Whites) (Parker et al., 2019). One central question for developmental science is "How do these beliefs play out in the lives of diverse children, youth, and their families in the United States and beyond, when those who hold power view them as a threat?" Therefore, it is evident that in order for our research findings to have ecological validity, we must integrate the unique cultural values, beliefs, norms, and traditions of all groups as they shape the psychological, educational, and health outcomes we seek to uncover. Equally important is the need for more research that examines the ramifications of systems of oppression, power, prejudice, and racism that are evident in the xenophobic and racist beliefs endorsed by a large portion of populations worldwide on core developmental processes.

Color-Evasive Developmental Science

Our science has been largely founded on the fallacy that with "rigorous" methodology, experimental designs, and objective science, we can uncover universal "truths" of human development that are applicable to all (Else-Quest & Hyde, 2016) despite the fact that the majority of our scientific inquiry has been based on a small, non-generalizable sliver of the world's population (e.g., Nielsen et al., 2017). The go-to solution has been that we must diversify our samples—whether within the United States or across the globe. This initiative to diversify our samples, albeit limited, has been pushed along by funders requiring recruitment of more diverse populations and journals developing recommendations for more rigorous sampling descriptions (e.g., APA, 2019). Yet many published psychology studies do not even meet this bar, with a recent review of psychology articles published from 2015 to 2016 in 11 journals finding that 73% did not report participants' race (DeJesus et al., 2019). Nevertheless, the effort to diversify and report the racial/ethnic makeup of the sample is only one small step in the right direction. Without the consideration of the unique cultural and societal processes that influence psychological outcomes across and within diverse populations, we ignore important unstudied processes and produce an incomplete understanding of development that does not have universal applicability. A diverse sample with different racial or ethnic groups represented should be a *bare minimum* for generalizability but is simply not enough. As scientists, we need to consider the lived experiences of race and ethnicity that inherently shape every environment, every interaction, and every psychological process as these are embedded within a larger developmental context (García Coll et al., 1996; Helms et al., 2005; Spencer, 2006).

This consideration does not necessarily mean that we measure or capture every aspect of the lived experience of all groups in every study. Such a feat would be impossible. However, we must *consider* the significance of such lived experiences and their potential influences on what we are studying, and how it is influenced by cultural socialization practices associated with race and ethnicity. Methodological considerations and decisions would then need to be made based on an integration of theories of race and ethnicity with specific theories and questions that have been developed as "culturally universal" (e.g., attachment, emotion socialization, peer processes). Unfortunately, most developmental scientific programs have trained multiple generations of scientists without the skill-set to do this. The training of most of these scientists has left them with a weak foundational knowledge of conceptual frameworks and methodologies best suited to understand the diversity in their samples and ensure that their scientific inquiries consider these processes. Instead, a paradoxical thing occurred: Due to reporting requirements by funders or journals, scientists describe their sample in the method section—with different levels of granularity to be sure—or indicate the lack of their diversity consideration as a limitation in the discussion, but the remaining sections of the article are often devoid of consideration of diversity. That is, the research questions, research designs, analyses of the data, and interpretations of the findings often ignore this diversity and instead are "color-evasive," meaning that the authors proceed to conduct "science as usual" without acknowledging the potentially different racial, cultural, and societal socialization practices and experiences of these groups (Syed et al., 2018). This oversight is the scientific equivalent to assertions of "I don't see race, just the person," which is now widely acknowledged to be harmful for communities of color.

To be clear, we are not proposing that every empirical endeavor in developmental science must directly examine ethnic, racial, or cultural processes. However, we are proposing that when our societies are diverse, as we described above, every developmental scientific inquiry needs to acknowledge the significance of this diversity and wrestle with the inevitable impact of ethnicity, race, power, oppression, racism, and culture on the developmental processes under study. For example, how can we understand the influence of peer processes, like peer relations, norms or bullying, on psychological outcomes without considering the ethnic-racial makeup of the sample and the racial power dynamics in the school (e.g., Graham & Echols, 2018; Kogachi & Graham, 2021)? Similarly, what is the meaning of findings on the development of prosocial behavior if the research questions, samples selected, and interpretations of findings are not guided by these principles we propose when cultural groups espouse different beliefs and practices in the integration of their toddlers in helping behaviors (e.g., Brownell et al., 2016; Streit et al., 2021)?

Fortunately, there is a legacy of scientific theory and research that can guide our field into integrating race and ethnicity in ways that honor the communities that we seek to understand, but that developmental scientists as a larger field have mostly viewed as distinct from their own work. The scientific inquiry about non-White populations was developed by scholars of color (primarily Black scholars and then by other scholars of color) in response to the deficit models that were used to study and characterize the development of children of color (García Coll et al., 1996). This intellectually rich work served to grow an entire discipline of diversity science that is vibrant, innovative, and evolving (Plaut, 2010; Miller et al., 2019) but has been largely ignored by "mainstream" developmental scientists. The result is a segregated developmental science that carries with it all the limitations that accompany residential and school segregation such that issues of culture, race, ethnicity, power, and oppression have been only studied by a few, been largely ignored by the flagship journals in the field or relegated to "special issues" or "special sections," been underfunded and poorly resourced, and viewed as falling outside "mainstream" developmental science (e.g., Syed et al., 2018). Scientific knowledge generated with diverse children that considers their lived experiences is still viewed as a distinct subdiscipline within developmental science rather than a cornerstone of all scientific inquiry. Indeed, psychology as a field has largely ignored issues of race across all sub-disciplines (Roberts et al., 2020). The unfortunate consequence is that much of the work on communities of color outside diversity science continues to espouse deficit-oriented perspectives that study populations of color without considering *processes* of historical and contemporary societal, systemic, and institutional oppression impacting these communities (García Coll et al., 1996, 2000).

This problem stems from the fact that our field was formed based on racist and colonist frameworks and methods (Syed et al., 2018). The American Psychological Association (APA) finally issued an apology *in 2021* for their role in the propagation and support of racism and oppression, although many feel this apology was insufficient (APA, 2021). While some scholars have, against all odds, created subfields within psychology that counter these racist origins, sustained efforts to truly center diversity science as psychological science require an entire overhaul of the field. As a field, we must contend with this racist legacy and directly grapple with how it has influenced the way we integrate marginalized populations into our science beyond inclusion in the sample. We agree with the assertion made so boldly and simply by Helms et al., in a 2005 *American Psychologist* paper that, although pivotal in the field of diversity science, is still not being heeded by the majority of developmental scientists: race is not an independent variable and race is not a control variable. Instead, *processes* of identity, cultural values, racism, discrimination, acculturation, etc., need to be included into our designs and models—not their proxies (i.e., demographic characteristics like racial group members, ethnic group membership, and immigrant status).

The overarching goal for this book is to provide developmental scientists tools to move beyond segregated science and consider cultural, racial, and ethnic processes as central to their scientific endeavors. In particular, some chapters challenge epistemological perspectives that are foundational to developmental science in how we engage with research and propose different approaches to the research *process* as much as to the research *questions* (see Chaps. 7 and 9). Other chapters focus on defining *key constructs* that are necessary to include in developmental inquiries beyond describing the race and ethnicity of the sample (e.g., discrimination and ethnic-racial identity) and how these need to be understood not just in communities of color but also in the developmental processes of White youth (see Chaps. 2, 3, and 10). In addition, given the contextualized experiences of racial-ethnic diverse children, youth and families, other chapters focus on specific developmental contexts (e.g., schools and neighborhoods; see Chap. 8). We now turn to some exemplars of how moving beyond a segregated science has been important to our field. To this end, we first highlight key theories can be used as a foundational starting point to consider the processes that should be included into studies of developmental science to advance our knowledge and make developmental science inherently a diversity science.

Theories of Developmental Diversity Science

Although a summary of all the pivotal theories of developmental diversity science is beyond the scope of this chapter, we want to highlight two key theories that can be used by all developmental scientists as they consider integrating cultural, ethnic, and racial processes into their work. In 1996, García Coll and colleagues developed a foundational model of diversity science that delineated how to move beyond race and ethnicity as categorical variables by first describing how these social position variables, through racism and segregation, shape the contexts in which families of color live. Second, it posited how processes in these contexts like schools and neighborhoods (i.e., inhibiting or promoting environments) interact with adaptive cultural and familial processes and child characteristics to influence the development of cognitive, social, emotional, and linguistic competencies. The model was groundbreaking, transformational, and generative—inspiring research on the dual processes of marginalization (Seaton et al., 2018) and adaptation (Perez-Brena et al., 2018; White et al., 2018) that were summarized in a special issue of *American Psychologist* in 2018. This rich body of literature provides a solid foundation for future studies that integrate these processes but also extend this to work for outcomes and processes that have not traditionally considered cultural and social position variables. In particular, we want to underline the importance of the joint study of marginalization and resilience processes in developmental science that unfortunately has mostly been inadequately captured using race or ethnicity as a variable. As so eloquently argued in this model (and by Helms et al., 2005), race or ethnicity are inadequate proxies for capturing these processes and should not be included into statistical models without explicit theoretically justified consideration of why it is included. We further highlight that few studies of ethnic-minoritized children have integrated child characteristics beyond age and gender even though García Coll et al. (1996) propose that it is important to consider other aspects of the child, including but not only biological factors such as temperamental and physical characteristics. Further, there are a host of social, cognitive, and emotional competencies that have not received significant research attention (Carlos et al., 2022).

Another pioneering and trailblazing theory that has been a guiding light in developmental diversity science is Spencer's phenomenological variant of ecological systems theory (PVEST) (2006). Although PVEST is not only applicable to the study of diversity processes (e.g., race, cultural identity) and can be applied to broader developmental phenomena as Spencer has done in her theoretical writings, this model is well-suited for the study of diverse and ethnic-racial minority populations and has been applied to intersectional, interlocking systems of inequality (Velez & Spencer, 2018). This model also considers the influence of social position with an explicit focus on systems of oppression and marginalization as a developmental context, but centers how youth in these contexts experience concomitant supports that synergistically balance with risk to create net vulnerability. Importantly, Spencer asserts that all populations experience vulnerability, and it is this balance between risk and protection that predicts multiple stage-salient coping outcomes, including both productive and unproductive ones (Spencer, 2006).

What also makes this theory so innovative is that it centers on the meaning making (i.e., phenomenology) processes that is essential to unpack our understanding of how oppression and support lead to behavioral, coping, and identity outcomes (Spencer, 2006). Thus, prominent in PVEST is the focus not just on the outcomes (i.e., the *what*) but also on the cognitive, affective, and interpretative processes (i.e., the *how*) that intersect with context, considering diverse individual-context interactions addressing the sources and pathways. PVEST has provided the theoretical foundational work for a multitude of studies, and in particular, due to its focus on phenomenology and integration of risk and protection across the life course, can be easily integrated to the study of social, cognitive, health, educational and other developmental processes across the lifespan.

We invite the reader to consider how to meaningfully integrate the PVEST and Integrative models into their work, first to define what is currently missing from their work and then to consult and collaborate with others with expertise in these topics (see Chaps. 9 and 14 that discuss positionality as research teams and how to grow in doing this type of work). We hope these models can serve as the kindling for new ideas and insights into how to best capture the experiences of diverse populations in rich, thoughtful, and innovative ways. To truly investigate meaning making processes, we need to break out of traditional research designs and methods and incorporate qualitative and mixed-methods work more explicitly (Lerner & Tolan, 2016; Yoshikawa et al., 2008).

In addition to these two models, we would also like to highlight other theoretical work that we believe can strengthen research that is informed by diversity science. Identity processes have been at the forefront of the study of diverse populations, with a particular focus on the understanding of racial and/or ethnic identity and the factors that fosters resilience and thriving in the face of oppression (see Umaña-Taylor et al., 2014 and the Ethnic and Racial Identity in the Twenty-First Century study group special section in Child Development). Given the centrality of identity processes to meaning making, we recommend that developmental scientists deeply consider the inclusion of identity processes into their work and point readers to Chap. 3 in this book. Racial/ethnic discrimination and its pernicious impact on developmental outcomes have also received significant research attention (see Benner et al., 2018, for a recent meta-analysis). Research teams examining experiences of stress and poverty should consider including measures of discrimination, as is discussed in Chap. 2 of this book. Multiple models unique to the impact of discrimination might be helpful for readers to consider including Clark et al. (1999), Harrell (2000), and especially Gee et al. (2012) who adopt a life course perspective.

Akin to the need to examine discrimination processes is the capturing of interlocking systems of oppression and how these factors and processes influence youth; here, theoretical writings on intersectionality are essential for guiding researchers to consider systems of classism, sexism, ableism, heterosexism, xenophobia, colorism (and other forms of oppression) and their intersection as a developmental context. Original thinking by Crenshaw (1989), steeped in a tradition of Black feminism, has led to a rich area of work in diversity science and in psychology (see Cole, 2009, in *American Psychologist*). Newer theoretical writings have applied intersectionality specifically to developmental science (see Ghavami et al., 2016, and the special section edited by Santos and Toomey (2018) on intersectionality and developmental science in *New Directions for Child and Adolescent Development*). Finally, we point readers to Neblett and colleagues' conceptual model (2012) that brings together multiple cultural resilience processes (cultural orientation, identity, and racial-ethnic socialization) with mediating processes like coping, self-concept, and attributions that synergistically contribute to well-being in the face of discrimination.

For the study of immigrant populations, there is a considerable body of theoretical writings and a large body of empirical work on the adaption and resilience of youth and their families with immigrant backgrounds across the globe. Suárez-Orozco and colleagues (2018) developed an integrative model that can be helpful in considering applying these processes to an increasingly global developmental science. Two pieces by Juang et al. (2018) and Titzmann and Lee (2018) highlight how integrating developmental concepts with principles of diversity science can be useful as well. Juang et al. apply an attachment perspective to the immigrant process, whereas Titzmann and Lee introduce dynamic concepts of tempo and timing borrowed from the pubertal timing literature to understanding acculturation. Two chapters in this book specifically consider acculturation and immigration processes from a methodological perspective (see Chaps. 5 and 12). Most important here is for researchers to move away from using the twin demographic proxies of generation status and language use or proficiency to better flesh out the processes under study.

As scientists, we can sometimes consistently rely on theories familiar to us or that are unique to our sub-discipline; this tendency is understandable to a certain degree as the field continues to grow exponentially. However, as a field, we cannot be complacent and continue to view diversity science as a sub-discipline; instead, we must consider these processes in every aspect of developmental science (Syed et al., 2018). Importantly, we need to train developmental scientists to consider these processes in their work in meaningful ways. Our re-education as a field is imperative—not only for the current rigorous endeavor but also for the ever-evolving science, and theories that center diverse lives and experiences should be as central to our curriculum as those of Piaget, Vygotsky, Erikson, or Bronfenbrenner to name a few. We now turn work that has countered the mainstream narrative by integrating racial, ethnic, and cultural processes into the study of development in innovative ways, leading to clarity and better generalizability in the study of developmental outcomes.

Leading the Way

There has been an emergence of work considering the integration of cultural processes into a variety of developmental science sub-disciplines and resulting empirical work that highlights the gained benefit of attending to these processes within "mainstream" developmental science. For example, Causadias (2013) has provided developmental psychopathologists with ways to consider the intersection of development, psychopathology, and culture that elucidates the constructs and processes important to integrate. In the same vein, Doane and colleagues (2018) proposed cultural neurobiology as the study of "the transactions among cultural processes and central and peripheral aspects of neurobiology" across development and multiple time frames in the first Handbook of Culture and Biology (p. 228).

Exemplifying this approach, Adam et al. (2020) went beyond documenting ethnic-racial disparities in cortisol levels (e.g., Deer et al., 2018) and tested the relation of both discrimination as a stressor and ethnic-racial identity as a protective factor across adolescence and early adulthood in Black and White participants. Consistent with their hypotheses informed by theories described above and past research on risk and resilience, these authors found that Black adolescent participants' adaptive ethnic-racial identity processes (e.g., stronger private and community regard) supported better cortisol functioning in early adulthood after controlling for racial-ethnic discrimination. A similar pattern was not evident among White participants suggesting different pathways and protections to cortisol dysregulation may be found for that group, which underscores the importance of examining culturally informed processes in our science. Clearly, by considering racial and ethnic processes like discrimination and identity, the study findings provide the field a more complete understanding of disparities (e.g., fueled by discrimination). Most importantly, this approach moves away from a deficit orientation that merely documents disparities and instead identifies community developed resilience factors (e.g., ethnic-racial identity). Similarly, another study also documented the protective cultural processes in Latinx families, whereby greater endorsement of Latinx cultural values (e.g., familism, *respeto*, religiosity) served to buffer the experiences of stress on diurnal cortisol rhythm for Latinx adolescents (Sladek et al., 2019).

Other researchers have focused on integrating *cultural processes* into the study "mainstream" psychological processes demonstrating how this inclusion can better refine our understanding of specific populations by bringing attention to the meaning making and contextualization of developmental processes. For example, Dunbar and colleagues (2017) articulated a model integrating ethnic-racial socialization and emotion socialization for African American families. This model highlighted how emotions like anger elicit different responses in the environment for African American youth, and thus, emotion socialization processes take on a different shape as parents consider how best to prepare their youth to thrive in that context. Without that meaning making component, parental practices may be mischaracterized and prediction to outcomes muddled. Demonstrating the critical need to integrate race into the study of emotion socialization and outcomes, for youth with higher biological sensitivity to context (measured by respiratory sinus arrhythmia [RSA]), parental suppression of negative emotion led to fewer externalizing behaviors in African American youth (5–6-year-old) only when parents discussed racism with their youth, but this protective effect was not evident when parents provided fewer preparation for bias messages (Dunbar et al., 2021). However, parental suppression messages were associated with greater internalizing symptoms for youth with high RSA regardless of preparation for bias messages. This type of theory and study shines a light on the information that can be gleaned about ethnic-racial differences when processes instead of proxies are included in study designs.

This approach can also be seen in applications of the family stress model that have explicitly incorporated racial and ethnic processes. The family stress model posits that perceptions of economic pressure lead to parental psychological distress that foments inter-parental conflict and disrupts protective parenting processes ultimately resulting in child adjustment problems (Conger et al., 2010). Tests of the family stress model in families of color have generally demonstrated support for the basic posited processes (Masarik & Conger, 2017), but the most innovative work that integrates diversity science has provided richer understanding of these processes within a culturally informed family stress model. For example, in examining these processes in Mexican American families, the links between parental psychological distress and interparental conflict were moderated by the endorsement of familism values such that mothers and fathers who had high levels of reciprocated family loyalty, obligation, and support were buffered against experiencing interparental conflict when psychologically distressed relative to those without this value endorsement (Martin et al., 2019). In a different sample of Mexican American families, maternal familism values served as a buffer in the link between economic pressure and maternal warmth further documenting the benefit of considering cultural protection (White et al., 2015). At the same time, neighborhood danger exacerbated the links between paternal economic pressure and paternal warmth noting the importance of considering contextual risk.

The family stress model has also been enhanced by considering parental experiences of racial-ethnic discrimination as predictors of disrupted parental and parenting processes beyond those related specifically to race and culture. In a sample of Chinese American families, parents' perceived discrimination was associated with parental depressive symptoms and marital and parent-child hostility that subsequently was associated with adolescent depressive symptoms and delinquent behaviors (Hou et al., 2017). Xue et al. (2022) revealed that maternal experiences of racial discrimination were associated with psychologically controlling parenting practices with their Chinese American preschoolers, through maternal depressive symptoms. Importantly, these indirect associations were further moderated by maternal acculturation toward the mainstream culture, where greater mainstream acculturation buffered mothers' mental health and subsequent engagement in psychologically controlling parenting against the negative effect of racial discrimination.

Similarly, Murry and colleagues (2021) demonstrated that parental discrimination experiences disrupt parent-child relationship quality in African American families, resulting in poorer self-control and emotion regulation in youth. Experiences of discrimination have also been tested as moderators of the links in the family stress model (like the studies above with familism) finding that the context of racism and oppression significantly exacerbates the links across the model between family stress and parental psychological functioning and then psychological functioning to partner and child relationship quality (Murry et al., 2001). We direct interested readers to Murry and colleagues' (2018) elegant integration of family stress theory with García Coll et al. (1996) model that fuses together more fully the collective cultural resilience of African American families, as well as the ways racism continues to serve as risk for impeding positive development.

Considering cultural values and practices may help to inform efforts to develop more culturally sensitive programs to facilitate youths' social-behavioral functioning in an increasingly diverse world. For example, adolescents' dyadic relationships with their mothers, fathers, and friends are influenced not only by each other but also by the broader relationship network and cultural context within which they exist. Most studies, however, neglect relationship networks as well as culture. Yet the consideration of both is necessary to understand universal and culture-specific relationship norms, as well as the extent to which different relationship constellations are more or less adaptive in different cultures. For example, a cross-cultural study identified how perceived support and negativity were similarly and differently manifested across youth relationships with mothers, fathers, and best friends, and how distinct relationship profiles were linked to adaptive and maladaptive functioning (aggression, anxious-withdrawal, prosocial behavior) in three countries—South Korea, Portugal, and the Unite States, which differ significantly in the degree to which their cultural norms (Oh et al., 2021). Such effort illuminated culturally common (high-quality; discordant) and culturally specific (e.g., only high-friendship quality; less involved father) relationship patterns in the relations between a high-quality relationship profile and adaptive social-behavioral functioning, as well as cultural specificity in the differential effects of distinct relationship profiles on youth outcomes. Together, these studies show developmental science has a lot to gain in integrating cultural and racial processes of risk and resilience into our models, and these gains will result in a richer science, a more complete understand of all families, and a more precise set of recommendations for policy and intervention in an increasingly diverse world.

Building a Collaborative, Diversity-Infused Developmental Science

The time has come for developmental scientists to embrace diversity science and integrate its principles to the very core of our discipline. As a field, we can no longer afford to be a science based on and relevant to only a subset of the global population. Research theories and designs must no longer ignore processes of racism, marginalization, and oppression. Whereas, simultaneously, developmental science must capture the resilience, spirit, and hope that have been the foundation of "thrival" (see Tynes et al., Chap. 11) and the development of competencies across communities of color. This book provides developmental scientists with tools to reimagine a different type of research agenda by taking advantage of new theories (see Chap. 11 by Tynes et al. on Afrofuturist Theory and Chap. 4 by Smith-Bynum on Racial-Ethnic Socialization), considering novel research approaches and methods (see Chap. 5 by Lorenzo Blanco et al. on Acculturation, Chap. 9 by Bañales et al. on Youth Voice, and Chap. 7 by Coppens & Coppinger on Ecologically Valid Developmental Science), integrating innovative measures of discrimination, identity, and critical civic engagement processes (see Chap. 2 by Morris et al. on Racial-Ethnic Discrimination, Chap. 3 by Rivas-Drake & Abi on Ethnic-Racial Identity, and Chap. 10 by Mathews et al. on Critical Consciousness), attending to populations that have been ignored (see Chap. 6 by Gabriel et al. on Multiracial Populations and Chap. 12 by Gulamhussein et al. on Transnational Immigrants), and considering context (see Chap. 8 on Neighborhoods and Schools by Witherspoon et al.). Next, chapters that consider how to build integrative, collaborative research teams that can facilitate this work (Chap. 14 by Caughy et al.) and inform practice and policy (Chap. 13 by Smith et al.) are presented. We conclude with key recommendations and takeaways (Witherspoon & Stein, Chap. 15). We hope that you heed our call and move the field into the twenty-first century to better reflect, represent, and serve all the youth of our changing world.

References

Adam, E. K., Hittner, E. F., Thomas, S. E., Villaume, S. C., & Nwafor, E. E. (2020). Racial discrimination and ethnic racial identity in adolescence as modulators of HPA axis activity. *Development and Psychopathology, 32*(5), 1669–1684. https://doi.org/10.1017/s095457942000111x

American Psychological Association. (2021). *Apology to people of color for APA's role in promoting, perpetuating, and failing to challenge racism, racial discrimination, and human hierarchy in U.S.* https://www.apa.org/about/policy/racism-apology

American Psychological Association, APA Task Force on Race and Ethnicity Guidelines in Psychology. (2019). *Race and ethnicity guidelines in psychology: Promoting responsiveness and equity*. Retrieved January 14, 2022, from https://www.apa.org/about/policy/guidelines-race-ethnicity.pdf

Australia Bureau of Statistics Census. (2017). *Cultural diversity in Australia, 2016*. Retrieved November 23, 2021, from https://www.abs.gov.au/ausstats/abs@.nsf/Lookup/by%20Subject/2071.0%7E2016%7EMain%20Features%7ECultural%20Diversity%20Article%7E60

Benner, A. D., Wang, Y., Shen, Y., Boyle, A. E., Polk, R., & Cheng, Y. P. (2018). Racial/ethnic discrimination and well-being during adolescence: A meta-analytic review. *American Psychologist, 73*(7), 855.

Brownell, C. A., & Early Social Development Research Lab. (2016). Prosocial behavior in infancy: The role of socialization. *Child Development Perspectives, 10*(4), 222–227. https://doi.org/10.1111/cdep.12189

Budiman, A. (2020, August 20). *Key findings about US immigrants*. Pew Research Center. https://www.pewresearch.org/fact-tank/2020/08/20/key-findings-about-u-s-immigrants/

Budiman, A., Tamir, C., Mora, L., & Noe-Bustamante, L. (2020). *Facts on US immigrants, 2018: Statistical portrait of the foreign-born population in the United States*. Pew Research Center. https://www.pewresearch.org/hispanic/2020/08/20/facts-on-u-s-immigrants/

Carlos, G., Cheah, C, S. L., Conejo Bolaños, L. D., & Cho, H.-S. (2022). The social development of immigrant children: A focus on Asian and Latinx children in the U.S. In P. K. Smith & C. H. Hart (Eds.), *Wiley-Blackwell handbook of childhood social development* (3rd ed.). Wiley-Blackwell Publishers.

Casey, A. E. (2021, October 21). *Who are children in immigrant families?* https://www.aecf.org/blog/who-are-the-children-in-immigrant-families

Causadias, J. M. (2013). A roadmap for the integration of culture into developmental psychopathology. *Development and Psychopathology, 25*(4pt2), 1375–1398. https://doi.org/10.1017/s0954579413000679

Child Trends. (2018). *Racial and ethnic composition of the child population*. Retrieved from https://www.childtrends.org/indicators/racial-and-ethnic-composition-of-the-child-population. Accessed, November 23, 2021.

Clark, R., Anderson, N. B., Clark, V. R., & Williams, D. R. (1999). Racism as a stressor for African Americans: A biopsychosocial model. *American Psychologist, 54*(10), 805–816. https://doi.org/10.1037//0003-066x.54.10.805

Cole, E. R. (2009). Intersectionality and research in psychology. *American Psychologist, 64*(3), 170–180. https://doi.org/10.1037/a0014564

Conger, R. D., Conger, K. J., & Martin, M. J. (2010). Socioeconomic status, family processes, and individual development. *Journal of Marriage and Family, 72*(3), 685–704.

Connor, P. (2016, December 15). *International migration: Key findings from the US, Europe and the world*. Pew Research Center. https://www.pewresearch.org/fact-tank/2016/12/15/international-migration-key-findings-from-the-u-s-europe-and-the-world/

Crenshaw, K. W. (1989). Demarginalizing the intersection of race and sex: A Black feminist critique of antidiscrimination doctrine, feminist theory and antiracist politics. *University of Chicago Legal Forum, 1989*, 139–167. https://chicagounbound.uchicago.edu/uclf/vol1989/iss1/8

Deer, L. K., Shields, G. S., Ivory, S. L., Hostinar, C. E., & Telzer, E. H. (2018). Racial/ethnic disparities in cortisol diurnal patterns and affect in adolescence. *Development and Psychopathology, 30*(5), 1977–1993. https://doi.org/10.1017/2FS0954579418001098

DeJesus, J. M., Callanan, M. A., Solis, G., & Gelman, S. A. (2019). Generic language in scientific communication. *Proceedings of the National Academy of Sciences, 116*(37), 18370–18377. https://doi.org/10.1073/pnas.1817706116

Doane, L. D., Sladek, M. R., & Adam, E. K. (2018). An introduction to cultural neurobiology: Evidence from physiological stress systems. In J. M. Causadias, E. H. Telzer, & N. A. Gonzales (Eds.), *The handbook of culture and biology* (pp. 227–254). Wiley.

Dunbar, A. S., Leerkes, E. M., Coard, S. I., Supple, A. J., & Calkins, S. (2017). An integrative conceptual model of parental racial/ethnic and emotion socialization and links to children's social-emotional development among African American families. *Child Development Perspectives, 11*(1), 16–22. https://doi.org/10.1111/cdep.12218

Dunbar, A. S., Zeytinoglu, S., & Leerkes, E. M. (2021). When is parental suppression of black children's negative emotions adaptive? The role of preparation for racial bias and children's resting cardiac vagal tone. *Research on Child and Adolescent Psychopathology*, 1–14. https://doi.org/10.1007/s10802-021-00779-z

Else-Quest, N. M., & Hyde, J. S. (2016). Intersectionality in quantitative psychological research: I. Theoretical and epistemological issues. *Psychology of Women Quarterly, 40*(2), 155–170. https://doi.org/10.1177/0361684316629797

García Coll, C. G., Crnic, K., Lamberty, G., Wasik, B. H., Jenkins, R., Garcia, H. V., & McAdoo, H. P. (1996). An integrative model for the study of developmental competencies in minority children. *Child Development, 67*(5), 1891–1914. https://doi.org/10.2307/1131600

García Coll, C. G., Akerman, A., & Cicchetti, D. (2000). Cultural influences on developmental processes and outcomes: Implications for the study of development and psychopathology. *Development and Psychopathology, 12*(3), 333–356. https://doi.org/10.1017/s0954579400003059

Gee, G. C., Walsemann, K. M., & Brondolo, E. (2012). A life course perspective on how racism may be related to health inequities. *American Journal of Public Health, 102*(5), 967–974. https://doi.org/10.2105/2FAJPH.2012.300666

Ghavami, N., Katsiaficas, D., & Rogers, L. O. (2016). Toward an intersectional approach in developmental science: The role of race, gender, sexual orientation, and immigrant status. *Advances in Child Development and Behavior, 50*, 31–73. https://doi.org/10.1016/bs.acdb.2015.12.001

Gonzalez-Barrera, A., & Connor, P. (2019, March 14). *Around the world, more say immigrants are a strength than a burden*. Pew Research Center. https://www.pewresearch.org/global/2019/03/14/around-the-world-more-say-immigrants-are-a-strength-than-a-burden/

Graham, S., & Echols, L. (2018). Race and ethnicity in peer relations research. In W. M. Bukowski, B. Laursen, & K. H. Rubin (Eds.), *Handbook of peer interactions, relationships, and groups* (pp. 590–614). The Guilford Press.

Harrell, S. P. (2000). A multidimensional conceptualization of racism-related stress: Implications for the well-being of people of color. *American Journal of Orthopsychiatry, 70*(1), 42–57. https://doi.org/10.1037/h0087722

Helms, J. E., Jernigan, M., & Mascher, J. (2005). The meaning of race in psychology and how to change it: A methodological perspective. *American Psychologist, 60*(1), 27.

Hou, Y., Kim, S. Y., Hazen, N., & Benner, A. D. (2017). Parents' perceived discrimination and adolescent adjustment in Chinese American families: Mediating family processes. *Child Development, 88*(1), 317–331. https://doi.org/10.1111/cdev.12603

Jensen, E., Jones, N., Rabe, M., Pratt, B., Medina, L., Orozco, K., & Spell, L. (2021). *The chance that two people chosen at random are of different race or ethnicity groups has increased since 2010*. United States Census Bureau. https://www.census.gov/library/stories/2021/08/2020-united-states-population-more-racially-ethnically-diverse-than-2010.html#:~:text=Using%20the%20same%20Diversity%20Index,2020%20from%2054.9%25%20in%202010

Jones, N., Marks, R., Ramirez, R., & Ríos-Vargas, M. (2021). 2020 census illuminates racial and ethnic composition of the country. United States Census Bureau. https://www.census.gov/library/stories/2021/08/improved-race-ethnicity-measures-reveal-united-states-population-much-more-multiracial.html

Juang, L. P., Simpson, J. A., Lee, R. M., Rothman, A. J., Titzmann, P. F., Schachner, M. K., et al. (2018). Using attachment and relational perspectives to understand adaptation and resilience among immigrant and refugee youth. *American Psychologist, 73*(6), 797–811. https://doi.apa.org/doi/10.1037/amp0000286

Kogachi, K., & Graham, S. (2021). Same-race friendship preference across the middle school years: The role of school racial/ethnic context. *Developmental Psychology, 57*(12), 2134–2149. https://doi.apa.org/doi/10.1037/dev0001263

Lerner, R. M., & Tolan, P. H. (2016). On the qualitative transformation of developmental science: The contributions of qualitative methods. *Qualitative Psychology, 3*(1), 120–124. https://doi.apa.org/doi/10.1037/qup0000052

Martin, M. J., Conger, R. D., & Robins, R. W. (2019). Family stress processes and drug and alcohol use by Mexican American adolescents. *Developmental Psychology, 55*(1), 170–183. https://doi.org/10.1037/dev0000629

Masarik, A. S., & Conger, R. D. (2017). Stress and child development: A review of the Family Stress Model. *Current Opinion in Psychology, 13*, 85–90. https://doi.org/10.1016/j.copsyc.2016.05.008

Miller, A. L., Stern, C., & Neville, H. (2019). Forging diversity-science-informed guidelines for research on race and racism in psychological science. *Journal of Social Issues, 75*(4), 1240–1261. https://doi.org/10.1111/josi.2019.75.issue-4/issuetoc

Murry, V. M., Brown, P. A., Brody, G. H., Cutrona, C. E., & Simons, R. L. (2001). Racial discrimination as a moderator of the links among stress, maternal psychological functioning, and family relationships. *Journal of Marriage and Family, 63*(4), 915–926. https://doi.org/10.1111/j.1741-3737.2001.00915.x

Murry, V. M., Butler-Barnes, S. T., Mayo-Gamble, T. L., & Inniss-Thompson, M. N. (2018). Excavating new constructs for family stress theories in the context of everyday life experiences of Black American families. *Journal of Family Theory & Review, 10*(2), 384–405. https://doi.org/10.1111/jftr.12256

Murry, V. M., Gonzalez, C. M., Hanebutt, R. A., Bulgin, D., Coates, E. E., Inniss-Thompson, M. N., et al. (2021). Longitudinal study of the cascading effects of racial discrimination on parenting and adjustment among African American youth. *Attachment & Human Development*, 1–17. https://doi.org/10.1080/14616734.2021.1976926

Neblett, E. W., Jr., Rivas-Drake, D., & Umaña-Taylor, A. J. (2012). The promise of racial and ethnic protective factors in promoting ethnic minority youth development. *Child Development Perspectives, 6*(3), 295–303. https://doi.org/10.1111/j.1750-8606.2012.00239.x

Nielsen, M., Haun, D., Kärtner, J., & Legare, C. H. (2017). The persistent sampling bias in developmental psychology: A call to action. *Journal of Experimental Child Psychology, 162*, 31–38. https://doi.org/10.1016/j.jecp.2017.04.017

Oh, W., Bowker, J., Kim, H., Santos, A., Ribeiro, O., Guedes, M., Freitas, M., Song, S., & Rubin, K. H. (2021). Distinct profiles of relationships with mothers, fathers, and best friends and socio-emotional functioning in early adolescence: A cross-cultural study. *Child Development, 92*(6), 1154–1170. https://doi.org/10.1111/cdev.13610

Parker, K., Morin, R., & Horowitz, J. M. (2019, March 21). *Looking to the future, public sees an America in decline on many fronts*. Pew Research Center. https://www.pewresearch.org/social-trends/2019/03/21/public-sees-an-america-in-decline-on-many-fronts/

Perez-Brena, N. J., Rivas-Drake, D., Toomey, R. B., & Umaña-Taylor, A. J. (2018). Contributions of the integrative model for the study of developmental competencies in minority children: What have we learned about adaptive culture? *American Psychologist, 73*(6), 713–726. https://doi.org/10.1037/amp0000292

Pew Research Center. (2019, January 30). *International migrants by country*. https://www.pewresearch.org/global/interactives/international-migrants-by-country/

Plaut, V. C. (2010). Diversity science: Why and how difference makes a difference. *Psychological Inquiry, 21*(2), 77–99. https://doi.org/10.1080/10478401003676501

Poushter, J., & Fetterolf, J. (2019, April 22). *A changing world: Global views on diversity, gender equality, family life and the importance of religion*. Pew Research Center. https://www.pewresearch.org/global/2019/04/22/a-changing-world-global-views-on-diversity-gender-equality-family-life-and-the-importance-of-religion/

Roberts, S. O., Bareket-Shavit, C., Dollins, F. A., Goldie, P. D., & Mortenson, E. (2020). Racial inequality in psychological research: Trends of the past and recommendations for the future. *Perspectives on Psychological Science, 15*(6), 1295–1309. https://doi.org/10.1177/2F1745691620927709

Santos, C. E., & Toomey, R. B. (2018). Integrating an intersectionality lens in theory and research in developmental science. In C. E. Santos & R. B. Toomey (Eds.), *Envisioning the integration of an intersectional lens in developmental science. New Directions for Child and Adolescent Development* (Vol. 161, pp. 7–15). https://doi.org/10.1002/cad.20245

Seaton, E. K., Gee, G. C., Neblett, E., & Spanierman, L. (2018). New directions for racial discrimination research as inspired by the integrative model. *American Psychologist, 73*(6), 768–780. https://doi.org/10.1037/amp0000315

Sladek, M. R., Doane, L. D., Gonzales, N. A., Grimm, K. J., & Luecken, L. J. (2019). Latino adolescents' cultural values associated with diurnal cortisol activity. *Psychoneuroendocrinology, 109*, 104403. https://doi.org/10.1016/j.psyneuen.2019.104403

Spencer, M. B. (2006). Phenomenology and ecological systems theory: Development of diverse groups. In R. M. Lerner & W. Damon (Eds.), *Handbook of child psychology: Theoretical models of human development* (Vol. 1, 6th ed., pp. 829–893). Wiley.

Statistics Canada. (2013). *Immigration and ethnocultural diversity in Canada*. Retrieved November 23, 2021, from https://www12.statcan.gc.ca/nhs-enm/2011/as-sa/99-010-x/99-010-x2011001-eng.pdf

Streit, C., Carlo, G., Knight, G. P., White, R. M., & Maiya, S. (2021). Relations among parenting, culture, and prosocial behaviors in U.S. Mexican youth: An integrative socialization approach. *Child Development, 92*(4), 383–397. https://doi.org/10.1111/cdev.13550

Suárez-Orozco, C., Motti-Stefanidi, F., Marks, A., & Katsiaficas, D. (2018). An integrative risk and resilience model for understanding the adaptation of immigrant-origin children and youth. *American Psychologist, 73*(6), 781–796. https://doi.org/10.1037/amp0000265

Syed, M., Santos, C., Yoo, H. C., & Juang, L. P. (2018). Invisibility of racial/ethnic minorities in developmental science: Implications for research and institutional practices. *American Psychologist, 73*(6), 812–826. https://doi.org/10.1037/amp0000294

Tamir, C. (2021, March 25). *The growing diversity of Black America*. Pew Research Center. https://www.pewresearch.org/social-trends/2021/03/25/the-growing-diversity-of-black-america/

Tamir, C. (2022, January 27). *Key findings about Black immigrants in the U.S.* Pew Research Center. https://www.pewresearch.org/fact-tank/2022/01/27/key-findings-about-black-immigrants-in-the-u-s/

Tamir, C., & Anderson, M. (2022, January 20). *One-in-ten Black people living in the U.S. are Immigrants.* Pew Research Center. https://www.pewresearch.org/race-ethnicity/2022/01/20/the-caribbean-is-the-largest-origin-source-of-black-immigrants-but-fastest-growth-is-among-african-immigrants/

Titzmann, P. F., & Lee, R. M. (2018). Adaptation of young immigrants. *European Psychologist, 23*(1), 72–82. https://doi.org/10.1027/1016-9040/a000313

Umaña-Taylor, A. J., Quintana, S. M., Lee, R. M., Cross, W. E., Jr., Rivas-Drake, D., Schwartz, S. J., et al. (2014). Ethnic and racial identity during adolescence and into young adulthood: An integrated conceptualization. *Child Development, 85*(1), 21–39. https://doi.org/10.1111/2Fcdev.12196

Velez, G., & Spencer, M. B. (2018). Phenomenology and intersectionality: Using PVEST as a frame for adolescent identity formation amid intersecting ecological systems of inequality. In C. E. Santos & R. B. Toomey (Eds.), *Envisioning the integration of an intersectional lens in developmental science. New Directions for Child and Adolescent Development* (Vol. 161, pp. 75–90). https://doi.org/10.1002/cad.20247

Vespa, J., Armstrong, D. M., & Medina, L. (2018). *Demographic turning points for the United States: Population projections for 2020 to 2060.* US Department of Commerce. Retrieved November 23, 2021, from https://www.census.gov/content/dam/Census/library/publications/2020/demo/p25-1144.pdf

White, R., Liu, Y., Nair, R. L., & Tein, J. Y. (2015). Longitudinal and integrative tests of family stress model effects on Mexican origin adolescents. *Developmental Psychology, 51*(5), 649–662. https://doi.org/10.1037/2Fa0038993

White, R., Nair, R. L., & Bradley, R. H. (2018). Theorizing the benefits and costs of adaptive cultures for development. *American Psychologist, 73*(6), 727–739. https://doi.org/10.1037/amp0000237

Xue, X., Cheah, C. S. L., & Hart, C. H. (2022). Risk and protective processes in the link between racial discrimination and Chinese American mothers' psychologically controlling parenting. Cultural Diversity and Ethnic Minority Psychology. Advance online publication. https://doi.org/10.1037/cdp0000545

Yoshikawa, H., Weisner, T. S., Kalil, A., & Way, N. (2008). Mixing qualitative and quantitative research in developmental science: Uses and methodological choices. *Developmental Psychology, 44*(2), 344–354. https://doi.org/10.1037/0012-1649.44.2.344

Chapter 2
Level Up: Recommendations for Measuring Racial Discrimination

Kamryn S. Morris, Jayley A. Janssen, and Eleanor K. Seaton

Racism is systematic and includes dominance, power and privilege based on racial group designations (Harrell, 2000). Racism is also multilevel and includes individual racism, cultural racism, and institutional racism (Jones, 1997; Tynes et al., 2015). Children and adolescents identifying as Black, Indigenous, and People of Color (BIPOC) experience racism directly and vicariously through family members and friends (Heard-Garris et al., 2018). Racial discrimination, or individual racism, is the behavioral component of racism and includes dominant racial group members' actions that have differential and negative effects on subordinate ethnic-racial groups (Williams et al., 2003). Benner and colleagues (2018) conducted a meta-analysis among Black American, Latinx, and Asian American early adolescents and reported that racial discrimination experiences were positively associated with socioemotional distress (e.g., internalizing symptoms, depression) and risky health behaviors (e.g., substance use, risky sexual behaviors), and negatively associated with academic outcomes (e.g., achievement, engagement). Thus, racial discrimination is linked to negative outcomes among BIPOC children and adolescents. This chapter recommends incorporation of novel methods among BIPOC youth, and the incorporation of White youth for future research.

Novel Methods

Scientific understanding of racial discrimination among BIPOC youth is limited by the current methodologies used in the field. Acknowledging that developmental stages, demographic factors, and contextual considerations contribute to

K. S. Morris (✉) · J. A. Janssen · E. K. Seaton
School of Social and Family Dynamics, Arizona State University, Tempe, AZ, USA
e-mail: ksmorr10@asu.edu

© The Author(s), under exclusive license to Springer Nature Switzerland AG 2023

D. P. Witherspoon, G. L. Stein (eds.), *Diversity and Developmental Science*,
https://doi.org/10.1007/978-3-031-23163-6_2

experiences of racial discrimination calls for critical reflection of the field's current methodologies. Utilization of multiple, novel methods to investigate youths' experiences of racial discrimination will allow researchers to triangulate scientific understanding of the nuances that surround the prevalence of racism in the United States and experiences of racial discrimination. The aggregate knowledge that emerges as a result of the use of multiple methods will offer a holistic understanding of racial discrimination among youth and delineate how racism systematically disadvantages BIPOC youth and privileges White youth. Specifically, we underscore the importance and utility of daily diary, qualitative, experimental, and mixed methods in investigating racial discrimination among BIPOC youth and posit recommendations for moving the field forward.

Daily Diary Method

The use of surveys in the field of racial discrimination research has substantiated it as a common and pervasive stressor among BIPOC youth. Reviews of self-report measures of racism and racial discrimination can be found in the psychometric work conducted by Utsey (1998) and more recently Atkins (2014), both of which highlight the need for reliable, valid, and quality measures. Forms of racial discrimination include educational, vicarious, and online discrimination; accordingly, there are extensive measures developed to assess frequency, perpetrators, and contexts. For example, Gale and Dorsey (2020) examined in school discrimination using a 7-item measure developed by researchers from the Maryland Adolescent Development in Context study which assessed the frequency of events and indicated that higher scores reflected more experiences of discrimination. Huynh and Fuligni (2010) focused on the perpetrators of racial discrimination, specifically examining frequencies of adult and peer discrimination using two 7-item scales developed by Greene et al. (2006). Moreover, Morris et al. (2020) utilized the 18-item Daily Life Experiences of Racism (DLE) to assess racial discrimination stress, that is, negative emotional responses following perceiving racial discrimination. Acknowledging how racial discrimination is measured in studies on BIPOC youth is foundational to daily diary approaches which employ these tools and provide in-depth examinations on how racial discrimination relates to various outcomes. The daily diary method, or ecological momentary assessment (EMA), includes repeated sampling of participants over the course of days or months (Moskowitz & Young, 2006). This method has been increasingly utilized in racial discrimination research (English et al., 2020; Seaton & Iida, 2019; Zeiders, 2017). Diary methods examine momentary experiences and eliminate retrospective bias due to the repeated sampling of participants (English et al., 2020; Santiago et al., 2017; Seaton & Iida, 2019). In addition to being considered more reliable and accurate for examining racial discrimination, daily diary techniques are also better suited to detect daily fluctuations (Santiago et al., 2017).

Daily diary studies document the habitual occurrence of racial discrimination. Although the core of daily diary methods lies in its repeated sampling, there is variation among the length of studies using this approach. For example, some daily diary studies span 14 days (e.g., English et al., 2020; Wang & Yip, 2019), whereas others span 21 days (e.g., Cheeks et al., 2020; Dunbar et al., 2017). Given that racial discrimination is a normal, daily occurrence for BIPOC youth, the length of daily diary studies is particularly important for highlighting typical hassles youth encounter and generalizing about experiences *within* and *across* youth. That is, studies spanning longer periods provide greater depth into the prevalence of racial discrimination and its association with critical developmental outcomes. Among Black youth, studies spanning two-week periods reported between 2.4 racially discriminatory experiences over the course of the study (Seaton & Iida, 2019) and 5.2 experiences per day (English et al., 2020). As such, this technique allows for examination of how racial discrimination is related to short-term processes like increased same-day depressive symptoms (English et al., 2020; Seaton & Iida, 2019) and same-day sleep (Yip et al., 2020) in addition to long-term processes like increased stress across 14 days (Huynh & Fuligni, 2010) and decreased school engagement 6 months post survey (Dunbar et al., 2017). Researchers have used this method for investigation of the associations between racial discrimination and sleep disturbances (Wang & Yip, 2019), the association between racial discrimination and socialization messages in relation to psychological affect (Cheeks et al., 2020), and the frequency and psychological effect of forms of racial discrimination (e.g., teasing, online) in relation to depressive symptoms (English et al., 2020). Additionally, the contexts which youth navigate may vary and become conflated by the timing, people, and environments they interact with on a particular day. Wang and Yip (2019) accounted for week-day and week-end differences which they found contributed to missingness in data on racial discrimination. Specifically, opportunities for daily diary studies examining racial discrimination include accounting for fluctuations related to the time of the day, school and neighborhood contexts, and previous day discrimination.

Importantly, we encourage researchers to understand developmental differences using the daily diary approach in two ways. First, by investigating the frequency of racial discrimination in middle childhood. For example, racial discrimination experiences increase as BIPOC youth mature through adolescence (Greene et al., 2006). Second, by using multiple assessments within a day, which can detect contexts and perpetrators associated with fluctuations in discrimination, in addition to creating a more nuanced understanding of short- and long-term outcomes, like well-being, mental health, and academic trajectories (Dunbar et al., 2017; Wang & Yip, 2019). Would this same finding be replicated if daily diaries were administered across years? These methods have opportunity to cultivate knowledge on *interindividual* and *intraindividual* processes, which should further be explored through longitudinal designs spanning multiple years. The multiple occurrences and contexts which advance racism and racial discrimination should be investigated frequently and daily to understand the long-term repercussions across key developmental transitions such as adolescence and adulthood.

Qualitative Methods

Qualitative research is the exploration and discovery of social phenomena using non-numerical data (Austin & Sutton, 2014). Qualitative methods may include interviews, focus groups, and case studies. Examination of experiences of racial discrimination through participants' own words and perspectives has yielded a rich scientific understanding (Tynes et al., 2015). The goal of qualitative research is not generalizability (Austin & Sutton, 2014) but to form an in-depth understanding of lived experiences and capture nuances that remain unexamined in quantitative analyses. Yoon et al. (2017) highlight the lack of qualitative research conducted among Asian American adolescents, particularly regarding their navigation of ethnic identity and racialized contexts (i.e., racism, stereotypes), and address how current disparities in minority participants and methodologies make qualitative methods critical for addressing knowledge gaps.

Ethnographies

Ethnography is the collective use of qualitative methods to directly observe social practices and interactions in a natural environment (Morgan-Trimmer & Wood, 2016). Ethnography allows for the unique discovery of how contextual factors may conflict or complement the articulated experiences of youth (Duncan, 2002). A strength of ethnographic research is the ability to use aggregate forms of data to detail an inclusive narrative (Estrada & Hondagneu-Sotelo, 2011). This is crucial when researching sensitive topics, like racial discrimination, or sampling participants who are disproportionately excluded from empirical research. For example, Pollock's (2004) ethnographic study among multiracial youth describes the paradox of youth prescribing to simple racial categories while simultaneously fighting against simplistic racial binaries. Understanding the environments youth navigate is essential to gaining comprehensive accounts of racial discrimination among people of color living in a racialized country (Estrada & Hondagneu-Sotelo, 2011).

Considering the various forms of racial discrimination that BIPOC youth face (e.g., direct, vicarious, institutional), investigation of their personal accounts, in addition to the perpetrators and settings that contribute to their experiences, is critically needed. A key example of this is Hardie and Tyson's (2013) nine month ethnographic study of a high school in the Southern region of the United States where observers documented clear racial divisions during lunch. Specifically, students segregated by race and often chose to socialize with same-race peers within and outside school; further, narratives from in-school informants supported these observations by detailing the core social groups located in the school (Hardie & Tyson, 2013). Additionally, despite observance of a race riot between students that occurred during the data collection period, teachers and administrators dismissed the incident by frequently emphasizing the family-like nature of the school (Hardie & Tyson, 2013). This study demonstrates the ability of ethnographic research for aggregation of data regarding victims, perpetrators, and the contexts which shape racism and racial discrimination. Extending our emphasis on the importance of contexts and settings, ethnographic methods ensure that researchers are equipped to not only observe the settings and contexts alone but also understand how youth navigate these contexts (Ispa-Landa, 2013). For example, Ispa-Landa (2013) explored Black students' experiences in a racial integration program to understand racial stereotypes when navigating a majority White suburban school. Ispa-Landa (2013) detailed the stark contrast between the neighborhoods students lived in compared to the affluent neighborhoods where their schools were located. Black students were victims of several racial stereotypes within their schools and felt pressure to adapt to the expectations prescribed in their school settings (Ispa-Landa, 2013). Specifically, boys described feeling the need to be tough and street smart, whereas girls had to denounce being "ghetto" to be perceived as ladylike (Ispa-Landa, 2013). Ethnographic methods, particularly through fieldwork, can be utilized to examine how macro- and micro-level forces construct the environments youth are embedded in (Morgan-Trimmer & Wood, 2016) and help illuminate the numerous factors which contribute to racial discrimination experiences such as perpetrators, peers and community members, and environments (Seaton & Douglass, 2014).

Given that fieldwork spans a longer period, ethnographic methods provide breadth and depth of knowledge, which equips researchers with rich observational data that enables them to construct a detailed understanding of the contexts youth navigate (Duncan, 2002). This is a key distinction, given the growing research on how context shapes youths' experiences, especially with mitigating or exacerbating racial discrimination (Gale & Dorsey, 2020; Morris et al., 2020). Understandably, information may get lost when utilizing certain methodologies; events may go unreported, youth may be hesitant to discuss, or key experiences can be forgotten when employing commonly used data collection methods (i.e., surveys, interviews). The confluence of observational and interview data may situate narratives within a context, thus resulting in more comprehensive data analysis.

Interviews, Focus Groups, and Narratives

Racial discrimination has been documented as harmful to the physical, mental, emotional, and academic well-being of BIPOC youth (Benner et al., 2018). However, through qualitative research, the field has begun to understand variability in encounters and the processing of incidences of racial discrimination (Wegmann, 2017). Qualitative methods can be a powerful tool for giving BIPOC youth the space to assert their experiences with racial discrimination. For example, Yoon et al. (2017) conducted interviews and reported that harmful stereotypes became deeply engrained among their sample of East Asian youth, which subsequently influenced their ethnic-racial identity and the academic demands imposed by themselves and others. Córdova and Cervantes (2010) conducted focus groups and found that Latinx youth experienced intergroup and within-group discrimination, and these experiences varied by nativity status and geographic location. These findings highlight that within BIPOC groups, the experiences of racial discrimination can differ among youth. Qualitative methods have been foundational in reaffirming the notion that there is no universal experience of racial discrimination within or across groups.

Differential racialized experiences and perceptions of racial discrimination contribute to the understanding of BIPOC youths' navigation of their marginalized identities. Intersectionality is a framework for understanding how systems of power, which can include but are not limited to racism, sexism, and ableism, create overlapping and interconnected experiences of oppression, discrimination, and disadvantage (Cho et al., 2013). Meaning making of these systems is made possible through qualitative methods, which are central to models of development like the Phenomenological Variant of Ecological Systems Theory (PVEST; Spencer et al., 1997) and the Integrative Model for the Study of Developmental Competencies in Minority Children (García Coll et al., 1996). There may be considerable variation in youths' subjective meaning making of racial discrimination across developmental periods, social positionality, and environmental contexts. As youth cognitively mature, they are more likely to navigate, engage with, and understand the systems of power which construct their perceptions of racial discrimination (Velez & Spencer, 2018). Importantly, we note the crystallization of awareness surrounding racial discrimination across developmental periods, with evidence suggesting that youth as young as seven are able to identify stereotypes (Wegmann, 2017). Further, during adolescence youth are exposed to racial discrimination in the media and peer interactions; however, due to their social positionality, or level of vulnerability, they may feel inadequately prepared to combat these messages (Romero et al., 2015). A recent study used qualitative techniques to examine developmental changes in ethnic-racial identity among a sample of diverse BIPOC children (Rogers et al., 2021). The results indicated that the importance of racial identity increased over a two-year period for Black and Multiracial children, with some children referencing racial inequities related to the Black Lives Matter movement (Rogers et al., 2021). These findings highlight youths' awareness regarding racial discrimination and the process through which they comprehend and combat negative messages about their race. Researchers must acknowledge the complexity among BIPOC youth, the racial climate of their environments and broader society, and the systems which exacerbate and mitigate risk (Huynh & Fuligni, 2010; Tobler et al., 2013).

Experimental Methods

Experimental methods have been used unethically among BIPOC groups (see Brazier, 2008; Leiter & Herman, 2015) and used to disenfranchise alternative methodologies (i.e., survey, qualitative, mixed-methods works). We acknowledge this problematic history and support the use of experimental methods that are implemented judiciously and that incorporate BIPOC voices in method design. Experimental research on racial discrimination offers the possibility for causal inference via the implementation of treatment and controls (Reichardt, 2019).

However, race and racism represent a unique challenge to this method. First, one's race cannot be randomly assigned or experimentally modified. Further, randomly assigning treatment differences of racial discrimination is unethical. Despite these challenges, experimental methods examining racial discrimination have emerged in which actors, or fictitious situations, are introduced allowing researchers to measure both perpetrators and effects of racial discrimination.

Field Experiments

Historically, field experiments have been used to identify racism interpersonally and systemically by implementing a randomized research design in a field setting (Gaddis, 2018). Audit experiments are a type of field experiment in which researchers select, train, and match individuals (e.g., testers), or construct fictitious, matched email correspondence to present equally qualified individuals to play the position of interest (i.e., a job-seeker; Pager & Shepherd, 2008). The testers are designed to differ in specific domains (e.g., race-ethnicity) allowing researchers to objectively assess treatment differences (e.g., racial discrimination).

Audit experiments are unique in that they are not subject to social desirability, given the participants are blind to their participation. Audit experiments have traditionally been implemented in employment, real estate, and healthcare settings (see Auspurg et al., 2019). However, such experiments have recently been used to examine racial discrimination in BIPOC youth's educational contexts (Giulietti et al., 2019; Janssen et al., 2022). Both Giulietti and colleagues (2019) and Janssen and colleagues (2022) examined disparities in school email communications with BIPOC families and White families. Results revealed that racial disparities existed in school administrators and guidance counselors email responses to BIPOC families when compared to White families (Giulietti et al., 2019; Janssen et al., 2022). School administrators and guidance counselors sent fewer responses to BIPOC versus White inquiries. Further, school administrator and guidance counselor's email responses to BIPOC families revealed a less cordial tone and a lack of detailed information when compared to responses to White families (Giulietti et al., 2019; Janssen et al., 2022).

Evidence from audit experiments directly identify if and from whom racial discrimination is occurring in BIPOC youth's contexts. The application of audit experiments in educational contexts has provided the opportunity to objectively identify how school personnel perpetrate racism to the detriment of BIPOC youth. This experimental evidence not only supports survey and qualitative work documenting racial discrimination but also provides greater understanding for how interpersonal racism intersects to form a system of racism. That is, interpersonal racial discrimination at the school level (e.g., from a guidance counselor) intersects with racial discrimination at the district level (e.g., district administrators) to create a system that inhibits BIPOC youth's academic trajectories. We encourage the continued use of audit methodologies to probe racism and racial discrimination in youths' educational institutions and note the need to expand the use of audit experiments in youths' multiple contexts and at different ages to understand developmental differences. Specifically, we emphasize the need to measure BIPOC youth's experience of racial discrimination in social services, healthcare, the juvenile justice system, and youth-led organizations to understand how and if racism prohibits BIPOC youth's health, adjustment, and well-being. Further, we encourage the use of audit studies with youth populations to understand the developmental trajectory of racial discrimination.

Modified Audits/Vignette Experiments

Similar to the audit methodology, experimental vignettes present fictitious, matched narratives that are designed to differ in specific domains (Atzmüller & Steiner, 2010). However, in vignette experiments, recipients of narratives are often aware of their participation in research and typically understand that the individuals in the presented scenarios are fictitious (Atzmüller & Steiner, 2010). Though recipients' awareness of their participation in the study may lead to greater social desirability bias, additional participant data can be obtained to test for potential mechanisms and effects of racial discrimination.

Johnson and colleagues (2013) implemented a combined experimental vignette and survey study that revealed fictitious individuals described with more stereotypically Black features were more likely to be believed as the target of racism when compared to fictitious individuals with less stereotypic features. Survey data on participants' own discrimination expectations revealed that experimental effects were not present among participants with high discrimination expectations. Only participants with low discrimination expectations focused on the features of the target in deciding if racial discrimination occurred (Johnson et al., 2013). Gibbons and colleagues (2010, 2012) also implemented both experimental vignette and survey methods to examine the relation between experiencing racial discrimination and substance use. The results demonstrated that adolescent envisioning experiences of racial discrimination caused an increase in substance-related words in an association task and an increase in drug willingness (Gibbons et al., 2010, 2012). Further, the authors identified that parental support buffered, while adolescent negative affect exacerbated the effects of racial discrimination (Gibbons et al., 2010, 2012). The combination of experimental vignettes and survey methods in the summarized studies facilitated the examination of racial discrimination's effect on participants' characteristics that are undetectable in audit experiments (e.g., participant's own discrimination expectations, drug willingness, parental support).

The combination of experimental methods with survey and qualitative methods results in rigorous data that can triangulate understanding. Thus, experimental methods document participant actions, while surveys and qualitative methods provide understanding of participants' self-reported understanding and beliefs with past and future actions (Gaddis, 2018). We encourage researchers to combine experimental methods with survey and interview methods, given the resulting evidence could provide greater understanding of the mechanisms, intentions, and conditions under which racial discrimination is perpetrated and experienced. Both audit experiments and experimental vignettes offer a unique opportunity to examine intersecting systems of power and oppression. By implementing and manipulating multiple identities in a given experiment (i.e., race, gender, sexual orientation), researchers can probe the unique ways in which racial discrimination is experienced at the intersections of other systems of oppression. A given experiment could probe various intersections by creating fictitious profiles that vary on race, gender, sexuality, ability, etc. Though the complexity of this experimental design may be laborious, the resulting findings would accurately measure the cause and effect of racial discrimination as it is experienced by a person with multiple identities that are oppressed and/or privileged. Ultimately, experimental vignettes and audits incorporating multiple identities have the potential to move the field beyond examining racial discrimination experiences tied to a single system of oppression and allow the opportunity to obtain a more comprehensive understanding of intersectional oppression.

Mixed Methods

Racial discrimination is a complex stressor with widespread consequences for BIPOC youth. Generally, mixed-method approaches cross-validate data through various sources (Berkel et al., 2009). This approach permits insights into deviation among data and provides breadth and depth of experiences (Berkel et al., 2009). Quantitative data may be used for obtaining a general understanding, whereas qualitative data may identify inconsistencies and offer additional insight into constructs (Gibson et al., 2019). An advantage of this methodology is that it permits an understanding of the frequency and attribution of the events in addition to the distinct impact it makes on individuals (Gibson et al., 2019), thus, bridging the gap between methodologies through data triangulation to facilitate understanding of racial discrimination.

Johnston-Goodstar and VeLure Roholt (2017) examined racial discrimination in relation to the push-out of Native American youth in school settings. Qualitative results provided accounts of three categories of microaggressions (i.e., microassaults, microinsults, microinvalidations), and quantitative results supported these findings through the disproportionality in suspension rates and perceptions of treatment from school personnel (Johnston-Goodstar & VeLure Roholt, 2017). Mixed methods research allows for a deeper understanding of experiences of racial discrimination (Seaton & Tyson, 2019). Interlocking systems of oppression are complex; therefore, it may be difficult to capture subtleties with a singular approach. In addition to variation in developmental processes among youth, their interpretation of challenges related to their development and race leads to further nuance in their navigation of their social positionality created by systems of power.

Mixed methods research is valuable for acquiring evidence surrounding frequency, perceptions, and stress related to racism and racial discrimination. However, racism is not the only system of power BIPOC youth must navigate. There are numerous underlying and conflicting social structures that shape youths' reports and processing of racial discrimination (Cho et al., 2013; Seaton & Tyson, 2019). While the goal of research is to cultivate models and theories that are generalizable, we must not forget the variation that lies *within* groups. Acknowledging that there is variability among and within BIPOC youth underscores the importance of mixed methods for measurement development (Sangalang & Gee, 2015) and ensuring that those measures are intersectional. Scholars should consider intersectional mixed method approaches to investigating racial discrimination, as the foundation of this method is the knowledge and experiences of marginalized people. In addition, it is practical for moving beyond examining categorical identities (e.g., race, gender, sexual orientation) toward addressing structural and systemic inequality (Bailey et al., 2019). An example of an intersectional measurement instrument can be found in Roberts et al.'s (2018) adaption of the Experiences of Discrimination (EOD) measure in their study which examined gender and racial discrimination in relation to teens exposed to dating violence. One of the strongest ways to push the field forward is by developing reliable and valid intersectional measures.

Whiteness

Research examining racial discrimination has largely focused on BIPOC experiences, ignoring that the perpetrators of racism in the United States are White individuals. Given that racism is rooted in White supremacy (Harrell, 2000), the vein of research measuring Whiteness, specifically Whiteness among US youth, must be expanded. Innocence has long been associated with both Whiteness and White children (Bernstein, 2011), and hesitance to study racism among White children likely rests at the intersection of Whiteness and innocence. Yet scientific evidence of the development, prevalence, and sustenance of White youths' racial attitudes and racially discriminatory behavior are integral in dismantling racist structures that undermine BIPOC youth's development. Further, research demonstrates that early developmental periods (i.e., 3–7 years of age) mark not only the formation but also the crystallization of White children's racial biases (Raabe & Beelmann, 2011), which persist throughout adolescence and influence peer interactions (Hagerman, 2016, 2020). Thus, it is essential that the field examine White youths' development and expression of racist ideas to identify opportunities for prevention. Specifically, we highlight opportunities for qualitative, quantitative, and experimental methods to investigate components of Whiteness.

Qualitative Methods

Given investigations of White youth's racial attitudes and discriminatory behaviors are rare, we place qualitative methods at the forefront of our recommendations. Results from qualitative examinations are instrumental in establishing scientific understanding because they illustrate process (Maxwell, 2012), and the same is true for establishing scientific understanding of Whiteness among youth populations. Qualitative probes of White youths' racial attitudes and behaviors can offer rich, in-depth data that can disentangle nuances associated with the components of Whiteness, racism, and White supremacy.

Ethnographies

We highlight the utility of ethnographies in examining White youth. Evidence from Hagerman's (2016) two-year ethnographic study with White, affluent families revealed evidence that children reproduce and rework colorblind ideologies, suggesting adolescent's own agency and responsibility for racial attitudes. In addition, ethnographic examinations in educational institutions provide evidence for how White youth exhibit racial privilege and discriminatory behaviors toward BIPOC peers in classrooms and on the playground (Lewis, 2004; Hagerman, 2016, 2020). Further, ethnographic and observational research has identified nuances *where* White youth display such attitudes and behaviors such that White children and adolescents are aware that discussing race and exhibiting racial bias are taboo. White youth exhibit their racial attitudes and biased behaviors in the peer context but hide and tame such attitudes and behaviors in the presence of authority figures (e.g., teachers and parents; Hagerman, 2016, 2020; Lewis, 2004).

Ethnographic examinations of how race and racism operate in White youths' contexts have the ability to identify nuances not possible in other quantitative or qualitative methods (Heath, 1982). Given White youths' skill to mask their racial biases, ethnographic examinations provide the ability to observe and document uncontrollable incidents of racism and privilege as youth experience or exhibit them. Further, ethnographies provide the ability to observe the unknown. Given the limited scientific understanding of Whiteness among youth populations, ethnographies provide the ability to gather understanding for research questions not yet conceived. We stress the importance of continuing to ethnographically examine Whiteness in educational institutions and encourage the expansion of conducting ethnographies in other youth contexts. Specifically, we highlight youth organizations, extra-curricular activities, and sports teams as contexts that may be particularly important, given White youth's relationships with adults and peers are distinct from the educational context (e.g., coach and athlete).

Interviews, Focus Groups, and Narratives

In addition to the importance of ethnographic data, we also note the importance of in-depth interviews, focus groups, and narratives, given their potential to probe White youths' own conceptualizations of Whiteness, racism, and White supremacy. For example, Thomann and Suyemoto (2018) interviewed White seventh and eighth graders and reported that White youth needed to be introduced to the notion of racism at a structural level and then be required to (a) reflect, (b) develop sympathy and empathy, and (c) negotiate their own Whiteness. Further, Miller and colleagues (2018) and Tanner (2016) utilized theater-based interview and focus-group techniques to explore White youths' racialization of social hierarchies (e.g., castles; Miller et al., 2018) and to facilitate students' location, definition, and articulation of their own Whiteness (Tanner, 2016). Such studies offer extensive details for how youth conceptualize issues pertaining to their own Whiteness and privilege and how they understand racism both systemically and interpersonally. More qualitative work is needed across age groups and settings to accurately understand how White youth develop, understand, and negotiate Whiteness and racism. Specifically, opportunities for qualitative investigations include documenting the ways in which White youth exhibit or are complicit in racial discrimination in their immediate contexts (i.e., homes, schools, religious organizations, neighborhoods), and how White youth express and conceptualize components of Whiteness, such as colorblindness, racial privilege, xenophobia, and White supremacy.

Quantitative Methods

Survey research assessing White youth racial attitudes or behaviors is less common and relies solely on measures validated among college and adult populations. Results reveal evidence of the components of Whiteness among White adolescents: racially discriminatory attitudes, colorblind ideologies, and White privilege (Maxwell & Schulte, 2018; Weinberg, 2006). However, given their creation and validation with adult samples, the utility of these measures is restricted to assessing White adolescents. Further, scholars have suggested that existing measures of Whiteness are missing important themes such as xenophobia, White supremacy, and activism (Schooley et al., 2019). Thus, considerable gaps exist for the types of research questions that can be answered about Whiteness via survey research, especially during middle childhood and early adolescence. To fill these gaps, increased survey development and validation are necessary. Specifically, we emphasize the need for (a) the translation of current adult measures [e.g., the cost of racism for Whites (Spanierman & Heppner, 2004), cognitive dimensions of color-blind racial attitudes (Neville et al., 2000)] and (b) the development of multiple measures to document the longitudinal development of White supremacy over time (e.g., measures to assess attitudes relating to White supremacy across child, youth, and adolescent samples) and identify its correlates. Importantly, this recommendation is secondary to our previous emphasis on qualitative research. Themes and nuances discovered in qualitative examinations should be used to support Whiteness measurement development.

Experimental Methods

The experimental examination of Whiteness in youth may be particularly useful in uncovering White youths' concealed racist attitudes and behaviors (Hagerman, 2016, 2020; Lewis, 2004). Experimental probes of Whiteness in youth largely come from the developmental literature. Such work originates in examining (a) youth preference for racial in-group members and fictitious in-group members (e.g., yellow shirt versus red shirt) among all youth (Brown & Bigler, 2002; McGuire et al., 2019; Mulvey & Killen, 2017) and (b) how all youth identify and understand stereotyping and racism in their environments (Crystal et al., 2008; Wegmann, 2017; Brown, 2006). From this literature, a vein of experimental research with a distinct focus on White youth, only, is necessary, given their racial biases are corroborated by societal supremacy and power. Research by Gal-Szabo (2020) represents the novel implemented of the experimental "street task" with an all-White sample of children. Results revealed that young White children, aged 5–12 years old, placed photographs of White families in houses most proximal to their own and placed BIPOC families in houses further away (Gal-Szabo, 2020). Experimental findings of White youths' racial bias contribute to understanding of the development of racism. We recommend the implementation of experimental tasks and advise the use of realistic scenarios (e.g., a "desk" task where White children assign peers, varying by race, to desks around their classroom). In addition, we encourage that experimental tasks be paired with qualitative probes to uncover White youths' own reasoning for exhibiting racial bias.

Recommendations

We conclude this chapter with specific recommendations for future racial discrimination research on children and adolescents.

Recommendations for Daily Diary Methods

We note the utility of daily diary methods for examining daily occurrences of racial discrimination among BIPOC youth. However, a severely underutilized strength of this method is its capacity for examination of within-person variability. We urge the field to examine intra-individual fluctuation when employing daily diary methods.

Future studies implementing daily diary methods must give renewed investigation to the cumulative effect of multiple forms and perpetrators of racial discrimination (i.e., online discrimination, teacher discrimination) in addition to using multiple assessments within a day. This will allow for an in-depth investigation of fluctuations in racial discrimination and related outcomes that span across time (e.g., morning, night) and contexts (e.g., bus ride to school, local neighborhood store). Researchers should utilize daily diary methods to investigate day-to-day variation in levels of racial discrimination over the weekdays, when youth are more likely to attend school and interact with teachers, in comparison to levels of racial discrimination over the weekend. Daily diary methods must be implemented in longitudinal designs to examine how racial discrimination influences immediate and distal outcomes across critical developmental periods.

Recommendations for Qualitative Methods

We stress that intersectionality must drive the field's use of qualitative methods. BIPOC youths' perceptions of racial discrimination represent the collective experience of multiple intersectional systems of powers. Their accounts are confounded by power structures that disadvantage girls, sexual/gender minorities, and individuals with disabilities. We recommend that researchers address and investigate the systems of power documented in youths' accounts to create a comprehensive understanding of how marginalized identities interact and create nuances among youth who may belong to the same ethnic-racial group yet encounter different forms of oppression. We emphasize ethnographic methods as particularly advantageous for triangulation of data. Specifically, investigations including school administrative staff and teachers, extracurricular activities, and community gathering spaces can be especially informative.

Recommendations for Experimental Methods

We recommend an increase in the use of experimental methods to examine racial discrimination in youth contexts. Both audit and vignette experiments have a particular advantage when examining how discrimination occurs in contexts noted as marginalizing to BIPOC youth. The field should focus on experimentally probing youth experiences within the education system (e.g., teachers, staff, and administration), social and health services (e.g., social workers, therapists, and counselors), and the criminal justice system (e.g., police, correctional treatment specialists). Further, researchers should consider constructing complex fictitious testers, varying both by race and other identities (e.g., sexual orientation, ability) to probe the intersections of racism and other oppressive US systems (e.g., heterosexism, ableism).

Recommendations for Mixed Methods

We highlight how crucial measurement development is for the advancement of research on racial discrimination. To promote measurement development, we recommend increased mixed methods and intersectional research. The field's understanding of racial discrimination should not be limited to deconstructing the harmful implications of racism. Researchers should utilize such experimental methods with survey and qualitative methods to facilitate varied scientific understanding of processes surrounding racial discrimination (i.e., to document racially discriminatory actions and obtain self-reported beliefs). Employing mixed methods, researchers must investigate the multiple systems of power that create racialized experiences for BIPOC youth, to construct valid and reliable measures of the forms and repercussions of racial discrimination.

Recommendations for Research on Whiteness

We stress the importance of examining White youths' development of racist ideologies and internalization of White supremacy. Primarily, we recommend the field's increased use of qualitative (e.g., ethnographies, interviews, focus groups) methods to understand how White youth express and conceptualize the following ideologies: colorblindness, White privilege, xenophobia, White supremacy. Further, we encourage the investigation of White youths' narratives about the ways in which they have perpetrated or witnessed racial discrimination in their immediate contexts. We note the extreme limitations associated with current surveys assessing Whiteness among youth populations and recommend both the translation of current adult measures (e.g., the cost of racism for Whites; Spanierman & Heppner, 2004) and the development of new measures to document the development of White supremacy from childhood to adolescence. We highlight the utility of experimental tasks, especially to assess White children's racial attitudes and biased behaviors toward BIPOC youth. The development and implementation of antiracist interventions is critical in eradicating racism. Importantly, interventions encouraging prosocial behaviors, empathic concern, equality attributions and inter-racial contact, only, are not sufficient. Antiracist interventions must be rooted in understanding the differential power and privilege afforded to Whites and the systemic nature of racism.

Recommendations for Research Capturing Developmental Trends

Throughout these recommendations, we encourage developmental scientists to assess developmental trends in racism-related experiences. Although the empirical research is expansive among adolescent populations, more empirical research is needed among child populations. Black and Latinx youth initially experience racism in utero (Alhusen et al., 2016), and Black children experience racism in preschool settings (US Department of Education, 2014). Yet, we know little about how the quantity and quality of racism-related experiences change throughout early childhood into late adolescence among BIPOC youth. Similarly, we know little about how White supremacy develops throughout early childhood into late adolescence among White youth. Longitudinal research is needed to capture these developmental trends across developmental periods.

References

Alhusen, J. L., Bower, K. M., Epstein, E., & Sharps, P. (2016). Racial discrimination and adverse birth outcomes: An integrative review. *Journal of Midwifery & Women's Health, 61*, 707–720. https://doi.org/10.1111/jmwh.12490

Atkins, R. (2014). Instruments measuring perceived racism/racial discrimination: Review and critique of factor analytic techniques. *International Journal of Health Services, 44*(4), 711–734. https://doi.org/10.2190/HS.44.4.c

Atzmüller, C., & Steiner, P. M. (2010). Experimental vignette studies in survey research. *Methodology, 6*(3), 128–138. https://doi.org/10.1027/1614-2241/a000014

Auspurg, K., Schneck, A., & Hinz, T. (2019). Closed doors everywhere? A meta-analysis of field experiments on ethnic discrimination in rental housing markets. *Journal of Ethnic and Migration Studies, 45*(1), 95–114. https://doi.org/10.1080/1369183X.2018.1489223

Austin, Z., & Sutton, J. (2014). Qualitative research: Getting started. *The Canadian Journal of Hospital Pharmacy, 67*(6), 436. https://doi.org/10.4212/cjhp.v67i6.1406

Bailey, J., Steeves, V., Burkell, J., Shade, L. R., Ruparelia, R., & Regan, P. (2019). Getting at equality: Research methods informed by the lessons of intersectionality. *International Journal of Qualitative Methods, 18*, 1–13. https://doi.org/10.1177/1609406919846753

Benner, A. D., Wang, Y., Shen, Y., Boyle, A. E., Polk, R., & Cheng, Y.-P. (2018). Racial/ethnic discrimination and well-being during adolescence: A meta-analytic review. *American Psychologist, 73*, 855–883. https://doi.org/10.1037/amp0000204

Berkel, C., Murry, V. M., Hurt, T. R., Chen, Y., Brody, G. H., Simons, R. L., Cutrona, C., & Gibbons, F. X. (2009). It takes a village: Protecting rural African American youth in the context of racism. *Journal of Youth and Adolescence, 38*(2), 175–188. https://doi.org/10.1007/s10964-008-9346-z

Bernstein, R. (2011). *Racial innocence: Performing American childhood from slavery to civil rights*. New York University Press.

Brazier, M. (2008). Exploitation and enrichment: The paradox of medical experimentation. *Journal of Medical Ethics, 34*(3), 180–183. https://doi.org/10.1136/jme.2007.020990

Brown, C. S. (2006). Bias at school: Perceptions of racial/ethnic discrimination among Latino and European American children. *Cognitive Development, 21*(4), 401–419. https://doi.org/10.1016/j.cogdev.2006.06.006

Brown, C. S., & Bigler, R. S. (2002). Effects of minority status in the classroom on children's intergroup attitudes. *Journal of Experimental Child Psychology, 83*(2), 77–110. https://doi.org/10.1016/S0022-0965(02)00123-6

Cheeks, B. L., Chavous, T. M., & Sellers, R. M. (2020). A daily examination of African American adolescents' racial discrimination, parental racial socialization, and psychological affect. *Child Development, 91*(6), 2123–2140. https://doi.org/10.1111/cdev.13416

Cho, S., Crenshaw, K. W., & McCall, L. (2013). Toward a field of intersectionality studies: Theory, application and praxis. *Signs: Journal of Women in Culture and Society, 38*(4), 785–810. https://doi.org/10.1086/669608

Córdova, D., & Cervantes, R. C. (2010). Intergroup and within-group perceived discrimination among U.S.-born and foreign-born Latino youth. *Hispanic Journal of Behavioral Sciences, 32*(2), 259–274. https://doi.org/10.1177/0739986310362371

Crystal, D. S., Killen, M., & Ruck, M. (2008). It is who you know that counts: Intergroup contact and judgments about race-based exclusion. *British Journal of Developmental Psychology, 26*(1), 51–70. https://doi.org/10.1348/026151007X198910

Dunbar, M., Mirpuri, S., & Yip, T. (2017). Ethnic/racial discrimination moderates the effect of sleep quality on school engagement across high school. *Cultural Diversity and Ethnic Minority Psychology, 23*(4), 527–540. https://doi.org/10.1037/cdp0000146

Duncan, G. A. (2002). Beyond love: A critical race ethnography of the schooling of adolescent black males. *Equity & Excellence in Education, 35*(2), 131–143. https://doi.org/10.1080/713845286

English, D., Lambert, S. F., Tynes, B. M., Bowleg, L., Zea, M. C., & Howard, L. C. (2020). Daily multidimensional racial discrimination among Black U.S. American adolescents. *Journal of Applied Developmental Psychology, 66*, 101068. https://doi.org/10.1016/j.appdev.2019.101068

Estrada, E., & Hondagneu-Sotelo, P. (2011). Intersectional dignities: Latino immigrant street vendor youth in Los Angeles. *Journal of Contemporary Ethnography, 40*(1), 102–131. https://doi.org/10.1177/0891241610387926

Gaddis, S. M. (Ed.). (2018). *Audit studies: Behind the scenes with theory, method, and nuance* (Vol. 14). https://doi.org/10.1007/978-3-319-71153-9

Gale, A., & Dorsey, M. (2020). Does the context of racial discrimination matter for adolescent school outcomes?: The impact of in-school racial discrimination and general racial discrimination on Black adolescents' outcomes. *Race and Social Problems, 12*(2), 171–185. https://doi.org/10.1007/s12552-020-09286-0

Gal-Szabo, D. (2020). *White parents' color-blind racial ideology and implicit white preference as predictors of children's racial attitudes* (Publication No. 27837213). (Doctoral dissertation, Arizona State University), ProQuest Dissertations Publishing.

García Coll, C., Crnic, K., Lamberty, G., Wasik, B. H., Jenkins, R., Garcia, H. V., & McAdoo, H. P. (1996). An integrative model for the study of developmental competencies in minority children. *Child Development, 67*(5), 1891–1914. https://doi.org/10.2307/1131600

Gibbons, F. X., Etcheverry, P. E., Stock, M. L., Gerrard, M., Weng, C.-Y., Kiviniemi, M., & O'Hara, R. E. (2010). Exploring the link between racial discrimination and substance use: What mediates? What buffers? *Journal of Personality and Social Psychology, 99*(5), 785–801. https://doi.org/10.1037/a0019880

Gibbons, F. X., O'Hara, R. E., Stock, M. L., Gerrard, M., Weng, C.-Y., & Wills, T. A. (2012). The erosive effects of racism: Reduced self-control mediates the relation between perceived racial discrimination and substance use in African American adolescents. *Journal of Personality and Social Psychology, 102*(5), 1089–1104. https://doi.org/10.1037/a0027404

Gibson, P., Haight, W., Cho, M., Nashandi, N. J. C., & Yoon, Y. J. (2019). A mixed methods study of Black Girls' vulnerability to out-of-school suspensions: The intersection of race and gender. *Children and Youth Services Review, 102*, 169–176. https://doi.org/10.1016/j.childyouth.2019.05.011

Giulietti, C., Tonin, M., & Vlassopoulos, M. (2019). Racial discrimination in local public services: A field experiment in the United States. *Journal of the European Economic Association, 17*(1), 165–204. https://doi.org/10.1093/jeea/jvx045

Greene, M. L., Way, N., & Pahl, K. (2006). Trajectories of perceived adult and peer discrimination among Black, Latino, and Asian American adolescents: Patterns and psychological correlates. *Developmental Psychology, 42*, 218–236. https://doi.org/dkwfj8

Hagerman, M. A. (2016). Reproducing and reworking colorblind racial ideology: Acknowledging children's agency in the white habitus. *Sociology of Race and Ethnicity, 2*(1), 58–71. https://doi.org/10.1177/2332649215594817

Hagerman, M. A. (2020). Racial ideology and white youth: From middle childhood to adolescence. *Sociology of Race and Ethnicity, 6*(3), 319–332. https://doi.org/10.1177/2332649219853309

Hardie, J. H., & Tyson, K. (2013). Other people's racism: Race, rednecks, and riots in a southern high school. *Sociology of Education, 86*(1), 83–102. https://doi.org/10.1177/0038040712456554

Harrell, S. P. (2000). A multidimensional conceptualization of racism-related stress: Implications for the well-being of people of color. *American Journal of Orthopsychiatry, 70*, 42–57. https://doi.org/10.1037/h0087722

Heard-Garris, N. J., Cale, M., Camaj, L., Hamati, M. C., & Dominguez, T. P. (2018). Transmitting trauma: A systematic review of vicarious racism and child health. *Social Science & Medicine, 199*, 230–240. https://doi.org/10.1016/j.socscimed.2017.04.018

Heath, S. B. (1982). *Ethnography in education: Defining the essentials.*

Huynh, V. W., & Fuligni, A. J. (2010). Discrimination hurts: The academic, psychological, and physical well-being of adolescents. *Journal of Research on Adolescence, 20*(4), 916–941. https://doi.org/10.1111/j.1532-7795.2010.00670.x

Ispa-Landa, S. (2013). Gender, race, and justifications for group exclusion: Urban Black students bussed to affluent suburban schools. *Sociology of Education, 86*(3), 218–233. https://doi.org/10.1177/0038040712472912

Janssen, J., Seaton, E., Jager, J., & Miller, C. F. (2022). Guidance or gatekeeping: An audit examination of racial discrimination in leading STEM high schools. *Journal of Research on Adolescence, 32*, 625. https://doi.org/10.1111/jora.12722

Johnson, J. D., Ashburn-Nardo, L., & Lecci, L. (2013). Individual differences in discrimination expectations moderate the impact of target stereotypically Black physical features on racism-related responses in Blacks. *Journal of Black Psychology, 39*(6), 560–584. https://doi.org/10.1177/0095798412469229

Johnston-Goodstar, K., & VeLure Roholt, R. (2017). "Our kids aren't dropping out; they're being pushed out": Native American students and racial microaggressions in schools. *Journal of Ethnic & Cultural Diversity in Social Work, 26*(1–2), 30–47. https://doi.org/10.1080/15313204.2016.1263818

Jones, J. M. (1997). *Prejudice and racism* (2nd ed.). McGraw Hill.

Leiter, V., & Herman, S. (2015). Guinea pig kids: Myths or modern Tuskegees? *Sociological Spectrum, 35*(1), 26–45. https://doi.org/10.1080/02732173.2014.978429

Lewis, A. E. (2004). "What group?" studying Whites and Whiteness in the era of "color-blindness". *Sociological Theory, 22*(4), 623–646. https://doi.org/10.1111/j.0735-2751.2004.00237.x

Maxwell, J. A. (2012). The importance of qualitative research for causal explanation in education. *Qualitative Inquiry, 18*(8), 655–661. https://doi.org/10.1177/1077800412452856

Maxwell, A., & Schulte, S. R. (2018). Racial resentment attitudes among White millennial youth: The influence of parents and media*: Racial resentment attitudes among White millennials. *Social Science Quarterly, 99*(3), 1183–1199. https://doi.org/10.1111/ssqu.12488

McGuire, L., Elenbaas, L., Killen, M., & Rutland, A. (2019). The role of in-group norms and group status in children's and adolescents' decisions to rectify resource inequalities. *British Journal of Developmental Psychology, 37*(3), 309–322. https://doi.org/10.1111/bjdp.12274

Miller, E. T., Tanner, S. J., & Murray, T. E. (2018). Castle"ing" Whiteness: White youth and the racial imagination. *Journal of Language and Literacy Education, 14*(2), 1–23. https://eric.ed.gov/?id=EJ1196740

Morgan-Trimmer, S., & Wood, F. (2016). Ethnographic methods for process evaluations of complex health behaviour interventions. *Trials, 17*(1), 232. https://doi.org/10.1186/s13063-016-1340-2

Morris, K. S., Seaton, E. K., Iida, M., & Lindstrom Johnson, S. (2020). Racial discrimination stress, school belonging, and school racial composition on academic attitudes and beliefs among Black youth. *Social Sciences, 9*(11), 191. https://doi.org/10.3390/socsci9110191

Moskowitz, D. S., & Young, S. N. (2006). Ecological momentary assessment: What it is and why it is a method of the future in clinical psychopharmacology. *Journal of Psychiatry & Neuroscience, 31*(1), 13–20.

Mulvey, K. L., & Killen, M. (2017). Children's and adolescents' expectations about challenging unfair group norms. *Journal of Youth and Adolescence, 46*(10), 2241–2253. https://doi.org/10.1007/s10964-017-0671-y

Neville, H. A., Lilly, R. L., Duran, G., Lee, R. M., & Browne, L. (2000). Construction and initial validation of the color-blind racial attitudes scale (CoBRAS). *Journal of Counseling Psychology, 47*(1), 59. https://psycnet.apa.org/doi/10.1037/0022-0167.47.1.59

Pager, D., & Shepherd, H. (2008). The sociology of discrimination: Racial discrimination in employment, housing, credit, and consumer markets. *Annual Review of Sociology, 34*(1), 181–209. https://doi.org/10.1146/annurev.soc.33.040406.131740

Pollock, M. (2004). Race bending: “Mixed” youth practicing strategic racialization in California. *Anthropology Education Quarterly, 35*(1), 30–52. https://doi.org/10.1525/aeq.2004.35.1.30

Raabe, T., & Beelmann, A. (2011). Development of ethnic, racial, and national prejudice in childhood and adolescence: A multinational meta-analysis of age differences: Age differences in prejudice. *Child Development, 82*(6), 1715–1737. https://doi.org/10.1111/j.1467-8624.2011.01668.x

Reichardt, C. S. (2019). *Quasi-experimentation: A guide to design and analysis*. Guilford Press.

Roberts, L., Tamene, M., & Orta, O. R. (2018). The intersectionality of racial and gender discrimination among teens exposed to dating violence. *Ethnicity & Disease, 28*(Suppl 1), 253. https://doi.org/10.18865/ed.28.S1.253

Rogers, L. O., Rosario, R. J., Padilla, D., & Foo, C. (2021). “[I]t’s hard because it’s the cops that are killing us for stupid stuff”: Racial identity in the sociopolitical context of black lives matter. *Developmental Psychology, 57*, 87–1010. https://doi.org/10.1037/dev0001130

Romero, A. J., Gonzalez, H., & Smith, B. A. (2015). Qualitative exploration of adolescent discrimination: Experiences and responses of Mexican-American parents and teens. *Journal of Child and Family Studies, 24*(6), 1531–1543. https://doi.org/10.1007/s10826-014-9957-9

Sangalang, C. C., & Gee, G. C. (2015). Racial discrimination and depressive symptoms among Cambodian American adolescents: The role of gender. *Journal of Community Psychology, 43*(4), 447–465. https://doi.org/10.1002/jcop.21696

Santiago, C. D., Brewer, S. K., Fuller, A. K., Torres, S. A., Papadakis, J. L., & Ros, A. M. (2017). Stress, coping, and mood among Latino adolescents: A daily diary study. *Journal of Research on Adolescence, 27*(3), 566–580. https://doi.org/10.1111/jora.12294

Schooley, R. C., Lee, D. L., & Spanierman, L. B. (2019). Measuring Whiteness: A systematic review of instruments and call to action. *The Counseling Psychologist, 47*(4), 530–565. https://doi.org/10.1177/0011000019883261

Seaton, E. K. & Douglass, S. (2014). School diversity and racial discrimination among African-American adolescents. *Cultural Diversity and Ethnic Minority Psychology, 20*(2) 156–165. https://doi.org/10.1037/a0035322

Seaton, E. K., & Iida, M. (2019). Racial discrimination and racial identity: Daily moderation among Black youth. *American Psychologist, 74*(1), 117–127. https://doi.org/10.1037/amp0000367

Seaton, E. K., & Tyson, K. (2019). The intersection of race and gender among Black American adolescents. *Child Development, 90*(1), 62–70. https://doi.org/10.1111/cdev.13093

Spanierman, L. B., & Heppner, M. J. (2004). Psychosocial Costs of Racism to Whites Scale (PCRW): Construction and initial validation. *Journal of Counseling Psychology, 51*(2), 249. https://doi.org/10.1037/0022-0167.51.2.249

Spencer, M. B., Dupree, D., & Hartmann, T. (1997). A Phenomenological Variant of Ecological Systems Theory (PVEST): A self-organization perspective in context. *Development and Psychopathology, 9*, 817–833. https://doi.org/10.1017/S0954579497001454

Tanner, S. J. (2016). Accounting for whiteness through collaborative fiction. *Research in Drama Education: The Journal of Applied Theatre and Performance, 21*(2), 183–195. https://doi.org/10.1080/13569783.2016.1155404

Thomann, C. R. B., & Suyemoto, K. L. (2018). Developing an antiracist stance: How White youth understand structural racism. *The Journal of Early Adolescence, 38*(6), 745–771. https://doi.org/10.1177/0272431617692443

Tobler, A. L., Maldonado-Molina, M. M., Staras, S. A. S., O’Mara, R. J., Livingston, M. D., & Komro, K. A. (2013). Perceived racial/ethnic discrimination, problem behaviors, and mental health among minority urban youth. *Ethnicity & Health, 18*(4), 337–349. https://doi.org/10.1080/13557858.2012.730609

Tynes, B. M., Del Toro, J., & Lozada, F. T. (2015). An unwelcomed digital visitor in the classroom: The longitudinal impact of online racial discrimination on academic motivation. *School Psychology Review, 44*(4), 407–424. https://doi.org/10.17105/SPR-15-0095.1

U.S. Department of Education Office for Civil Rights. (2014). *Civil rights data collection data snapshot: School discipline* (Issue Brief No. 1). https://www2.ed.gov/about/offices/list/ocr/docs/crdc-discipline-snapshot.pdf

Utsey, S. O. (1998). Assessing the stressful effects of racism: A review of instrumentation. *Journal of Black Psychology, 24*(3), 269–288. https://doi.org/10.1177/00957984980243001

Velez, G., & Spencer, M. B. (2018). Phenomenology and intersectionality: Using PVEST as a frame for adolescent identity formation amid intersecting ecological systems of inequality. In C. E. Santos & R. B. Toomey (Eds.), *Envisioning the integration of an intersectional lens in developmental science. New Directions for Child and Adolescent Development* (Vol. 161, pp. 75–90). https://doi.org/10.1002/cad.20247

Wang, Y., & Yip, T. (2019). Sleep facilitates coping: Moderated mediation of daily sleep, ethnic/racial discrimination, stress responses, and adolescent well-being. *Child Development, 91*(4), e833–e852. https://doi.org/10.1111/cdev.13324

Wegmann, K. M. (2017). "His skin doesn't match what he wants to do": Children's perceptions of stereotype threat. *American Journal of Orthopsychiatry, 87*(6), 615–625. https://doi.org/10.1037/ort0000238

Weinberg, L. B. (2006). *An investigation of the correlates of white racial identity attitudes among white adolescents in a multiracial urban environment*. New York University. https://search.proquest.com/openview/98bec00ae5aa59e71ca1861547a8a4c4/1?pq-origsite=gscholar&cbl=18750&diss=y

Williams, D. R., Neighbors, H. W., & Jackson, J. S. (2003). Racial/ethnic discrimination and health: Findings from community studies. *American Journal of Public Health, 93*, 200–208. https://doi.org/10.2105/AJPH.93.2.200

Yip, T., Cheon, Y. M., Wang, Y., Cham, H., Tryon, W., & El-Sheikh, M. (2020). Racial disparities in sleep: Associations with discrimination among ethnic/racial minority adolescents. *Child Development, 91*(3), 914–931. https://doi.org/10.1111/cdev.13234

Yoon, E., Adams, K., Clawson, A., Chang, H., Surya, S., & Jérémie-Brink, G. (2017). East Asian adolescents' ethnic identity development and cultural integration: A qualitative investigation. *Journal of Counseling Psychology, 64*(1), 65–79. https://doi.org/10.1037/cou0000181

Zeiders, K. H. (2017). Discrimination, daily stress, sleep, and Mexican-origin adolescents' internalizing symptoms. *Cultural Diversity and Ethnic Minority Psychology, 23*(4), 570–575. https://doi.org/10.1037/cdp0000159

Chapter 3
Canaries and Bellwethers: What Can We Learn About Racial Justice from Studying Ethnic-Racial Identity Within and Across Groups?

Deborah Rivas-Drake, Jessica Montoro, and Abunya Agi

In popular culture and in developmental science, we hear a lot about the changing US demographics, especially among the youth population. We know that the changing landscape of people means we can look forward to a future in which the youngest, and most diverse, segment of our society will be parents, neighbors, employees, and voters of a place foreign to many of us (adults) (Rivas-Drake & Umaña-Taylor, 2019). As the title of this volume suggests, well-known projections show that in just a little over two decades, we can expect the White population to decline with the concurrent emergence of a majority composed of Black, Latinx, Asian, and multiracial youth (Colby & Ortman, 2015). Although significant, these changes are not the whole, or even most important part, of the history we are living.

An exclusive focus on the narrative of the future of the United States that highlights our growing diversity unquestionably risks overlooking the clear warning signs of racism in the air we breathe every day, that is evident in the daily lives of marginalized youth of color, and that will not simply disappear with greater racial diversity (Tatum, 2017). The very same youth who are contributing to the new emerging majority of the United States are growing up in a world that exposes them to racism time and again over the course of their lives. Indeed, their exposure to racism begins very early on when they are subjected to disciplinary disproportionalities in the early childhood classroom (U.S. Department of Education Office of Civil Rights, 2014), policies and practices meant to reject who they are (e.g., English only policies) and even their bodies as such (e.g., hair policies in schools and sports), and

For inclusion in the following volume: D. P. Witherspoon & G. L. Stein, (Eds.), *Conceptualizing and measuring culture, context, race and ethnicity: A focus on science, ethics, and collaboration in the spirit of 2044.*

D. Rivas-Drake (✉) · J. Montoro · A. Agi
University of Michigan, Ann Arbor, MI, USA
e-mail: drivas@umich.edu

© The Author(s), under exclusive license to Springer Nature Switzerland AG 2023

D. P. Witherspoon, G. L. Stein (eds.), *Diversity and Developmental Science*,
https://doi.org/10.1007/978-3-031-23163-6_3

expressions of political precarity and threats meant to inflict psychological and even physical harm (Unite the Right rally; Trump tweets; COVID-related racism; e.g., Ruiz et al., 2020). No town or city is exempt from or immune to these realities, and the global protests in support of Black Lives Matter suggest that young people are growing up keenly aware of them (Rogers et al., 2020b). One way to attend to the racism with which young people are growing up, as a field, is to intentionally orient ourselves to the pursuit of racial justice, or actions that foster the equality of status, opportunity, health, civic life, and well-being of racially marginalized people while recognizing their inherent humanity and worth.

In this chapter, we argue that the study of ethnic-racial identity (ERI) in youth can benefit from an explicit orientation to racial justice. ERI involves the beliefs and feelings young people have about their ethnic-racial group, their understanding of the role of race and ethnicity in their lives, and the processes by which they arrive at such beliefs, feelings, and understanding (Sellers et al., 1998; Umaña-Taylor et al., 2014). Supporting ERI development among Black, Indigenous, or other People of Color (BIPOC) is a way to recognize their worth by encouraging a sense of connection to their ethnic-racial group and rejecting the erasure of their experiences and histories. We can consider how the harms of racism may interfere with ERI development; in this way, we might use ERI as a "canary-in-the-coalmine" gauge of exposure to the toxicity of racism. Under certain conditions, moreover, ERI is also an aspect of youths' social development that may help them understand and work toward racial justice (Mathews et al., 2019). In this sense, ERI can also be used as a "bellwether" to gauge those conditions—that is, policies and practices—that aim to or are the result of social change. Thus, to the extent that ERI can be leveraged to support the equal status, opportunity, health, civic life, and well-being of BIPOC, specifically, then it can be a lever for racial justice. This is especially the case as certain aspects of ERI are associated with better mental health, well-being, and greater engagement in civic life among these communities.

A growing body of literature suggests ERI promotes positive development across youth of color, and protects these youth from negative environmental forces, such as discrimination (Neblett et al., 2012; Rivas-Drake et al., 2014b; Yip et al., 2019). Reviews of the literature have found that youth of color with a strong and positive connection to their ethnic-racial group report better academic adjustment (i.e., achievement and attitudes), better psychosocial functioning, and less negative adjustment (e.g., fewer internalizing and externalizing symptoms) compared to youth of color with a weaker or more negative connection (Miller-Cotto & Byrnes, 2016; Neblett et al., 2012; Rivas-Drake et al., 2014b). When ERI is assessed as a single composite of the Multi-Ethnic Identity Measure (MEIM) that includes exploration, achievement, positive affect, and/or group belonging, it is related to higher academic achievement and mitigates the negative role of discrimination on youth's mental health (Miller-Cotto & Byrnes, 2016; Yip et al., 2019). Although not the focus of the literature to date, an emerging body of work also suggests that ERI is positively linked to civic and political engagement and dispositions among diverse youth of color (Bañales et al., 2020; Fish et al., 2021; Hope et al., 2019; Pinetta et al., 2020; Wray-Lake et al., 2017).

The following sections proceed as follows. First, we illuminate the affordances for advancing racial justice provided by commonly used ERI theory and research with youth populations. In particular, we consider ERI as a psychological, social/contextual, and temporal phenomenon in order to describe facets that are common and those that are unique across groups. We then discuss how different measures have attempted to capture the psychological, social/contextual, and temporal spaces inhabited by youth across various groups, and we highlight conceptual issues that arise in ERI measurement in single- and multi-group studies. Following this discussion, we highlight findings that indicate for whom and how ERI can be leveraged to support the academic adjustment, psychological health, and civic engagement of racially marginalized youth. We conclude by providing recommendations for how to attend to these complexities when integrating ERI concepts into new and ongoing empirical studies.

Affordances of Common Approaches to Conceptualizing ERI in Studies of Youth

In the United States, the developmental science of ERI was founded on models that focus on the experiences of particular ethnic-racial groups, especially Black Americans (e.g., Cross, 1991), but it has been *popularized* with the study of this phenomenon across multiple groups (e.g., Phinney, 1990). That is, a developmental perspective on ERI did not originate with Phinney's work; scholars such as Cross (1991) theorized about the nature and quality of change in racial identity among Black people, primarily adults. However, Phinney's conceptualizations (1989, 1990, 1992; Phinney & Ong, 2007) of ethnic identity form the basis of the majority of research on ERI in *youth* (see Rivas-Drake et al., 2014a; Yip et al., 2019). This perspective advanced the idea that ethnic identity is a part of the normative development of any youth who are ethnic-racial minorities in the United States due to their marginalization (Phinney, 1990). She also conceptualized the developmental aspect by emphasizing the universal importance of arriving at an ERI through one's own exploration and meaning-making (for a fuller overview of this approach, see Rivas-Drake & Umaña-Taylor, 2019). Importantly, Phinney's conceptualization—and later elaborations by scholars such as Umaña-Taylor et al. (2004)—has served to advance the phenomenon of ERI as a worthy subject of scholarly inquiry in mainstream developmental science and to highlight a key ethnic-racial asset in child development.[1] Thus, this work sets a foundation for scholars of ERI to claim "air-time" in the developmental science of diverse ethnic-racial minority children—figuratively and literally, in terms of the presence of this research in our peer-reviewed journals.

[1] See also Sellers et al. (1998) for a discussion of *mainstream* versus *underground* approaches to racial identity at that time.

A group-specific approach, by contrast, is grounded in the unique experiences of particular ethnic-racial groups. This approach has operated to advance racial justice in multiple ways. First, group-specific models introduce ideas and phenomena that attend explicitly and consciously to the reality that such groups must contend with racism at the societal and interpersonal levels in daily life, along with the recognition of individual variability in the extent to which they are aware of these issues (Sellers et al., 1998). Further, from this perspective, ERI development not only occurs against the backdrop of such racism but also has unique elements rooted in the sociopolitical and historical contexts of a particular ethnic or racial group. In addition, some of the rationale for the development of these models involve a conscious rejection of a color-evasive and racism-avoidant approach to the psychological study of particular groups. For example, Cross (1991) and Sellers et al. (1997) proffered conceptualizations of racial identity that sought to *center* the experiences of diverse Black or African American individuals (i.e., adults) as well as the value of theorizing about those experiences due to the inherent worth and dignity of this group, in particular. These perspectives have also challenged the idea that there is a singular "optimal" racial identity, as different understandings of racial identity could be adaptive depending on the realities of individuals' proximal contexts.

Yet, over time, aspects of Sellers et al.'s (1997, 1998) Multidimensional Model of Racial Identity (MMRI) have become widely used in developmental science of ERI focused on *youth*, generally, and among non-African American youth populations, in particular (see Rivas-Drake et al., 2014a; Yip et al., 2019). For this reason, and for the purpose of this chapter and volume, it is important to clarify here as has been done elsewhere (e.g., Schwartz et al., 2014) that a subset of components of the MMRI has been applied to the study of ERI with diverse populations. The justification for this has been that, with the exception of the ideology dimension—which refers to individuals' beliefs about how other group members should think and behave (Sellers et al., 1998)—the dimensions of salience, centrality, and regard articulated in the MMRI were drawn from theories that were grounded in universal experiences and transcended the unique experiences of African Americans (e.g., social identity theory; see also Schwartz et al., 2014 for additional discussion). Thus, as discussed further below, these dimensions have been studied in diverse populations, ranging from Chinese American college students (e.g., salience; Yip, 2005) to Dominican sixth graders (e.g., regard; Rivas-Drake et al., 2009). This translational use of MMRI concepts has allowed for the representation of ERI content among youth of color in ways that do not necessarily imply a single optimal identity, which is consistent in principle with the original model.

ERI As a Psychological Place, In Context, and Over Time

To describe how ERI may operate across groups, it may be instructive to consider that at a broader level, beyond the perspective reflected in any given model, we can conceptualize ERI as a psychological (i.e., what Cross et al. (2017) describe as the "being, feeling, and knowing" aspect), contextual (i.e., shaped by the environment) (Cross et al., 2017; Verkuyten, 2016), and temporal, or time-dependent, phenomenon (Umaña-Taylor et al., 2014). As scholars have noted, an individual's ERI is an ever-changing psychological place that an individual inhabits, carries with them, and is continuously influenced by the environment (Cross et al., 2017; Verkuyten, 2016). A number of theoretical frameworks—for example, social identity theory (Verkuyten, 2016), Boykin's (1983) Triple Quandary framework, Berry's (1997) acculturation model and Cross et al.'s (2017) identity enactments model—have argued that social milieus position individuals so they must negotiate internal (intrapersonal) and external (interpersonal/intergroup) facets of their ERI. For example, Boykin's (1983) Triple Quandary framework delineated how racially marginalized people must maneuver through three different cultural realms: the mainstream, the oppressed minority, and the ethnic. Berry's (1997) acculturation model similarly theorizes that immigrant or immigrant-origin individuals must navigate their ERI across multiple realms including the non-dominant as well as the mainstream and dominant. Navigating both the mainstream and dominant culture has distinct implications for youths' ERI. For example, Latinx adolescents who experienced more acculturation-related conflict with their parents reported less ERI *private regard*, or perceived their ethnic-racial group less positively, when faced with discrimination from peers (Huq et al., 2016). Indeed, the strategies of biculturalism and *code-switching*, both of which involve oscillating between multiple languages or modes of identity expression, exemplify how people regulate their ERI across various cultural spheres (e.g., Birnie-Smith, 2016; Cross et al., 2017). Cross et al.' (2017) ERI enactments model draws from all of the aforementioned perspectives to provide an integrative framework of how ERI is enacted across different *intra*- and *inter*group contexts. Findings from Birnie-Smith (2016) illustrate the context-dependent nature of ERI expression. Chinese Indonesian young adults altered their language to differentially express or de-emphasize aspects of their ERI on distinct social network sites (i.e., posted in English or in Indonesian). More specifically, their language use was contingent on what social factors they encountered on those social network sites, such as the ERIs of their friends on that social network (Birnie-Smith, 2016). Taken together, these perspectives thus illuminate that ERI resides not only in the intrapsychological affect and meaning one ascribes to one's ethnic-racial group membership but also the places one inhabits in response to dominant societal views and expectations for members of that group relative to others.

Further, youth make meaning of their ethnicity and race amidst landscapes marked by psychological, symbolic, and physical borders. The complex placemaking involved in navigating one's ERI amidst opposing cultural realms or "borders" is illustrated by Anzaldua's (1999) *Borderlands.* Anzaldua captures the stigmatizing experiences of inhabiting and shifting between the physical borderlands or *fronteras* of Texas and Mexico as "swimming" in the often contradicting psychological and linguistic landscapes of American and Mexican culture (Anzaldua, 1999). Wilson et al. (2014) expand on Anzaldua's *fronteras* and describe the educational borderlands traversed by Latinx immigrant youth as "physical and/or contextual landscapes where one must negotiate notions of cultural difference" (Wilson et al., 2014,

p. 3). For some ethnic-minority youth, borderlands manifest in the disconnect between lived experiences at home and in school contexts. For example, scholars have found that Native American youth must traverse sociolinguistic borderlands as they negotiate their Indigenous heritage, multiple languages, and exclusionary school environments that oppress and erase their Native American upbringings (McCarty, 2014). Montoya (2020) notes how transformative educators can openly discuss dynamic issues such as racism, xenophobia, legality, and current immigration policy to encourage Latinx youth to embody, embrace, and even challenge their *borderlands identity* in the classroom. Moreover, the ways in which specific ethnic-racial groups are allowed to navigate such obstacles vary greatly and are context-dependent.

For youth of color, the xenophobic, racist, and oppressive sociopolitical climate that characterizes their experiences determines both the literal and metaphorical boundaries they are encouraged or allowed to cross (Huq et al., 2016; Seaton et al., 2009). Restrictive anti-immigrant and xenophobic US policies enforced by the Trump administration inhibit the physical borders immigrants can cross and have led to thousands of forced family separations, resulting in detrimental consequences for migrant families' and youth's well-being (Dreby, 2015; Enchautegui & Menjívar, 2015). Furthermore, historical events that are widely understood as racialized, such as endemic anti-Black police brutality (Rogers et al., 2020b), the 2008 Presidential election (Fuller-Rowell et al., 2011), and the 2016 Presidential election (Wray-Lake et al., 2018) can contribute to shifts in ERI among children and adolescents. Youth must thus make sense of what it means to be part of their ethnic-racial groups amidst the ever-present boundaries of racial stratification, discriminatory stereotypes, and oppression (Yip, 2018; Seaton et al., 2009). Indeed, throughout the Trump administration's implementation of anti-immigrant policy and xenophobic rhetoric, Latinx youth described having low ERI *public regard*, or more negative views of others' perceptions of their ethnic-racial group, as well as fear and anxiety around their ethnic-racial group affiliation (Wray-Lake et al., 2018). Ultimately, identifying the societal and psychological "borders" experienced by various ethnic-racial groups can aid scholars in understanding the shared, distinct, and dynamic racialized realities across groups.

A Temporal Phenomenon

ERI is not a collection of individual static characteristics but rather consists of constellations of content (i.e., significance and meaning) obtained or constructed through processes that unfold over time (Umaña-Taylor et al., 2014; Wang et al., 2017). The significance and meaning of ERI can encompass beliefs and attitudes such as how positively one feels about one's ethnic-racial group (i.e., private regard). The processes by which ERI develops, by contrast, underscore that it is a temporal phenomenon that unfolds and fluctuates over time (Rivas-Drake et al. 2014a, b; Wang et al., 2017). Age-dependent factors such as cognitive abilities and environmental factors such as historical events can inform when, how, and what aspects of ERI develop throughout the lifespan (Williams et al., 2020). Through a developmental lens, how a person identifies with their ethnic-racial group in early childhood differs drastically from how they do so in adolescence, middle adulthood, and so on (Rogers et al., 2020a; Williams et al., 2020). Whereas the extent of an infant's ERI may be a visual preference for members of their ethnic-racial group (Liu et al., 2015), an adolescent's ERI involves more agency, such as actively exploring activities and traditions valued within their ethnic-racial group (Rivas-Drake et al., 2014a; Umaña-Taylor et al., 2014). Even more, during middle childhood, youth have early understandings of social hierarchies across racial groups (e.g., Rogers et al., 2012). For example, an ethnically and racially diverse sample of second and fourth graders described more perceptual aspects of ERI such as language or physical appearance, as well as more complex and less observable dimensions such as culture and relative social position (Rogers et al., 2012).

Although the field has predominantly relied on cross-sectional, retrospective self-reports of ERI dimensions and development (Rivas-Drake et al., 2014a), scholars have looked to daily diary and longitudinal methods to capture daily fluctuations and trajectories of ERI (Wang et al., 2017). Chinese American adolescents' daily involvement in ethnic behaviors such as speaking Chinese or celebrating a Chinese holiday was positively associated with ethnic salience, regardless of their overall ERI (Yip & Fuligni, 2002). Other work shows that Black youth's daily decreases in public regard exacerbated depressive symptoms linked to previous-day experiences of racial discrimination (Seaton & Iida, 2019). Longitudinal studies have also captured the time-varying nature of ERI. Findings from Douglass and Umaña-Taylor (2016) suggest the positive association between Latinx adolescents' family ethnic socialization and ERI exploration strengthen from middle-to-late adolescence, whereas the positive link between family ethnic socialization and ERI resolution weakens. Other findings from Wang et al. (2017) suggest that over time, adolescents' ERI commitment predicted greater daily ethnic-racial salience, whereas greater variability in daily ethnic-racial salience prospectively predicted greater ERI exploration.

However, the distinction between environmental influences and the time-dependent nature of ERI is not always clear. Environmental influences such as the current ethnic-racial makeup of peers (Douglass et al., 2017), ethnic language proficiency (Phinney & Ong, 2007), exposure to ethnic-racial discrimination (Yip, 2018) and interracial contact (Wang & Yip, 2020), as well as the diversity of one's school environment can contribute to shifts in ERI over short periods of time. For instance, Yip (2005) found that among Chinese American college students, ethnic composition, predominant spoken language, and presence of family in their immediate environment were associated with higher ethnic centrality. In this same sample, when students were in the presence of family, increased ERI centrality was associated with greater ethnic salience (Yip, 2005). Kiang and Fuligni (2009) found similar contextual variations among a racially and ethnically diverse sample of young adults. ERI belonging and exploration were the highest when youth were in the presence of parents, decreased when surrounded by co-ethnic peers, and were at their lowest with different-ethnic peers (Kiang & Fuligni, 2009). Compared to their Asian American, Filipino American, and White counterparts, Latinx youth showed the highest levels of ethnic exploration and belonging with family (Kiang & Fuligni, 2009). Ultimately, acknowledging ERI as a temporal phenomenon also requires that we recognize the role of context in shaping youth's ERI.

A Social Contextual Phenomenon

ERI is a dynamic phenomenon influenced by social contexts. Indeed, much of the situational variations in ERI have been demonstrated across various ethnic-racial groups and throughout the lifespan (Yip & Fuligni, 2002). As posited by social identity theory, we conceptualize ERI as a social, contextual phenomenon in that the salience of one's identity varies based on that ethnic-racial group's representation and status in any given context (e.g., Syed et al., 2018; Verkuyten, 2016; Yip, 2005). Via an ecological/transactional lens, environmental changes within *microsystems*, or settings most proximal to youth, are often associated with ethnic-racial identity shifts among youth (Bronfenbrenner, 1979; Hughes et al., 2016). To fully capture the complexity of ERI, Syed et al. (2018) emphasize the need to assess distinct environmental characteristics within ethnic-racial settings: *objective/subjective perspective*, *differentiation*, *heterogeneity*, and *proximity*. Measuring objective perspectives of youth's environments such as the ethnic-racial composition of neighborhoods, schools, and peer networks is important in assessing their opportunities to interact with members of other ethnic-racial groups (Syed et al., 2018). The ethnic-racial differentiation and heterogeneity aspects of the environment also matter for youth ERI. *Differentiation*, or how groups are defined within a setting, as well as *heterogeneity*, or the ethnic-racial diversity of an environment, informs how youth make sense of their ethnic-racial group membership (Syed et al., 2018).

Black youth who live in primarily Black neighborhoods report fewer instances of racial discrimination and greater racial centrality, or a stronger sense of their ethnic-racial group membership (Stevenson & Arrington, 2009). For BIPOC youth, living in a predominantly White neighborhood may result in greater exposure to discrimination, racial profiling and racial microaggressions, all of which inform youths' ERI (Douglass & Umaña-Taylor, 2016; Stewart et al., 2009). Experiences of racial discrimination also relate to how youth of color perceive their own ethnic-racial groups. Perceptions of racial discrimination have been negatively associated with perceptions of group status and opportunity among Latinx (e.g., Constante et al., 2021), African American (Seaton et al., 2009), and diverse youth of color (Del Toro et al., 2020). Further, Latinx adolescents who resided in neighborhoods with higher percentages of Latinx neighbors showed greater ERI affirmation, or positive feelings about their ethnic-racial group (Supple et al., 2006). For multiracial youth, the environment in which they are embedded informs their racial identity development

(Csizmadia et al., 2012). Multiracial youth who identified as both White and Latinx identified more often as White when in a predominantly White social setting (Herman, 2008).

Proximity, or how close a person is to another person within a setting, also informs youth ERI. Friend and peer networks capture youths' proximity to different ethnic-racial groups and have implications for ERI formation (Rivas-Drake et al., 2017; Santos et al., 2017). For example, Kiang and Johnson (2013) found that Asian adolescents' ethnic identity labels when surrounded by Asian peers differed from when they were surrounded by non-Asian peers. Adolescents who reported a greater proportion of same ethnic-racial friends also felt their ethnicity/race was important to their identity 6 months later (Douglass et al., 2017). However, Douglass et al. (2017) also found that this association was inverse for adolescents who reported fewer same ethnic-racial peers at school, further underscoring the context-dependent nature of ERI.

Importantly, youths' subjective perspective of ethnic-racial settings can shed light on their understanding of racial hierarchies and power disparities between groups (Syed et al., 2018). For example, ethnic-racial minority youth report more perceived racial injustice within the criminal system compared to White youth (Hagan et al., 2005). Moreover, the filters through which youth interpret their experiences of racial marginalization are also continuously changing, as Seider et al. (2019) have shown that Black and Latinx adolescents' awareness of the links between structural factors and racism increase over time. An awareness of one's marginalized social position informs how youth learn about and relate to their ethnic-racial group. For example, stereotypes of Black racial identities as criminal, dangerous, unprofessional, or damaging are examples of stressors Black youth contend with on a daily basis (Adams-Bass et al., 2014; Zirkel & Johnson, 2016). From early childhood, Black youth receive more parental messages that prepare them for navigating and coping with negative stereotypes and racial bias (Neblett et al., 2009). Preparation for bias messages such as these has been linked to great exploration of one's ethnic-racial group membership (Hughes & Johnson, 2001).

Further, understanding that ERI is a contextual/societal phenomenon also requires we conceptualize it as a bidirectional one (Rogers et al., 2020a; Yip, 2018). Just as the environment shapes ERI, ERI can provide a lens through which youth acknowledge and navigate social inequities, racial injustice, and ultimately change their environment (Aldana et al., 2012; Rogers et al., 2020b; Szymanski & Lewis, 2015). Youths' experiences of oppression and racial marginalization inform their ERI, motivate them to join social movements, and contribute to their critical consciousness (Fish et al., 2021; Mathews et al., 2019). Other work by Szymanki and Lewis (2015) and Hope et al. (2019, 2020) suggests associations between dimensions of Black youths' racial identity and their activism. Taken together, understanding the reciprocal relationship between youth ERI and the environment is paramount in understanding ERI as a tool for racial justice. Next, we discuss common conceptual and measurement issues that arise when studying ERI within and across various ethnic-racial groups.

Measurement Issues: Single- Versus Multiple-Group Research

Whether to examine ERI in a specific ethnic-racial group or across multiple ethnic-racial groups depends on the research question under investigation (Schwartz et al., 2014). Given that ERI is a context-driven phenomenon, single-group research addresses a particular group's sociocultural history, which may not be applicable to other groups. Empirical research with ethnically and racially diverse samples supports the notion that aspects of ERI content are group specific (Hughes et al., 2017; Phinney & Ong, 2007). For example, Black, Dominican and Chinese adolescents reported greater ERI exploration than White youth, which parallels prior research that shows BIPOC youth navigate ethnic-racial experiences more frequently than non-BIPOC youth (Csizmadia et al., 2012; Hughes et al., 2017). Black and Dominican youth also reported higher ERI commitment compared to White adolescents. Additionally, Black youth reported the lowest public regard, whereas White youth reported the highest public regard (Hughes et al., 2017). These findings reflect theories of Black racial identity that have focused on Black people's status and awareness of their stigmatization within the U.S and the identities that form within this context (Sellers et al., 1998). The Multidimensional Inventory of Black Identity (MIBI) is a measure that captures these beliefs, in addition to various ideologies specific to African Americans (Sellers et al., 1997). Additional measures, such as Cross's Racial Identity Scale (Vandiver et al., 2002), capture not only ERI content specific to African Americans but also developmental processes specific to African Americans, and thus are not appropriate for use with other groups.

However, Schwartz et al. (2014) note limitations with the single-group approach—namely, that defining group boundaries is messy. They argue it is not always clear where to draw the line of what constitutes a group. For instance, some studies may examine Asian Americans as a panethnic category and others may examine Korean Americans more specifically. Thus, racial and ethnic labels are important when creating and selecting measures that define particular groups of individuals. Among college students of African origin, those who used the self-label Black also reported a less Afrocentric racial identity (Anglin & Whaley, 2006). Students who self-labeled as African American also reported more assimilated racial identities, whereas those who self-labeled as West Indian also endorsed more negative stereotypes of African-origin people (Anglin & Whaley, 2006). Further, due to histories of migration, a person may identify as belonging to multiple, overlapping ethnic-racial groups. Although some groups classified as a singular group, such as Latinx, are often considered one homogenous group, they are in fact racially diverse, with wide variations in European, Native American, and African ancestry (Bryc et al., 2015; Umaña-Taylor & Fine, 2001). Such variation within the Latinx population has been shown to have implications for ERI content and development (Itzigsohn & Dore-Cabral, 2000). Among a sample of Latinx adolescents, the MEIM has shown the strongest internal reliability for Puerto Rican and Nicaraguan adolescents, and the lowest internal reliability for Guatemalan and Honduran youth (Umaña-Taylor & Fine, 2001).

Further, some youth have a parent who does not share the same ethnicity or race as the other parent, and thus might identify as biracial or multiracial. Multiracial BIPOC youth remain an understudied group in the field and questions remain as to the best way to categorize them and capture their experiences (Rockquemore et al., 2009; Shih & Sanchez, 2009). Rockquemore et al. (2009) have suggested that in order to fully understand multiracial youth's ERI development, scholars must differentiate between their *racial category* (the racial identity chosen in specific environments), their *racial identity* (their own understanding of their race) and their *racial identification* (how others group them). Consequently, the limited research on multiracial youth supports the notion that the extent to which they identify with a racial group is context-dependent (Renn, 2000). Importantly, considering where these constructs are similar or overlap, and where they differ is crucial in understanding the multiracial experience of BIPOC youth (Rockquemore et al., 2009; see also Abigail et al., Chap. 7 in this volume).

Additionally, ethnic-racial experiences may vary within singular groups due to geographic region and immigration status (Agi & Rivas-Drake, 2021; Wiley et al., 2008). As noted by Schwartz et al., (2014), "groups have collective histories that influence what it means to identify with the group" (p. 63), potentially resulting in different meanings attached to one's ERI, both across panethnic categories and within. One study showed that Dominican immigrants predominantly self-identified as Black when in the Dominican Republic and as Latinx or Hispanic when in the U.S (Itzigsohn & Dore-Cabral, 2000). Among a separate group of Black immigrant college students of West-Indian origin, second-generation immigrants showed less favorable ERI public regard when compared to first-generation immigrants (Wiley et al., 2008). Contrastingly, findings from the same study showed that second-generation Latinx students showed higher public regard when compared to first-generation students (Wiley et al., 2008). Further, among White first-generation immigrants, ERI importance was linked to both private and public regard, but unrelated to public regard among second-generation immigrants (Wiley et al., 2008). The generational and regional differences found in these studies underscore the variation within singular groups.

Rather than a single-group approach, researchers might examine ERI content and processes across multiple groups. Whereas research with African American samples tends to focus on racial identity and research with Latinx or Asian American samples tends to focus on ethnic identity, previous researchers have argued these groups negotiate both ethnicity and race in their lived experiences (Johnston-Guerrero, 2016; Umaña-Taylor et al., 2014). The multiple-groups approach allows researchers to examine aspects of ERI that are theoretically universal and make cross-group comparisons, contributing to our general understanding of ERI (Phinney & Ong, 2007). For instance, a large body of literature has found both universal processes of developing one's ERI (e.g., exploration, resolution), and specific ERI content dimensions of meaning and importance (e.g., centrality, private regard, public regard) that apply across ethnic-racial groups, including Black, Asian, and Latinx youth (see meta-analysis by Miller-Cotto & Byrnes, 2016).

Examining multiple groups in a single study comes with its own challenges. Namely, the extent to which survey measures function similarly across ethnic-racial groups varies. Some studies have found the full and brief versions of the Ethnic Identity Scale (EIS), which are composed of exploration, resolution, and affirmation, demonstrate construct validity across ethnically and racially diverse adolescents and college students (Douglass & Umaña-Taylor, 2015; Umaña-Taylor et al., 2004). Other studies have found theoretically universal aspects of the MIBI, such as centrality and public regard, are not fully invariant across groups, with item means varying systematically across ethnic-racial groups as well as geographic regions (Sladek et al., 2020). Given different groups have different stereotypes applied to them, public regard measures may be particularly sensitive to group context. Indeed, the item means of public regard are systematically lower for Black American adolescents compared to Asian American adolescents, and Asian American adolescents vary depending on geographic region (Sladek et al., 2020). Public regard measures the extent to which a person believes other people view their own group positively or negatively, including perceptions of intelligence (Scottham et al., 2008). With longstanding stereotypes that paint people of African-descent as being less intelligent than others, and people of Asian-descent as being more intelligent than others, current ways of measuring public regard may not necessarily capture Asian American experiences (McGee & Martin, 2011; Thompson et al., 2016). For instance, when youth are asked to report whether they believe other people think their ethnic group is as smart as people from other ethnic groups, the prevailing stereotype that Asians are smarter than other groups (Thompson et al., 2016) cannot be captured by either agreeing or disagreeing with the prompt. Ultimately, researchers must ensure their research questions and measures are theoretically driven, whether they opt for a single-group approach or a multiple groups approach.

Recommendations for Research Seeking to Incorporate ERI

In this section, we offer recommendations for scholars seeking to incorporate ERI in their research. Our first recommendation is for researchers to be clear about their epistemological orientation. When studying a phenomenon with underlying ERI questions, scholars should identify whether the aim is to a) shed light on unique experiences of single groups situated within particular sociopolitical, historical, and geographical contexts or b) capture more generalized perspectives of multiple groups. Establishing whether the approach is a single-group or rather a multiple-group approach can distinguish whether a particular ethnic-racial group is granted or denied "airtime" within the developmental science literature (Schwartz et al., 2014). Furthermore, particularly if the aim is to center the ERI experiences of a marginalized group, scholars should acknowledge the backdrop of racism, as well as the sociopolitical and historical landscapes in which ERI develops and fluctuates (Rogers et al., 2020b; Zepeda-Millán & Wallace, 2013). Without doing so, researchers risk dismissing or inaccurately conveying the daily realities and narratives of entire groups that are often overlooked, understudied, and underestimated within developmental research.

A related question to ask is, what is the conceptual role of ERI in the phenomenon being studied? How central is ERI to the primary research objectives? Considering the scope of focus on ERI as well as the level of nuance can inform scholars on how to best empirically measure and examine the construct (Rivas-Drake et al., 2014a). As scholars have previously shown, ethnic-racial experiences vary within and across ethnic-racial groups (Csizmada et al., 2012; Del Toro et al., 2020; Seaton et al., 2009; Stevenson & Arrington, 2009). Further research should seek to understand the distinct influences of specific ERI dimensions, rather than relying on composite ERI measures (Rivas-Drake et al., 2014a). When possible, indicating measurement invariance both between and within groups is essential to ensure instruments are theoretically and psychometrically sound for both the population and phenomenon of interest (Douglass & Umaña-Taylor, 2015; McGee & Martin, 2011; Sladek et al., 2020).

As with any measurement tool, existing ERI measures such as the MIBI and the EIS have both strengths and limitations (Phinney & Ong, 2007; Schwartz et al., 2014) that scholars must be aware of prior to designing their studies. Whereas the EIS captures both ERI development (i.e., exploration) and content (i.e., affirmation) and is grounded in a universal, multigroup approach, the MIBI was designed to capture only ERI content, and in the original version, beliefs and ideologies that would be pertinent to African Americans. Despite the EIS capturing both ERI process and content, however, scholars have often focused on the process component of the scale (Schwartz et al., 2014), and the affirmation items are also all negatively worded, which may not accurately capture youths' positive feelings about their group (Meca et al., 2021). A shared limitation of both the MIBI and the EIS is that they have not been substantively updated or revised to potentially reflect more contemporary aspects of process and content that may be relevant to youths' ERI (e.g., intersectionality).

In that vein, authors taking a single-group approach should be mindful of what constitutes borders and boundaries when defining groups. When selecting and creating measures, ethnic and panethnic categories must be considered carefully (Anglin & Whaley, 2006; Yoshikawa et al., 2016). Additionally, acknowledging and addressing the tension between how researchers categorize groups versus how groups categorize themselves are crucial. Ethnic-racial categories as presented by researchers may not accurately capture youth's ERI. For instance, multiracial adolescents were four times more likely to change their racial category when presented with forced-choice racial categories at two different timepoints (Hitlin et al., 2006). Findings suggest that forced-choice, racial self-categorization is more limiting for multiracial youth when compared to monoracial youth (Hitlin et al., 2006). In addition, scholars have answered calls to the ERI field for more longitudinal, multivariate studies (Rivas-Drake et al., 2014a, b; Schwartz et al., 2014), interdisciplinary approaches (Neblett et al., 2019; Syed et al., 2011) as well as intersectional frameworks (Rogers et al., 2012, 2020a). Regarding interdisciplinary approaches, Syed and colleagues (2011) present prejudice and stereotype threat, social support, and identity development as three distinct areas of ERI and education research that present opportunities for collaboration across social science disciplines. Seaton et al. (2010) relied on an intersectional approach to examine how age as well as gender and ethnic identity moderated the link between discrimination and Black youth's psychological well-being. Going forward, scholars must determine whether existing measures and frameworks are sufficient, or if it is the case that new measures and approaches are needed to address their research aims.

Regardless of what theoretical orientation or methodological approach is taken, scholars looking to incorporate ERI in their work must consider that the context for ERI development is not equally contentious for all youth. The nature of the contention differs and is contingent upon their group's social position in society, which informs what identities are granted airtime and space (Coll et al., 1996). Revisiting our "canary-in-a coalmine" analogy, social hierarchies determine which identities breathe comfortable or healthily amidst the smog of racism and oppression. As such, ERI not only fosters BIPOC youth's connections to their ethnic-racial group but also promotes well-being, civic and political engagement as well as socioemotional and academic adjustment. Most importantly, ERI can be a catalyst for the true endpoint of our research: social justice. ERI may be conceptualized as a mechanism for youths' growing awareness and understanding of racial hierarchies and power disparities within and between BIPOC groups (Rivas-Drake & Bañales, 2018; Fish et al., 2021; Syed et al., 2018). Consequently, such shared experiences of oppression and marginalization can result in cross-racial coalitions, political activism, allyship and ultimately, solidarity among BIPOC groups (Fish et al., 2021; Hope et al., 2020). The study of ERI sets a stage for scholars to shed light on and gain a deeper understanding and appreciation of narratives and histories too often deemed unimportant and unworthy. Only by doing so can scholars actively work to dismantle racism. Ultimately, to take a racial justice approach to ERI research is to understand its potential for promoting equality and inclusivity in terms of opportunity, health, and well-being of youth of color.

References

Adams-Bass, V. N., Stevenson, H. C., & Kotzin, D. S. (2014). Measuring the meaning of Black media stereotypes and their relationship to the racial identity, Black history knowledge, and racial socialization of African American youth. *Journal of Black Studies, 45*(5), 367–395. https://doi.org/10.1177/0021934714530396

Agi, A., & Rivas-Drake, D. (2021). "Other people's battles"? Toward an integrative model of ethnic-racial identity for immigrant-origin Black youth. *Identity*, 1–15. https://doi.org/10.1080/15283488.2021.1919518

Aldana, A., Rowley, S. J., Checkoway, B., & Richards-Schuster, K. (2012). Raising ethnic-racial consciousness: The relationship between intergroup dialogues and adolescents' ethnic-racial identity and racism awareness. *Equity & Excellence in Education, 45*(1), 120–137. https://doi.org/10.1080/10665684.2012.641863

Anglin, D. M., & Whaley, A. L. (2006). Racial/ethnic self-labeling in relation to group socialization and identity in African-descended individuals. *Journal of Language and Social Psychology, 25*(4), 457–463. https://doi.org/10.1177/0261927X06292770

Anzaldúa, G. (1999). *Borderlands/la frontera: The new mestiza*. Aunt Lute.

Bañales, J., Hoffman, A. J., Rivas-Drake, D., & Jagers, R. J. (2020). The development of ethnic-racial identity process and its relation to civic beliefs among Latinx and Black American adolescents. *Journal of Youth and Adolescence, 49*(1), 2495–2508. https://doi.org/10.1007/s10964-020-01254-6

Berry, J. W. (1997). Immigration, acculturation, and adaptation. *Applied Psychology, 46*(1), 5–34. https://doi.org/10.1111/j.1464-0597.1997.tb01087.x

Birnie-Smith, J. R. (2016). Ethnic identity and language choice across online forums. *International Journal of Multilingualism, 13*(2), 165–183. https://doi.org/10.1080/14790718.2015.1078806

Boykin, A. W. (1983). The academic performance of Afro American children. In *Achievement and achievement motives* (pp. 321–371). W. Freeman.

Bronfenbrenner, U. (1979). Contexts of child rearing: Problems and prospects. *American Psychologist, 34*(10), 844–850. https://doi.org/10.1037/0003-066X.34.10.844

Bryc, K., Durand, E. Y., Macpherson, J. M., Reich, D., & Mountain, J. L. (2015). The genetic ancestry of African Americans, Latinos, and European Americans across the United States. *The American Journal of Human Genetics, 96*(1), 37–53. https://doi.org/10.1016/j.ajhg.2014.11.010

Colby, S. L., & Ortman, J. M. (2015). *Projections of the size and composition of the U Spopulation: 2014 to 2060*. US Census Bureau. https://census.gov/content/dam/Census/library/publications/2015/demo/p25-1143.pdf

Coll, C. G., Crnic, K., Lamberty, G., Wasik, B. H., Jenkins, R., Garcia, H. V., & McAdoo, H. P. (1996). An integrative model for the study of developmental competencies in minority children. *Child Development, 67*(5), 1891–1914. https://doi.org/10.2307/1131600

Constante, K., Piceno, V., Medina, M., & Rivas-Drake, D. (2021). Latinx adolescents' ethnic-racial discrimination and expected future barriers over time. *Developmental Psychology, 57*, 991–999.

Cross, W. E. (1991). *Shades of Black: Diversity in African American identity*. Temple University Press.

Cross, W. E., Seaton, E., Yip, T., Lee, R. M., Rivas, D., Gee, G. C., Roth, W., & Ngo, B. (2017). Identity work: Enactment of racial-ethnic identity in everyday life. *Identity, 17*(1), 1–12. https://doi.org/10.1080/15283488.2016.1268535

Csizmadia, A., Brunsma, D. L., & Cooney, T. M. (2012). Racial identification and developmental outcomes among Black–White multiracial youth: A review from a life course perspective. *Advances in Life Course Research, 17*(1), 34–44. https://doi.org/10.1016/j.alcr.2011.10.003

Del Toro, J., Hughes, D., & Way, N. (2020). Inter-relations between ethnic-racial discrimination and ethnic racial identity among early adolescents. *Child Development, 92*(1), e106–e125. https://doi.org/10.1111/cdev.13424

Douglass, S., & Umaña-Taylor, A. J. (2015). A brief form of the Ethnic Identity Scale: Development and empirical validation. *Identity: An International Journal of Theory and Research, 15*(1), 48–65. https://doi.org/10.1080/15283488.2014.989442

Douglass, S., & Umaña-Taylor, A. J. (2016). Examining discrimination, ethnic-racial identity status, and youth public regard among Black, Latino, and White adolescents. *Journal of Research on Adolescence, 27*(1), 155–172. https://doi.org/10.1111/jora.12262

Douglass, S., Mirpuri, S., & Yip, T. (2017). Considering friends within the context of peers in school for the development of ethnic-racial identity. *Journal of Youth and Adolescence, 46*(2), 300–316. https://doi.org/10.1007/s10964-016-0532-0

Dreby, J. (2015). US immigration policy and family separation: The consequences for children's well-being. *Social Science & Medicine, 132*, 245–251. https://doi.org/10.1016/j.socscimed.2014.08.041

Enchautegui, M. E., & Menjívar, C. (2015). Paradoxes of family immigration policy: Separation, reorganization, and reunification of families under current immigration laws. *Law & Policy, 37*(1–2), 32–60. https://doi.org/10.1111/lapo.12030

Fish, J., Aguilera, R., Ogbeide, I. E., Ruzzicone, D. J., & Syed, M. (2021). When the personal is political: Ethnic identity, ally identity, and political engagement among Indigenous people and people of color. *Cultural Diversity and Ethnic Minority Psychology, 27*(1), 18–36. https://doi.org/10.1037/cdp0000341

Fuller-Rowell, T. E., Burrow, A. L., & Ong, A. D. (2011). Changes in racial identity among African American college students following the election of Barack Obama. *Developmental Psychology, 47*(6), 1608–1618. https://doi.org/10.1037/a0025284

Hagan, J., Shedd, C., & Payne, M. R. (2005). Race, ethnicity, and youth perceptions of criminal injustice. *American Sociological Review, 70*(3), 381–407. https://doi.org/10.1177/000312240507000302

Herman, M. R. (2008). Racial identification among multiracial youth: Implications for adjustment. In S. Quintana & C. McKown (Eds.), *Handbook of race, racism, and the developing child* (pp. 203–225). Wiley and Sons.

Hitlin, S., Brown, J. S., & Elder, G. H., Jr. (2006). Racial self-categorization in adolescence: Multiracial development and social pathways. *Child Development, 77*(5), 1298–1308. https://doi.org/10.1111/j.1467-8624.2006.00935.x

Hope, E. C., Gugwor, R., Riddick, K. N., & Pender, K. N. (2019). Engaged against the machine: Institutional and cultural racial discrimination and racial identity as predictors of activism orientation among Black youth. *American Journal of Community Psychology, 63*(1–2), 61–72. https://doi.org/10.1002/ajcp.12303

Hope, E. C., Cryer-Coupet, Q. R., & Stokes, M. N. (2020). Race-related stress, racial identity, and activism among young Black men: A person-centered approach. *Developmental Psychology, 56*(8), 1484–1495. https://doi.org/10.1037/dev0000836

Hughes, D., & Johnson, D. (2001). Correlates in children's experiences of parents' racial socialization behaviors. *Journal of Marriage and Family, 63*(4), 981–995. https://doi.org/10.1111/j.1741-3737.2001.00981.x

Hughes, D. L., Watford, J. A., & Del Toro, J. (2016). A transactional/ecological perspective on ethnic-racial identity, socialization, and discrimination. *Advances in Child Development and Behavior, 51*, 1–41. https://doi.org/10.1016/bs.acdb.2016.05.001

Hughes, D. L., Del Toro, J., & Way, N. (2017). Interrelations among dimensions of ethnic-racial identity during adolescence. *Developmental Psychology, 53*(11), 2139–2153. https://doi.org/10.1037/dev0000401

Huq, N., Stein, G. L., & Gonzalez, L. M. (2016). Acculturation conflict among Latino youth: Discrimination, ethnic identity, and depressive symptoms. *Cultural Diversity and Ethnic Minority Psychology, 22*(3), 377–385. https://doi.org/10.1037/cdp0000070

Itzigsohn, J., & Dore-Cabral, C. (2000). Competing identities? Race, ethnicity and panethnicity among Dominicans in the United States. *Sociological Forum, 15*(2), 225–247. https://doi.org/10.1023/A:1007517407493

Johnston-Guerrero, M. P. (2016). Embracing the messiness: Critical and diverse perspectives on racial and ethnic identity development. In E. S. Abes (Ed.), *Critical perspectives on student development theory* (pp. 43–55). Wiley Periodicals. https://doi.org/10.1002/ss.20174

Kiang, L., & Fuligni, A. J. (2009). Ethnic identity in context: Variations in ethnic exploration and belonging within parent, same-ethnic peer, and different-ethnic peer relationships. *Journal of Youth and Adolescence, 38*(5), 732–743. https://doi.org/10.1007/s10964-008-9278-7

Kiang, L., & Johnson, N. C. (2013). Relationship specific variability in adolescents' ethnic self-labeling preferences. *Journal of Adolescence, 36*(5), 815–823. https://doi.org/10.1016/j.adolescence.2013.07.002

Liu, S., Xiao, W. S., Xiao, N. G., Quinn, P. C., Zhang, Y., Chen, H., et al. (2015). Development of visual preference for own-versus other-race faces in infancy. *Developmental Psychology, 51*(4), 500–511. https://doi.org/10.1037/a0038835

Mathews, C. J., Medina, M. A., Bañales, J., Pinetta, B. J., Marchand, A. D., Agi, A. C., et al. (2019). Mapping the intersections of adolescents' ethnic-racial identity and critical consciousness. *Adolescent Research Review, 5*, 1–17. https://doi.org/10.1007/s40894-019-00122-0

McCarty, T. L. (2014). Negotiating sociolinguistic borderlands—Native youth language practices in space, time, and place. *Journal of Language, Identity & Education, 13*(4), 254–267. https://doi.org/10.1080/15348458.2014.939031

McGee, E. O., & Martin, D. B. (2011). "You would not believe what I have to go through to prove my intellectual value!" Stereotype management among academically successful Black mathematics and engineering students. *American Educational Research Journal, 48*(6), 1347–1389. https://doi.org/10.3102/0002831211423972

Meca, A., Gonzales-Backen, M., Rodil, J. C., Cowan, I., Sharma, S., Webb, T. N., & Hayes, T. (2021). The ethnic identity scale: Affirmation, really? *Cultural Diversity and Ethnic Minority Psychology*. https://doi.org/10.1037/cdp0000485. Advance online publication.

Miller-Cotto, D., & Byrnes, J. P. (2016). Ethnic/racial identity and academic achievement: A meta-analytic review. *Developmental Review, 41*, 51–70. https://doi.org/10.1016/j.dr.2016.06.003

Montoya, T. M. (2020). Unlearning racism: The classroom as a space for social transformation through borderlands pedagogy. *Contemporary Justice Review, 23*(2), 148–157. https://doi.org/10.1080/10282580.2020.1755844

Neblett, E. W., Jr., Smalls, C. P., Ford, K. R., Nguyen, H. X., & Sellers, R. M. (2009). Racial socialization and racial identity: African American parents' messages about race as precursors to identity. *Journal of Youth and Adolescence, 38*(2), 189–203. https://doi.org/10.1007/s10964-008-9359-7

Neblett, E. W., Jr., Rivas-Drake, D., & Umaña-Taylor, A. J. (2012). The promise of racial and ethnic protective factors in promoting ethnic minority youth development. *Child Development Perspectives, 6*(3), 295–303. https://doi.org/10.1111/j.1750-8606.2012.00239.x

Neblett, E. W., Jr., Roth, W. D., & Syed, M. (2019). Ethnic and racial identity development from an interdisciplinary perspective: Introduction to the special issue. *Emerging Adulthood, 7*(2), 79–84. https://doi.org/10.1177/2167696819834366

Phinney, J. S. (1989). Stages of ethnic identity development in minority adolescents. *Journal of Early Adolescence, 9*, 34–49. https://doi.org/10.1177/0272431689091004

Phinney, J. S. (1990). Ethnic identity in adolescents and adults: Review of research. *Psychological Bulletin, 108*, 499–514. https://doi.org/10.1037/0033-2909.108.3.499

Phinney, J. S. (1992). The multigroup ethnic identity measure: A new scale for use with diverse groups. *Journal of Adolescent Research, 7*(2), 156–176. https://doi.org/10.1177/074355489272003

Phinney, J. S., & Ong, A. D. (2007). Conceptualization and measurement of ethnic identity: Current status and future directions. *Journal of Counseling Psychology, 54*(3), 271–281. https://doi.org/10.1037/0022-0167.54.3.271

Pinetta, B. J., Blanco Martinez, S., Cross, F. L., & Rivas-Drake, D. (2020). Inherently political? Associations of parent ethnic-racial socialization and sociopolitical discussions with Latinx youths' emergent civic engagement. *American Journal of Community Psychology, 66*(1–2), 94–105. https://doi.org/10.1002/ajcp.12435

Renn, K. A. (2000). Patterns of situational identity among biracial and multiracial college students. *The Review of Higher Education, 23*(4), 399–420. https://doi.org/10.1353/rhe.2000.0019

Rivas-Drake, D., & Bañales, J. (2018, November 16). In the same boat? Developing cross-racial coalitions. *Psychology Today*. https://www.psychologytoday.com/us/blog/american-me-american-we/201811/in-the-same-boat-developing-cross-racial-coalitions

Rivas-Drake, D., & Umaña-Taylor, A. (2019). *Below the surface: Talking with teens about race, ethnicity, and identity*. Princeton University Press.

Rivas-Drake, D., Hughes, D., & Way, N. (2009). Public ethnic regard and perceived socioeconomic stratification: Associations with well-being among Dominican and Black American youth. *Journal of Early Adolescence, 29*(1), 122–141. https://doi.org/10.1177/0272431608324479

Rivas-Drake, D., Seaton, E. K., Markstrom, C., Quintana, S., Syed, M., Lee, R. M., et al. (2014a). Ethnic and racial identity in adolescence: Implications for psychosocial, academic, and health outcomes. *Child Development, 85*(1), 40–57. https://doi.org/10.1111/cdev.12200

Rivas-Drake, D., Syed, M., Umaña-Taylor, A., Markstrom, C., French, S., Schwartz, S. J., et al. (2014b). Feeling good, happy, and proud: A meta-analysis of positive ethnic-racial affect and adjustment. *Child Development, 85*(1), 77–102. https://doi.org/10.1111/cdev.12175

Rivas-Drake, D., Umaña-Taylor, A. J., Schaefer, D. R., & Medina, M. (2017). Ethnic-racial identity and friendships in early adolescence. *Child Development, 88*(3), 710–724. https://doi.org/10.1111/cdev.12790

Rockquemore, K. A., Brunsma, D. L., & Delgado, D. J. (2009). Racing to theory or retheorizing race? Understanding the struggle to build a multiracial identity theory. *Journal of Social Issues, 65*(1), 13–34. https://doi.org/10.1111/j.1540-4560.2008.01585.x

Rogers, L. O., Zosuls, K. M., Halim, M. L., Ruble, D., Hughes, D., & Fuligni, A. (2012). Meaning making in middle childhood: An exploration of the meaning of ethnic identity. *Cultural Diversity and Ethnic Minority Psychology, 18*(2), 99–108. https://doi.org/10.1037/a0027691

Rogers, L. O., Kiang, L., White, L., Calzada, E. J., Umaña-Taylor, A. J., Byrd, C., et al. (2020a). Persistent concerns: Questions for research on ethnic-racial identity development. *Research in Human Development, 17*(2–3), 1–24. https://doi.org/10.1080/15427609.2020.1831881

Rogers, L. O., Rosario, R. J., Padilla, D., & Foo, C. (2020b). "[I]t's hard because it's the cops that are killing us for stupid stuff": Racial identity in the sociopolitical context of Black Lives Matter. *Developmental Psychology, 57*(1), 87–101. https://doi.org/10.1037/dev0001130

Ruiz, N. G., Horowitz, J., & Tamir, C. (2020). *Many Black and Asian Americans say they have experienced discrimination amid the COVID-19 outbreak [Report].* Pew Research Center. https://www.pewsocialtrends.org/2020/07/01/many-black-and-asian-americans-saythey-have-experienced-discrimination-amid-the-covid-19-outbreak/

Santos, C., Kornienko, O., & Rivas-Drake, D. (2017). Peer network influence on ethnic-racial identity development: A multi-site investigation. *Child Development, 88*(3), 725–742. https://doi.org/10.1111/cdev.12789

Schwartz, S. J., Syed, M., Yip, T., Knight, G. P., Umaña-Taylor, A. J., Rivas-Drake, D., & Lee, R. M. (2014). Methodological issues in ethnic and racial identity research with ethnic minority populations: Theoretical precision, measurement issues, and research designs. *Child Development, 85*(1), 58–76. https://doi.org/10.1111/cdev.12201

Scottham, K. M., Sellers, R. M., & Nguyên, H. X. (2008). A measure of racial identity in African American adolescents: The development of the Multidimensional Inventory of Black Identity-Teen. *Cultural Diversity and Ethnic Minority Psychology, 14*(4), 297–306. https://doi.org/10.1037/1099-9809.14.4.297

Seaton, E. K., & Iida, M. (2019). Racial discrimination and racial identity: Daily moderation among Black youth. *American Psychologist, 74*(1), 117–127. https://doi.org/10.1037/amp0000367

Seaton, E. K., Yip, T., & Sellers, R. M. (2009). A longitudinal examination of racial identity and racial discrimination among African American adolescents. *Child Development, 80*(2), 406–417. https://doi.org/10.1111/j.1467-8624.2009.01268.x

Seaton, E. K., Caldwell, C. H., Sellers, R. M., & Jackson, J. S. (2010). An intersectional approach for understanding perceived discrimination and psychological well-being among African American and Caribbean Black youth. *Developmental Psychology, 46*(5), 1372–1379. https://doi.org/10.1037/a0019869

Seider, S., Clark, S., Graves, D., Kelly, L. L., Soutter, M., El-Amin, A., & Jennett, P. (2019). Black and Latinx adolescents' developing beliefs about poverty and associations with their awareness of racism. *Developmental Psychology, 55*(3), 509–524. https://doi.org/10.1037/dev0000585

Sellers, R. M., Rowley, S. A. J., Chavous, T. M., Shelton, J. N., & Smith, M. A. (1997). Multidimensional inventory of Black identity: A preliminary investigation of reliability and construct validity. *Journal of Personality and Social Psychology, 73*(4), 805–815. https://doi.org/10.1037/0022-3514.73.4.805

Sellers, R. M., Smith, M. A., Shelton, J. N., Rowley, S. A., & Chavous, T. M. (1998). Multidimensional model of racial identity: A reconceptualization of African American racial identity. *Personality and Social Psychology Review, 2*(1), 18–39. https://doi.org/10.1207/s15327957pspr0201_2

Shih, M., & Sanchez, D. T. (2009). When race becomes even more complex: Toward understanding the landscape of multiracial identity and experiences. *Journal of Social Issues, 65*(1), 1–11. https://doi.org/10.1111/j.1540-4560.2008.01584.x

Sladek, M. R., Umaña-Taylor, A. J., McDermott, E. R., Rivas-Drake, D., & Martinez-Fuentes, S. (2020). Testing invariance of ethnic-racial discrimination and identity measures for adolescents across ethnic-racial groups and contexts. *Psychological Assessment, 32*(6), 509–526. https://doi.org/10.1037/pas0000805

Stevenson, H. C., & Arrington, E. G. (2009). Racial/ethnic socialization mediates perceived racism and the racial identity of African American adolescents. *Cultural Diversity and Ethnic Minority Psychology, 15*(2), 125–136. https://doi.org/10.1037/a0015500

Stewart, E. A., Baumer, E. P., Brunson, R. K., & Simons, R. L. (2009). Neighborhood racial context and perceptions of police-based racial discrimination among black youth. *Criminology, 47*(3), 847–887. https://doi.org/10.1111/j.1745-9125.2009.00159.x

Supple, A. J., Ghazarian, S. R., Frabutt, J. M., Plunkett, S. W., & Sands, T. (2006). Contextual influences on Latino adolescent ethnic identity and academic outcomes. *Child Development, 77*(5), 1427–1433. https://doi.org/10.1111/j.1467-8624.2006.00945.x

Syed, M., Azmitia, M., & Cooper, C. R. (2011). Identity and academic success among underrepresented ethnic minorities: An interdisciplinary review and integration. *Journal of Social Issues, 67*(3), 442–468. https://doi.org/10.1111/j.1540-4560.2011.01709.x

Syed, M., Juang, L. P., & Svensson, Y. (2018). Toward a new understanding of ethnic-racial settings for ethnic-racial identity development. *Journal of Research on Adolescence, 28*(2), 262–276. https://doi.org/10.1111/jora.12387

Szymanski, D. M., & Lewis, J. A. (2015). Race-related stress and racial identity as predictors of African American activism. *Journal of Black Psychology, 41*(2), 170–191. https://doi.org/10.1177/0095798414520707

Tatum, B. D. (2017). *Why are all the Black kids sitting together in the cafeteria?: And other conversations about race*. Basic Books.

Thompson, T. L., Kiang, L., & Witkow, M. R. (2016). "You're Asian; You're supposed to be smart": Adolescents' experiences with the Model Minority Stereotype and longitudinal links with identity. *Asian American Journal of Psychology, 7*(2), 108–119. https://doi.org/10.1037/aap0000038

U.S. Department of Education Office of Civil Rights. (2014). *Civil rights data collection data snapshot: School discipline* (Issue Brief No. 1). https://www2.ed.gov/about/offices/list/ocr/docs/crdc-discipline-snapshot.pdf

Umaña-Taylor, A. J., & Fine, M. A. (2001). Methodological implications of grouping Latino adolescents into one collective ethnic group. *Hispanic Journal of Behavioral Sciences, 23*(4), 347–362. https://doi.org/10.1177/0739986301234001

Umaña-Taylor, A. J., Yazedjian, A., & Bámaca-Gómez, M. (2004). Developing the ethnic identity scale using Eriksonian and social identity perspectives. *Identity: An International Journal of Theory and Research, 4*(1), 9–38. https://doi.org/10.1207/S1532706XID0401_2

Umaña-Taylor, A. J., Quintana, S. M., Lee, R. M., Cross, W. E., Jr., Rivas-Drake, D., Schwartz, S. J., et al. (2014). Ethnic and racial identity during adolescence and into young adulthood: An integrated conceptualization. *Child Development, 85*(1), 21–39. https://doi.org/10.1111/cdev.12196

Vandiver, B. J., Cross, W. E., Jr., Worrell, F. C., & Fhagen-Smith, P. E. (2002). Validating the Cross racial identity scale. *Journal of Counseling Psychology, 49*(1), 71–85. https://doi.org/10.1037/0022-0167.49.1.71

Verkuyten, M. (2016). Further conceptualizing ethnic and racial identity research: The social identity approach and its dynamic model. *Child Development, 87*, 1796–1812. https://doi.org/10.1111/cdev.12555

Wang, Y., Douglass, S., & Yip, T. (2017). Longitudinal relations between ethnic-racial identity process and content: Exploration, commitment, and salience among diverse adolescents. *Developmental Psychology, 53*(11), 2154–2169. https://doi.org/10.1037/dev0000388

Wang, Y., & Yip, T. (2020). Parallel changes in ethnic/racial discrimination and identity in high school. *Journal of Youth and Adolescence, 49*, 1517–1530. https://doi.org/10.1007/s10964-019-01186-w

Wiley, S., Perkins, K., & Deaux, K. (2008). Through the looking glass: Ethnic and generational patterns of immigrant identity. *International Journal of Intercultural Relations, 32*(5), 385–398. https://doi.org/10.1016/j.ijintrel.2008.04.002

Williams, C. D., Byrd, C. M., Quintana, S. M., Anicama, C., Kiang, L., Umaña-Taylor, A. J., et al. (2020). A lifespan model of ethnic-racial identity. *Research in Human Development*, 1–31. https://doi.org/10.1080/15427609.2020.1831882

Wilson, C. M., Ek, L. D., & Douglas, T. R. M. (2014). Recasting border crossing politics and pedagogies to combat educational inequity: Experiences, identities, and perceptions of Latino/a immigrant youth. *The Urban Review, 46*(1), 1–24. https://doi.org/10.1007/s11256-013-0246-5

Wray-Lake, L., Tang, J., & Victorino, C. (2017). Are they political? Examining Asian American college students' civic engagement. *Asian American Journal of Psychology, 8*(1), 31–42. https://doi.org/10.1037/aap0000061

Wray-Lake, L., Wells, R., Alvis, L., Delgado, S., Syvertsen, A. K., & Metzger, A. (2018). Being a Latinx adolescent under a Trump presidency: Analysis of Latinx youth's reactions to immigration politics. *Children and Youth Services Review, 87*, 192–204. https://doi.org/10.1016/j.childyouth.2018.02.032

Yip, T. (2005). Sources of situational variation in ethnic identity and psychological well-being: A Palm Pilot study of Chinese American students. *Personality and Social Psychology Bulletin, 31*(12), 1603–1616. https://doi.org/10.1177/0146167205277094

Yip, T. (2018). Ethnic-racial identity—A double-edged sword? Associations with discrimination and psychological outcomes. *Current Directions in Psychological Science, 27*(3), 170–175. https://doi.org/10.1177/0963721417739348

Yip, T., & Fuligni, A. J. (2002). Daily variation in ethnic identity, ethnic behaviors, and psychological well–being among American adolescents of Chinese descent. *Child Development, 73*(5), 1557–1572. https://doi.org/10.1111/1467-8624.00490

Yip, T., Wang, Y., Mootoo, C., & Mirpuri, S. (2019). Moderating the association between discrimination and adjustment: A meta-analysis of ethnic-racial identity. *Developmental Psychology, 55*(6), 1274–1298. https://doi.org/10.1037/dev0000708

Yoshikawa, H., Mistry, R., & Wang, Y. (2016). Advancing methods in research on Asian American children and youth. *Child Development, 87*(4), 1033–1050. https://doi.org/10.1111/cdev.12576

Zepeda-Millán, C., & Wallace, S. J. (2013). Racialization in times of contention: How social movements influence Latino racial identity. *Politics, Groups, and Identities, 1*(4), 510–527. https://doi.org/10.1080/21565503.2013.842492

Zirkel, S., & Johnson, T. (2016). Mirror, mirror on the wall: A critical examination of the conceptualization of the study of Black racial identity in education. *Educational Researcher, 45*(5), 301–311. https://doi.org/10.3102/0013189X16656938

Chapter 4
The Theory of Racial Socialization in Action for Black Adolescents and Their Families

Mia A. Smith-Bynum

Racial socialization is fundamental to understanding the development of African American children in the United States (Anderson & Stevenson, 2019; Hughes et al., 2006; Wang et al., 2020a). A corpus of studies has established a relationship between messages and behavioral practices aimed at teaching African American children about racial issues and coping with racism and their academic and mental health outcomes. Landmark scholarship and theoretical work produced in the last 40 years outlined influential concepts and practices that comprise racial socialization processes (Anderson & Stevenson, 2019; Hughes et al., 2006; Neblett Jr. et al., 2012; Spencer 1983, Umaña-Taylor & Hill, 2020). Though variation in theoretical perspectives and constructs exists, research largely agrees that racial socialization consists of four key domains: cultural socialization, preparation for bias, egalitarianism, and cultural mistrust (Hughes et al., 2006).

Racial socialization research is maturing. As noted, major constructs have been identified. Powerful explanatory theories have gained currency and helped establish the importance of this field of inquiry to research on child development and family development (Umaña-Taylor & Hill, 2020).[1] There is also growing recognition that racial socialization among White children is essential to understanding the role of race and ethnicity in child and family development in the United States (Hagerman, 2018). These are long overdue developments given the US Census data indicating that children of color now constitute the majority of youth in United States (Vespa

[1] This chapter will focus on racial socialization specifically, but the author recognizes the broad use of ethnic-racial socialization for families of different racial and ethnic backgrounds. The author uses the terms African American and Black interchangeably to refer to children and families who identify with Black American racial and cultural experience in the United States.

M. A. Smith-Bynum (✉)
University of Maryland, College Park, MD, USA
e-mail: msbynum@umd.edu

© The Author(s), under exclusive license to Springer Nature Switzerland AG 2023

D. P. Witherspoon, G. L. Stein (eds.), *Diversity and Developmental Science*,
https://doi.org/10.1007/978-3-031-23163-6_4

et al., 2020). It is time to integrate racial socialization research more fully with the study of other aspects of child and family development in which the study of race has received limited attention (Smith-Bynum, 2022).

This chapter introduces the Theory of Racial Socialization in Action for African American families (TRSA). It begins with a brief overview of the TRSA, then provides a rationale for this expansion of seminal theoretical work in field. Research evidence that supports the theoretical tenets of the TRSA is also provided. Additionally, this chapter describes the (RSOTCS; Smith-Bynum) the measurement tool that informed its development. I conclude with implications for the future research regarding the role of race and ethnicity in child development.

Expanding Theory on Racial Socialization

The Theory of Racial Socialization in Action (TRSA) is an attempt to forge stronger connections between racial socialization and the broader theoretical and research literature parenting (Ahn et al., 2021; Smith-Bynum et al., 2016). Based on 12 years of research (Smith-Bynum, 2006, 2008, 2011a; Smith-Bynum & Usher, 2004), the TRSA is charting new ground by articulating a conceptual framework that unpacks the dyadic aspects of racial socialization between African American parents and their children. To that end, this chapter has several goals. The main goal is to describe the TRSA, the rationale for it, and the observational methodology I developed that undergirds it.

A challenge in articulating the TRSA at this time is the reality that the racial socialization literature is still evolving and doing so rapidly (Umaña-Taylor & Hill, 2020). Like all areas of research focused on the development of children of color, racial socialization research has been constrained by structural barriers that render it as less worthy of scholarly attention (Syed et al., 2018). Until the death of George Floyd and the Black Lives Matter movement, a persistence in the doubt of the relevance of racism to the lives of children permeated the scientific process (Buchanan et al., 2021). Charting new ground is necessary, even in the context of gaps in the literature, but hopefully it can spur the field to continue to advance. Thus, an ancillary goal of this chapter is to describe my intellectual journey that coincided with the development of the TRSA and other key turning points in the racial socialization literature. I hope that by taking some intellectual risks, others are encouraged to do so as well. This chapter concludes with future directions for racial socialization research with a focus on specific ways that diversity and developmental science within this domain can be integrated.

The Theory of Racial Socialization in Action (TRSA) rests on four empirical and theoretical foundations: (1) existing research on parenting and ethnic-racial socialization in Black families and developmental outcomes in Black youth; (2) seminal theories in the field, most notably the Theory of Parental Ethnic-Racial Socialization (TPRES; Hughes et al., 2006); (3) my observational research studies on African American adolescents and their parents; and (4) research and theory on parenting, parent-adolescent communication, and goodness-of-fit (Belsky, 1984; Darling & Steinberg, 1993; Telzer et al., 2014). Though the TRSA is based on my research with Black adolescents and their families, I have extrapolated from that work to theorize about racial socialization in Black children more broadly.

A Need for Assessment of Racial Socialization as a Bidirectional Processes

Racial socialization is one of the most studied topics in contemporary Black child development with multiple recent comprehensive reviews on the topic (Reynolds & Gonzales-Backen, 2017; Stein et al., 2018; Umaña-Taylor & Hill, 2020; Wang et al., 2020a, b; Washington et al., 2015). As such, only a brief overview of key constructs and issues informing the present work is provided here. The seminal and most influential theory in the field to date is the Theory of Parental Ethnic-Racial Socialization (TPERS; Hughes et al., 2006). The TPRES asserts that there are four domains of ethnic-racial socialization: cultural socialization, preparation for bias, cultural mistrust, and egalitarianism/silence about race. Briefly, *cultural socialization* is defined as parental teachings to children about their cultural background through exposure to the cultural traditions, rituals, social mores, language, and food. *Preparation for racial bias* consists of parental teachings about the realities of racism and discrimination and strategies for coping with those challenges. *Cultural mistrust* focuses on parental lessons focused on distrust of other groups due to the historical mistreatment of African Americans. *Egalitarianism* consists of parental messages focused on equality among all human beings. *Silence about race* refers to the parental decision to eschew direct communication, consciously or unconsciously, about racial issues with their children.

There is wide agreement that racial socialization is a bidirectional process undergirded by several key assumptions outlined by Hughes et al. (2006). We know that both parents and children shape the racial socialization process through their conversations about race-related and cultural events in their lives (Brown & Lesane-Brown, 2006; Caughy et al., 2002; Hughes & Chen, 1999; Riina & McHale, 2012; Stevenson et al., 2002; Thomas & Blackmon, 2015). Parents initiate the process with intentions to convey specific information about culture and racial realities. Events within or outside the family can trigger dyadic communication about racial dynamics in society that shape the development of African American children. Still the family process-oriented assumptions remain among the least investigated aspects of the TPRES.

The lack of attention to racial socialization bidirectional process along with the limited tools for measurement beyond self-report likely explains some of the mixed findings in the literature regarding the developmental impact of racial socialization on African American youth and adolescents (Smith-Bynum et al., 2016). Recently, researchers have asserted that the heavy reliance on self-report measures is one of the reasons that the conditions under which racial socialization impacts youth outcomes are not yet clearer. (Wang et al., 2020a, b). As Wang et al. (2020a) assert about racial socialization, "The global construct is multidimensional, but so [are] several of its constituent subtypes" (p. 17). Continued reliance on self-report measures limits our capacity to identify the multidimensional components of cultural socialization, preparation for bias, promotion of mistrust, and egalitarianism/silence about race. Observing dyadic conversations between parents and children gives researchers access to the granular aspects of the racial socialization process as outlined by Hughes et al. (2006).

Creation of the Racial Socialization Observational Task and Coding System

My effort to explore the interactive elements within the racial socialization process led to the development of the Racial Socialization Observational Task and Coding System (RSOTCS; Smith-Bynum, 2008; Smith-Bynum et al., 2005, 2016; Smith-Bynum & Usher, 2004). The RSOTCS is the culmination of 12 years of intensive, community-based, family-focused research involving video recordings of family conversations about racial issues. The Racial Socialization Observational Task (RSOT) consists of 5-minute discussions between parents and adolescents about one of five hypothetical situations involving Black teenagers coping with subtle, ambiguous forms of racial discrimination (*The Teacher*, *The Mall*, *The School Counselor*, *The Police*, *Soccer;* Smith-Bynum et al., 2005; Fisher et al., 2000; Nyborg & Curry, 2003; Smith-Bynum & Usher, 2004). Parent-adolescent dyads receive a prompt to discuss one of five vignettes as if the situation happened to their family. Researchers can elect to present multiple vignettes to participant families in a counterbalanced order to gain information about how the families respond to different situations involving subtle, ambiguous racial discrimination (Dovidio & Gaertner, 2004; Duckitt, 1992; Salvatore & Shelton, 2007; Sue, 2010). The RSOT has the added advantage of permitting observational assessment of the general features of communication and relationship quality between the parents and adolescents using coding systems designed to assess general family dynamics (e.g., Smetana & Abernethey, 1998).

Because contemporary racial prejudice and discrimination are often ambiguous (Dovidio & Gaertner, 2004; Duckitt, 1992; Salvatore & Shelton, 2007; Sue, 2010), the RSOT focuses on eliciting the reasoning process in Black family communication about racial issues that determine whether the events and treatment experienced constitute racial discrimination. Parents can facilitate mastery of this reasoning process through direct communication about racial matters, discussions about specific discriminatory incidents, and through role modeling strategies for coping with discriminatory situations (Anderson et al., 2020; Anderson & Stevenson, 2019). Consistent with the tenets of the TPRES, the RSOT is designed to capture spontaneous discussion about racial discrimination between Black adolescents and their parents not captured by existing self-report questionnaires. The five vignettes vary in racial ambiguity cues to elicit a range of dyadic responses to the situations in the spirit of the assumptions outlined by the TPRES (Hughes & Chen, 1999). These discussions are also recorded so that the actual process of communication can be observed, enabling granular assessments of verbal and nonverbal family communication about racial issues. Recordings also facilitated the opportunity for new constructs to be identified as research on racial socialization continues to evolve.

Through transcription and analysis of video discussions across two pilot studies and a larger, validation study, I identified at least 11 new behavioral strategies and dyadic patterns of problem-solving styles embedded in the racial socialization process (Smith-Bynum, 2008; Smith-Bynum et al., 2005; Smith-Bynum & Usher, 2004). This methodological approach moves the field beyond its heavy reliance on self-report and in so doing, reveals the various ways that parents adapt their racial socialization strategies to the various situations that children encounter. Observed family communication about racial issues also permits assessment of the parent-adolescent relationship (e.g., warmth, conflict, intensity of affect; Dunbar et al., 2022a, b; Smetana & Abernethy, 1998; Smith-Bynum et al., 2016). These data are the ideas on which the TRSA is based.

Centering Research on Racial Socialization within Developmental Science on Parenting

Pioneering studies in the racial socialization literature (e.g., Boykin & Toms, 1985; Sanders Thompson, 1994; Spencer, 1983; Stevenson, 1994) cut new ground *and* undertook an intellectual battle for the legitimacy for the work itself. Today, investigators are now positioned to consider ways to integrate racial socialization with existing theory and research on parenting and child development with greater depth (Huguley et al., 2019; Wang et al., 2020a, b). To date, there remain large areas of developmental science in which race and racial issues are rarely considered (Iruka et al., 2021; Lozada et al., 2022; Roberts et al., 2020; Syed et al., 2018). The Theory of Racial Socialization in Action incorporates three such developmental constructs into racial socialization research: (a) parental judgement and sensitivity, (b) features of child temperament, and (c) goodness-of-fit between parents and children in the racial socialization process. This field of inquiry is moving rapidly (Smith-Bynum, 2022). Consistent with the progress in the field, readers should consider this chapter the first iteration of the TRSA. The TRSA articulates four assumptions about the racial socialization process between Black parents and their children. Assumption 4.1 states that racial socialization is a social learning process in the tradition of Vygotsky. Assumption 4.2 states that the parent's capacity for perspective taking and empathy affects their deployment of racial socialization messages and behaviors. Assumption 4.3 states that the attributes of the child affect the receipt and uptake of racial socialization around coping with discrimination. Assumption 4.4 states that the combination/goodness-of-fit between parental judgement, sensitivity, empathy, and child attributes predict child developmental outcomes.

A Preliminary Conceptual Framework for Racial Socialization in Action

Figure 4.1 presents a conceptual framework for the TRSA. It is based on existing research on (1) racism, racial socialization, and racial identity in Black youth (e.g., García-Coll et al., 1996; Neblett Jr. et al., 2012) and (2) seminal developmental theory and notable research findings that cut across parenting experiences and child development (Belsky, 1984; Darling & Steinberg, 1993; Maccoby & Martin, 1983). The framework also draws upon my research data (Smith-Bynum et al., 2005, 2006; Smith-Bynum & Usher 2004), my review of observational data collected with an adapted version of the ROTCS that was used with a sample of Black and Latinx pre-adolescent youth and their caregivers (Osborne et al., 2022; Osborne et al., in press) and my clinical observations of the families who participated in my observational studies between 2002 and 2011.

As can be seen in Fig. 4.1, the TRSA conceptual framework accounts for and builds on existing established research findings related to the effects of racism and racial discrimination (path A_1), child temperament (path A_2), and child racial identity (path A_3) on mental health in Black youth (e.g., Davis Tribble et al., 2017; Lavner et al., 2022; Neblett Jr. et al., 2012; Rothbart, 2007). It also accounts for the

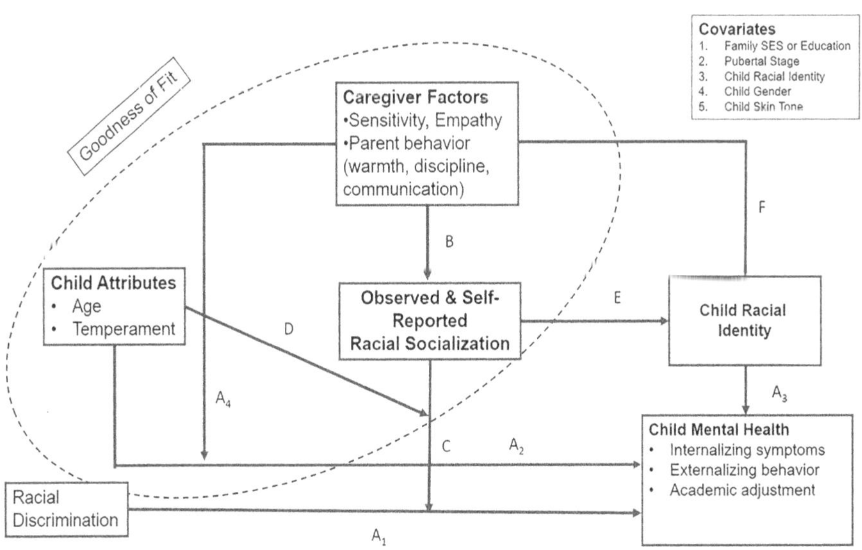

Fig. 4.1 Illustration of the Theory of Racial Socialization in Action (TRSA)

role of the goodness-of-fit between caregiver factors (i.e., sensitivity, empathy, warmth, discipline, communication quality) and youth temperament in shaping risk for internalizing and externalizing behaviors (path A_4; Sharp et al., 2006; Telzer et al., 2014; Williams et al., 2009). Path B proposes that caregiver factors inform the deployment of racial socialization communication detected through observational research. Path C proposes that observed (and self-reported) racial socialization moderates the relationship between racism and Black youth mental health (Smith-Bynum et al., 2019). Path D proposes a three-way interaction between child age, temperament, and observed racial socialization variables. Specifically, it is posited that the proposed moderating role for racial socialization for the discrimination—mental health link (Path C) also depends on the age and temperament of the child. Observed (and self-reported) racial socialization communication is expected to predict child racial identity (path E), which has been shown to predict mental health in Black youth (Davis Tribble et al., 2017; Huguley et al., 2019; Neblett Jr. et al., 2013).

This framework also presents several plausible covariates that should be considered for research questions using part of the conceptual framework to answer questions about racism, observed racial socialization, and Black youth development. Those are family socioeconomic status, child gender, and child racial identity (e.g., Davis Tribble et al., 2017; Smith-Bynum et al., 2016; Jones et al., 2021). Another factor likely to play a role in how racial socialization plays out is youth's pubertal status because of the ways pubertal status can exacerbate the impact of racism and discrimination on Black youth (Carter & Flewellen, 2022; Cook & Haberstadt, 2021; Goff et al., 2014). Furthermore, skin color and colorism have received limited attention in the racial socialization literature to date. Still, these factors are likely important covariates as well as possible moderators and mediators of racial socialization processes on Black youth mental health (Davis Tribble et al., 2019; Landor et al., 2013). More details about the research rationale for each assumption that undergird the TRSA conceptual framework are provided in the next sections.

The Four Assumptions of the TRSA

Assumption 4.1: Racial socialization is a social learning process in the tradition of Vygotsky Specifically, the parent or caregiver is the first teacher in racial issues as is the case with language development and emotion regulation (Eisenberg et al., 2004; Golinkoff et al., 2015; Rasmussen et al., 2017; Roque et al., 2013; Shai & Belsky, 2017). Effective racial socialization emanates from parents' sensitivity to specific individual attributes in their child and their own judgement about when and how to deploy teachings about racial socialization. Specifically, parents are positioned to judge their children's developmental capacity for acquiring new knowledge about coping with racial dilemmas. Furthermore, parents calibrate their teachings about racial issues (Anderson & Stevenson, 2019; Smith-Bynum et al., 2016). Consistent with previous research, it is posited that the dyadic process of communication about racial issues between parents and their children is developmentally graded (Hughes & Johnson, 2001; Hughes et al., 2006; Priest et al., 2014). That is, parents make conscious and unconscious judgements about when their children are ready to acquire new information about racism and discrimination and how to cope with racial dilemmas. Parents with greater sensitivity to their children's attributes in this regard are likely positioned to help foster better outcomes.

Vygotsky (1978) posited that the parent is the first teacher in helping their children acquire new skills such as speech, social rules of interaction, and cultural norms (Wertsch & Wertsch, 2009) even as children age (Gredler & Shields, 2008). The family is the primary context for socializing children regarding race, African American culture, racial identity, and racism. Parents choose neighborhoods, schools, after-school programs, places of worship, and other enrichment activities that shape children's ideas and notions about life for African Americans (Boykin & Toms, 1985). In so doing, parents create the tapestry of the contexts in which children learn to make sense of the sociocultural and racial landscape of their lives.

Racial socialization maps onto Vygotsky's (1978) Social Learning Theory in several ways. The situations involving racism and discrimination that Black children and families confront on a regular basis represent the institutional and sociocultural realities in their lives (Harrell, 2000; Harrison et al., 1990; Peters & Massey, 1983). In each situation, a child must problem-solve around these dilemmas, and it is the parent's task to assess whether the skillsets needed to solve the dilemma occurs within or outside of the child's *zone of proximal development* (Vygotsky, 1978). When these racial dilemmas occur within the zone of proximal development, children can master new knowledge with the support of their parents or other social teachers (See Fig. 4.2). Further, Vygotsky (1978) asserted that cognitive learning is facilitated directly by engagement with the social environment, and in this way, parents are often the first teachers that facilitate cognitive reasoning through various problems and tasks. Also, the cultural context is implicated in the developmental sequencing and course of social learning. To bring the application of Vygotsky's theory to racial socialization full circle, I also posit that the capacity to respond to certain racial dilemmas may fall outside the skillset of the target youth such that even with parental guidance and support, they are unable to respond to racial

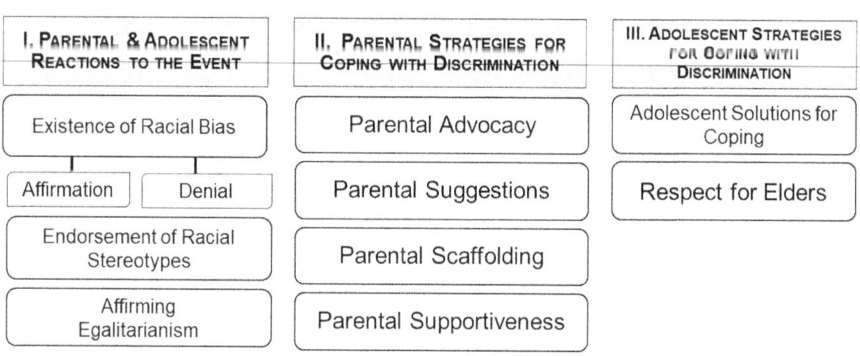

Fig. 4.2 The Racial Socialization Observational Task and Coding System (RSOTCS)

dilemmas effectively (Anderson, Heard-Garris et al. 2022). These racial dilemmas may involve several complex components such as subtle situational and environmental cues that signal the possibility of racial bias, power differentials between the target youth and other people involved, threat(s) to personal safety, and dilemmas involving multiple steps to address racial dilemmas effectively (Hoggard et al., 2016).

When applied to racial socialization, Social Development Theory would indicate that racial socialization leads to the development of effective racial coping skills (e.g., Johnson, 2005; Scott et al., 2020), but few studies have articulated the process of skill acquisition in this arena (Anderson & Stevenson, 2019). When done effectively, parents who scaffold adolescents' efforts to solve racial dilemmas will have adolescents with fewer behavioral difficulties (Mulvaney et al., 2006). Effective scaffolding of racial messages to African American adolescents is infused with parental emotional support for mastery of racial coping skills. Additionally, I assert that adolescents learn coping strategies through their efforts to develop solutions for responding to discriminatory encounters. It also includes acquisition of the capacity to display respect for elders as a general orientation towards adults and other family and community elders and when judging a power differential between themselves and adults in a discriminatory encounter.

In support of these ideas, parents exhibit four types of behaviors in support of acquisition of racial coping skills. The behaviors are *parental advocacy*, *parental suggestions*, *parental supportiveness*, and *parental scaffolding*. They have not been captured by self-report measures to date (Smith-Bynum & Usher, 2004; Smith-Bynum, 2006, 2008, 2011a, b; Smith-Bynum et al., 2016). The constructs will be defined later in this chapter. In this context, parents often present their children additional explanations and pointers to help adolescents grasp the nuances of the racial dilemma in the vignette. These explanations and pointers are also intended to help the child be better equipped to respond to similar dilemmas in the future.

Assumption 4.2: The parent's capacity for perspective taking and empathy affects their deployment of racial socialization messages and behaviors I have written previously about the importance of considering the parent-adolescent relationship in understanding the role of racial socialization in adolescents' developmental outcomes (Ahn et al., 2021; Smith-Bynum et al., 2016). Several studies suggest that parental warmth/involvement co-occurs with greater engagement regarding racial socialization (Cooper & McLoyd, 2011; Frabutt et al., 2002; Smalls, 2010; Umaña-Taylor & Hill, 2020). Why do parents who have warmer relationships with their children engage in more racial socialization? My anecdotal clinical observations of parents' behavior during the RSOT suggested that some parents were simply better at guiding their children than others in the study samples (Smith-Bynum et al., 2005; Osborne et al., 2022; Osborne et al., in press; Smith-Bynum, 2011a, b; Smith-Bynum & Usher, 2004). This led me to think about the underlying processes responsible for this correlation and other constructs that had not yet been considered in the racial socialization literature to date.

I posit that parents' judgment regarding when and how to deploy racial socialization derives from *parental sensitivity*. Parental sensitivity is awareness of their children's point of view and emotional experience (Camoirano, 2017; Cliffordson, 2001; Scopesi et al., 2015; Slade, 2005). It reflects the capacity for emotional perspective-taking needed to inform judgements about parenting behavior (Camoirano, 2017). I propose that parental sensitivity is a parenting trait from which effective judgment in deploying racial socialization emanates. As a parenting trait, parental sensitivity is superordinate to parental judgments about racial socialization, and thus is a predictor of racial socialization. Parents with high levels of sensitivity are posited to be more effective at judging the boundaries of their children's zone of proximal learning and calibrating their guidance about racial issues effectively. Furthermore, I propose that the goodness-of-fit between parental sensitivity and adolescent temperament are critical factors in understanding the protective benefits of racial socialization.

In this way, teaching children about race requires an adaptation of dominant models of general parenting (e.g., Darling & Steinberg, 1993) to include the parenting goal of racial socialization or parental teaching about racial issues. The video data reveal that some parents engage in contemplation of their actions and communications before speaking (Smith-Bynum, 2006, 2011a, b; Smith-Bynum & Usher, 2004). I propose that parents' emotional judgment and sensitivity are a precursor to parenting style. Healthy emotional judgment and sensitivity is fostered by a secure attachment history to their own parents (Dunbar et al., 2021; Stern et al., 2021). As such, like with any parenting task, effective socialization about race, particularly in teaching about racism and racial bias, requires astute judgements about children's tolerance for the frank realities of racism and use of age-appropriate language and teaching strategies.

Relatedly, I propose that parental sensitivity is critical to helping adolescents acquire useful skills for coping with racial discrimination. I further propose that the fit between parental attributes (i.e., sensitivity, warmth) and adolescent temperament may be systematically related to their response to racial socialization message content and process. My data involving video-recorded racial socialization behaviors suggest that parents vary in their sensitivity regarding how to deploy racial socialization. Such variation could be consequential for adolescents who differ on specific aspects of temperament (i.e., fearfulness, surgency, effortful control). Thus, the goodness-of-fit between parental attributes, racial socialization, and adolescent temperament can affect adolescents' receipt and assimilation of parents' racial socialization content.

In the racial socialization context, parental capacity for perspective taking and empathy can be made more challenging because of the threat that racism and discrimination pose (Anderson & Stevenson, 2019; Heard-Garris et al., 2018; Smith-Bynum et al., 2016). It is distressing for parents to hear about discriminatory incidents their children experience (Anderson & Stevenson, 2019; Heard-Garris et al., 2018; Peters & Massey, 1983). It is distressing for parents to be the cope with racial discrimination (Condon et al., 2022) and to process racist events happening to friends and family members or in the news (Anderson, Jones et al., 2020; Bor et al., 2018; Smith-Bynum et al., 2016). Parents' capacity to manage their own emotions to teach about race likely varies as a function of their own life experiences and the general overall stress they may be under in a given day (Coard, 2021; Murry et al., 2021). Indeed, exposure to racist life events render their own wear and tear on the body for parents and children alike (Jones et al., 2019; Trent et al., 2019). They remind parents of the ways American society continues to dehumanize African Americans (Eberhardt et al., 2004; Goff et al., 2008, 2014). These are enormous emotional demands on parents and exact their own psychological toll (Keating et al., 2022). Parents' own capacity to manage emotions and self-regulate matters for deployment of racial socialization. Thus, I posit that the parent-adolescent relationship quality and the parents' capacity for empathy and taking the perspective of the child (i.e., *parental sensitivity*) likely fosters better judgment about how to engage children and teach about racial dilemmas.

Just as in the way that parents' effective management and sensitivity to child temperament can curtail risk for certain psychological disorders, these parental skills can also bolster adolescent's coping with racial dilemmas that their children and the family encounter. Available evidence suggests that the interplay between (a) parent attributes, (b) the parent-adolescent relationship, and (c) child/adolescent temperament can affect the development of emotional and behavioral disorders in childhood (Hirvonen et al., 2018; Morales et al., 2016; Shiner et al., 2012; van der Voort et al., 2014). Parents who are better at mentalizing have children and adolescents with better emotional expressivity skills (Telzer et al., 2014) and better psychosocial adjustment (Sharp et al., 2006). I posit that parents who are more effective at mentalizing the emotional needs of their adolescents will also be better at executing the racial socialization process. I also predict that these parents will be more effective at considering their adolescent's temperament when they deploy racial socialization content and suggestions for racial coping strategies.

The Theory of Racial Socialization in Action presents two constructs critical to understanding children's mastery of racial coping: (a) parents' *emotional support* of adolescents' efforts to solve racial dilemmas and (b) adolescents' *solutions to racial dilemmas*. Data reveal that parental warmth is distinct from parents' emotional support for solving racial dilemmas (Smith-Bynum et al., 2016, 2021). In other words, children likely need targeted effective support for racial coping specifically. Merely having a positive relationship with the parent may be a necessary precursor to effective support, but insufficient for the acquisition of racial coping skills.

RECAST Theory (Anderson & Stevenson, 2019) introduced the concept of racial socialization competency, defined as the knowledge, skills, and confidence families possess to cope effectively with racial discrimination. Anderson and Stevenson (2019) asserted that through the process appraisal of discriminatory situations, parent-child dyadic coping, parents support children's acquisition of racial coping self-efficacy. Specifically, parents' skills with racial coping guide their instruction of you in the appraisal process, fostering children's capacity to evaluate a situation through a racial lens effectively. RECAST Theory is complementary to the TRSA in its articulation of the cognitive processes embedded in children's acquisition of racial coping through the racial socialization process. The TRSA emphasizes our understanding of the relationship between the parent-child relationship and racial socialization as an important ingredient to the racial socialization process. Additionally, the TRSA would posit that the goodness-of-fit between (1) the parents' capacity for perspective taking, parental judgement, and parenting skill and (2) children's own developmental capacity for learning likely shape the youth's assimilation of parents' lessons about race and ultimately, the acquisition of the children's racial coping skills (i.e., racial socialization competency).

Assumption 4.3: The attributes of the child affect impact, the receipt, and uptake of racial socialization around coping with discrimination Established research highlights the importance of child age and gender in the effects of racial socialization on African American youth (Wang et al., 2020a). Child temperament and emotional maturity likely serve as moderators of linkages between racial socialization practices and child outcomes (Dunbar et al., 2017). Existing theoretical models have not articulated the precise role of temperament in racial socialization processes to date (e.g., García Coll et al., 1996). The TRSA argues that selected constructs outlined in Rothbart's (2011) Theory of Temperament and Personality Development are most relevant to racial socialization: (1) fearfulness and negative emotionality, (2) surgency/novelty seeking, and (3) effortful control. Given the limited research in this area, I present several hypothetical ideas that might explain the role of temperament in this context.

Teaching their children to cope with racism and discrimination involves communication about racially stressful and traumatic information that affects children with different temperaments in different ways. Candid information can be overwhelming to youth in general, and especially troublesome for youth with temperaments that are high in fearfulness (Chae et al., 2009). Similarly, a child high in negative emotionality will likely have strong reactions to emotionally charged race-related information. Children with high *fearfulness and negative emotionality* may experience heightened stress response to already difficult, unfair situations such as those involving racism and discrimination. A problematic parent-child relationship with poor dyadic communication or high levels of alienation can diminish the effectiveness of adaptive racial socialization efforts (Darling & Steinberg, 1993). For example, if the parent dismisses the child's dispositional response, exhibits low warmth or emotional support, or has difficulty taking the perspective of the child, efforts to teach the child how to respond to the dilemma are likely to be ineffective. These parental behaviors may also be harmful to the uptake of racial socialization by youth in some situations.

Another aspect of temperament not yet investigated in the racial socialization literature is the child's novelty seeking behavior, which is also referred to as surgency or engagement (Rothbart, 2011). I hypothesize that parents may vary in how intentionally they respond to their children's preference for novelty seeking or engagement with racial dilemmas. The child may prefer an immediate, robust response to discriminatory treatment when the features of the situation warrant it; such a reaction may be justified under the circumstances as is the case with racial discrimination. However, the features of the racial discrimination dilemmas African Americans encounter daily are varied and complex (Fisher et al., 2000; Murphy et al., 2012; Salvatore & Shelton, 2007; Thomas & Blackmon, 2015). African American children *and* adults must think about potential risks to their safety and long-term well-being and weigh whether a more cautious approach is warranted.

Effortful control, defined as the capacity to inhibit a response in a given situation, has relevance for racial coping because there may be times when a child needs to inhibit a normal response in order to avoid placing themselves in danger (Dunbar et al., 2022a, b; Rothbart, 2011). African American children are often judged as being older than their chronological age (Goff et al., 2014). The realities of racial discrimination places greater developmental demands on African American children to be highly attuned to situational cues and see the big picture before responding. Children and adolescents with greater capacity for *effortful control* may be well positioned to receive and implement racial socialization messages and coping strategies from a dispositional standpoint. Adolescents high in surgency/novelty seeking and effortful control may be better able to judge the contours of a racial dilemma and levy a response. In contrast, adolescents classified as high in surgency/novelty seeking and low in effortful control may be more vulnerable and in greatest need of effective parental guidance.

Assumption 4.4: The combination/goodness-of-fit between parental judgement, sensitivity, empathy, and child attributes predict child developmental outcomes As noted, insufficient attention has been devoted to one of the most influential components of the TPRES; its assertion that racial socialization is a dyadic process between parents and children. Assumption 4.4 of the TRSA builds on this assertion by attempting to articulate some of the ways communication about racial issues unfolds. Specifically, goodness-of-fit and attachment conceptual frameworks should be applied to the racial socialization process (Dunbar et al., 2021; Stern et al., 2021). Goodness-of-fit models posit that variation in parents' capacity for perspective taking and empathy play a significant role in fostering youth self-regulation skills and capacity (Bañales et al., 2021; Camoirano, 2017; Scopesi et al., 2015; Wagers & Kiel, 2019).

Given that racial dilemmas are unpredictable, stressful, and perplexing for children (Fisher et al., 2000), the goodness-of-fit between parents' judgement, sensitivity, and empathy and the child's attributes (e.g., temperament, emotional maturity, cognitive understanding of race and racial dilemmas) is key to the dyadic racial socialization process. Parents' nuanced management of children's temperament factors in the face of spontaneous racial dilemmas (Dunbar et al., 2017; García Coll et al., 1996; Hughes & Chen, 1999) and capacity for managing their own emotions in the context of discrimination are posited to predict children's developmental outcomes (Lozada et al., 2022; Peters & Massey, 1983; Smith-Bynum et al., 2016). Goodness-of-fit likely facilitates children's cognitive understanding of racial dilemmas and their capacity to appraise them effectively (Anderson & Stevenson, 2019; Darling & Steinberg, 1993; Neitzel & Stright, 2003; Zimmer-Gembeck & Skinner, 2016).

Expansion of Racial Socialization Theory Through Observational Methods

I began observing African American parents and their teenaged children having discussions about hypothetical incidents involving potential racial discrimination in 2002. I relied heavily upon the early version of TPRES (Hughes & Chen, 1999). In developing the observational task and coding system (Smith-Bynum, 2011b), I created behavioral codes for the TPRES (Hughes & Chen, 1999; Hughes et al., 2006) seminal constructs: cultural socialization, preparation for bias, promotion of cultural mistrust, and silence/egalitarianism about race (Smith-Bynum & Usher, 2004). The discussions reflected various problem-solving strategies to address the dilemmas in the vignettes. Certain patterns of communication by the parents and the adolescents emerged across the videos that had not been identified at that time (Smith-Bynum & Usher, 2004). Against prediction, the TPRES behaviors rarely occurred as defined in the pilot studies (Smith-Bynum et al., 2005; Smith-Bynum, 2006; Smith-Bynum & Usher, 2004). Therefore, behavioral codes for newly identified behaviors were successfully developed to capture the variation across the dyads in a standardized manner (Smith-Bynum, 2008). These are listed in Fig. 4.2.

Based on pilot data from my early studies (Smith-Bynum, 2006, 2008; Smith-Bynum et al., 2005; Smith-Bynum & Usher, 2004), I posit that there are at least three superordinate areas of communication implicated in the parent-adolescent racial socialization process: (A) *Parent and Adolescent Reactions to the Events*; (B) *Parental Strategies for Coping with Racial Dilemmas*; (C) *Adolescent Strategies for Coping with Racial Dilemmas*. These communication areas can be observed in brief discussions as short as 5 minutes (Smith-Bynum, 2011b). In the sections to follow, I define the three major categories and the constructs within those categories.

The first category, *Parent and Adolescent Reactions to the Events,* consists of the wide array of responses family members have regarding the potential racism and bias in the vignettes. To date, I have identified at least four constructs under this category. The first, *Existence of Racial Bias*, involves the frequency of any verbal or nonverbal communication regarding parent or adolescent beliefs about the existence of racial bias in any form (e.g., racial discrimination, institutionalized racism, systemic racism, racial prejudice; see Chap. 2 in this book). *Existence of Racial Bias* consists of two subconstructs: *Affirmation of Racial Bias* and *Denial of Racial Bias*. Parents and adolescents each receive a score for the number of communications made that affirms or denies racial bias.

The following statements represent examples of affirmation of racial bias by parents and adolescents.

> "So really, he had a racist teacher" (Son, age 16, to mother, *The Teacher*)
>
> "Sounds to me like this one salesman is a straight bigot… I'm pretty sure this is not the first person he has been this blatantly unfair to," (Father to son, age 17, *The Mall*)
>
> "… racial profiling is going on everywhere" (Father to son, age 17, *The Mall*)

The construct, *Parent and Adolescent Denials of Racial Bias*, is defined as any direct statement indicating their belief denying the reality of racism in American society. These statements can be expressed in the context of the features of the vignette or in general. It can also occur in the form of an active disagreement with the statements made by a family member endorsing the existence of racism. Parents and adolescents receive separate ratings for this variable (Dunbar et al., 2022a; Smith-Bynum, 2011b). In my data collected to date, denials of racial bias are infrequent (Smith-Bynum, 2006, 2011a; Smith-Bynum & Usher, 2004). This is likely a function of the sample selection factors. Black families who enroll in studies focused on race are likely to already be actively engaged in racial socialization. Thus, when one or more family members express statements denying that racial bias exists, it is likely to be meaningful for youth adjustment (O'Brien Caughy et al., 2004). There is very limited scholarship to guide speculation about pathways of influence for these types of behaviors and communication; this is an area in need of further research attention.

The third construct, *Endorsement of Racial Stereotypes*, is defined as communication endorsing racial stereotypes about African Americans in a direct or indirect manner (Smith-Bynum et al., 2005; Smith-Bynum & Usher, 2004). Parents and adolescents receive separate ratings for statements reflecting beliefs about Black racial stereotypes. Endorsement of racial stereotypes likely emanates from belief systems based on internalized racism (Jones, 2000). Jones (2000) defines internalized racism as:

> "acceptance by members of the stigmatized races of negative messages about their own abilities and intrinsic worth. It is characterized by their not believing in others who look like them and not believing in themselves. It involves accepting limitations to one's own full humanity, including one's spectrum of dreams, one's right to self-determination, and one's range of allowable self-expression" (p. 1213).

Endorsement of Racial Stereotypes is a type of internalized racism expressed through racial socialization communication (Jones, 2000). It is similar to Stevenson et al.'s (2002) construct *of cultural endorsement of the mainstream. Cultural endorsement of the mainstream* is defined as racial socialization messages that emphasize access and involvement in White American cultural values and institutions as well as de-emphasizing the effects of racism on African Americans and the value of cultural connections to the African American community. In the validity study for the Teenager Experiences of Racial Socialization (TERS), Stevenson et al. (2002) identified a construct referred to as *cultural endorsement of the mainstream*. It is defined as "the relative importance of majority culture institutions and *values* [emphasis added] and the affective and educational benefits that African Americans can receive by being involved in those institutions" (p. 92).

Cultural endorsement of the mainstream was positively correlated with another TERS construct similar to preparation for bias (*cultural coping with antagonism*) and other constructs similar to cultural socialization (*cultural legacy appreciation, cultural pride reinforcement;* Stevenson et al., 2002). This small but important pattern of relationships among TERS constructs has not gained a great deal of attention in the literature. It likely means that a sizeable portion of Black parents consciously or unconsciously communicate mainstream or Western values to their children.

Furthermore, endorsement of racial stereotypes can co-occur with other racial socialization messages, thereby reducing the impact of positive messages about being Black. It may also explain some of the mixed findings in the literature regarding the impact of racial socialization more broadly. Although endorsement of racial stereotypes within Black families has received limited attention in the literature (Mouzon & McLean, 2017), the evidence demonstrates that racial stereotypes in general are harmful to Black people (Eberhardt et al., 2006; Griffith et al., 2019; Legewie, 2016; Smedley et al., 2009; Steele, 1997; Wyatt et al., 2003). As such, they are likely to have a harmful impact when they come from parents.

Examples of parental statements reflecting endorsement of racial stereotypes are:

> "That's true, it's [soccer] is a White people's sport, like rugby, and all that other stuff." (Son, age 16, to Mother, *Soccer*)
>
> "I don't know if it's boring, I just don't like it. It don't seem like a sport for Black people." (Mother to son, age 16, *Soccer*).

These statements reflect stereotypic beliefs about Black people and sports, resulting from the hypervisibility of Black people in certain sports (Turner, 2014). Given the widespread presence of racial stereotypes about African Americans in many arenas (Jean et al., 2022; Miles et al., 2020; Mouzon & McLean, 2017; Smith & Hope, 2020; Steele, 1997), it is not surprising that Black parents and children can internalize these stereotypes (Jones, 2000).

Observational data with the RSOTCS also revealed that some parents and adolescents actively reject racial stereotypes. For example, one parent stated the following in response to the Soccer vignette (Smith-Bynum & Usher, 2004).

> "What's a White sport and what's a Black sport?" (Daughter, age 15, to father, *Soccer*)
>
> "I would like to think that as a Black person, we would want to be global in a sense of there's no boundaries…Whatever it is you do, that's what you do, be it basketball or badminton, [race] shouldn't matter. (Father to daughter, age 15, *Soccer*)

Note that language used by the father intentionally disconnects racial group membership from participation in specific sports. Although limited research to date has examined this type of communication explicitly, communication focused on helping Black youth disentangle their humanity from the burden of racial stereotypes is likely to be beneficial to Black youth in many ways, including fostering development of a positive racial identity (Ahn et al., 2021; Nelson et al., in press; Oliver et al., 2017).

The fourth construct, *Affirming Egalitarianism*, is defined as parental communication that addresses the humanity of their children by addressing fundamental issues of equality in response to racial discriminatory incidents in the vignettes (Hughes & Chen, 1999; Hughes et al., 2006). *Affirming Egalitarianism* is distinct from general egalitarianism messages because these parental messages are meant to repair any harm done due to discriminatory experiences their children have encountered. General parental expressions of egalitarianism are often race-neutral in content and tone, reflecting broad American cultural narratives (e.g., "All men are created equal"; "We are all the same"). Affirming egalitarianism acknowledges the fact that African Americans still live with the impacts of harmful, dehumanizing racial stereotypes (Bonam et al., 2016; Goff et al., 2008, 2014; Quillian & Pager, 2001). Parents' *Affirming Egalitarianism* messages are intended to repair children's sense of self from the effects of these harmful stereotypes and dehumanizing situations they encounter throughout their daily lives.

Affirming egalitarianism is conceptually similar to Stevenson et al.'s (2002) construct *cultural coping with antagonism*. Cultural Coping with Antagonism focuses on the value of knowledge of Black history and spirituality as fuel for persistence in the face of racial barriers. Affirming egalitarianism parental messages help children develop psychological armor when confronting discriminatory situations and conditions. This psychological armor likely elevates Black children's developing racial centrality and contributes their positive racial identity (i.e., private regard; Davis Tribble et al., 2017; Huguley et al., 2019; Neblett Jr. et al., 2012).

Affirming egalitarianism are parental messages that validate the humanity of their adolescents in ways that help them persist in the face of racial discrimination (Smith-Bynum, 2006, 2008, 2011a, b; Smith-Bynum & Usher, 2004). Note that instructional guidance that parents provide their children is customized to the discriminatory situation in the vignette:

> "Well, first you got to recognize that you can't let somebody deter your dreams of what you want to do, you know? If it's something that you want to do in life, and if that's what you, it just takes hard work and dedication and commitment on your part to want to do it." (Mother to daughter, age 15, *The School Counselor*)
>
> "So just don't never let nobody ever tell you in life that you can't do something, you gotta rise above that, and just know in your heart, and you made up in your mind what you want to do, set out to achieve it. You may fail the first time, but you just still keep on trying." (Mother to daughter, age 15, *The School Counselor*)

These parental statements all reflect messages intended to deflect negative feedback or discouragement from naysayers and harmful situations. Also present in these messages are themes meant to reinforce or preserve children's self-efficacy despite discouragement and disillusionment borne from racial discrimination (Anderson et al., 2020; Anderson & Stevenson, 2019).

The second major category of racial socialization processes in the TRSA is *Parent Strategies for Coping with Discrimination*. This category consists of four communication strategies deployed by parents to their children to guide youth coping with situations involving complex racial dilemmas. As can be seen in Fig. 4.1, I have identified four strategies: parental advocacy, parental suggestions, parental scaffolding, and parental supportiveness. *Parental Advocacy* is defined as any strategy used by Black parents to equalize the power dynamic between a youth and a White adult. It involves parents stepping in on behalf of their child and engaging in problem -solving behavior to resolve the racial dilemma (Holman, 2012; Smith-Bynum et al., 2016).

The following statements are examples of parental advocacy by parents (Smith-Bynum et al., 2016; Smith-Bynum, 2011b; Smith-Bynum & Usher, 2004).

> "I'd leave work and go to the school and go to the Dean's office or the principal's office and ask to set up a meeting with the teacher" (Mother to daughter, age 16, *The Teacher*).

"You need an adult to intervene…That's an issue [for] the parent, the child, and the teacher." (Mother to daughter, age 15, *The Teacher*).

"I want a meeting between the principal, between the counselor and you and your parents, both parents. (Father to daughter, age 15, *The School Counselor*).

In contrast, *Parental Suggestions* involves specific strategies that parents share with their child to guide children in resolving the situation on their own. These solutions consist of practical, behavioral solutions as well as suggestions for regulating negative emotion in the face of potential discrimination (Dunbar et al., 2015, 2017; Lozada et al., 2022). The following statements are exemplars of parental suggestions in response to the hypothetical racial dilemmas (Smith-Bynum & Usher, 2004).

"Try to stay calm. I know it's hard." (Mother to daughter, age 16, *The Teacher*)

"I wouldn't stop at the principal. I would go to the school administrator, I would tell my parents, I would write a letter. You being 16, you have all types of resources available to you. And it's hard for anyone to listen to a 16-year-old black youth, but someone out there will. You have to find the right person and keep being relentless till someone calls you or sits down with you and says, 'Hey, I understand your problem. Let's see what we can talk about solving it.' (Father to son, age 17, *The Teacher*)

"… We talk about [it] in Bible study all the time. People not always going to receive us the way they should. What you know, we have, we still accountable to act as certain way. You know that don't mean go into his store and cuss and get mad and throw things or nothing else. If you walk into a store and somebody don't want you to spend your money in that store, you just walk out of the store. (Mother to daughter, age 15, *The Mall*)

The second parental communication strategy is *Parental Scaffolding*. Consistent with a Vygotskian perspective on racial socialization, *Parental Scaffolding* is the parents' efforts to support the mastery of their child's skills. Effective parental scaffolding requires that the parent be astute in identifying their child's zone of proximal learning (Vygotsky, 1978). They must have keen observational skills to assess the verbal and nonverbal indicators of their child's learning capacity. They must be able to ask questions effectively to assess understanding of racial dilemmas. They must also explain complex racial dilemmas to facilitate learning. Skilled scaffolders also affirm their children's efforts to solve racial dilemmas and offer additional explanatory guidance to their children in age-appropriate language.

Parental Scaffolding as defined in the TRSA is akin to the concept of racial socialization competency as outlined in by Anderson and Stevenson (2019) in the Racial Encounter Coping Appraisal and Socialization Theory (RECAST). In these excerpted statements, one participant father of a 15-year-old daughter affirmed her problem-solving effort and expanded on her effort with additional actions she could implement in the situation. The instruction or expansion in the problem solving is provided to his daughter and his affirmation after she expresses agreement and understanding about the strategy for solving the problem (Habermas et al., 2010).

Father: "Listen to me sweetheart, there's a protocol to everything…you should treat everyone fairly. The principal is the manager of that school. The principal needs to have an opportunity to chastise and to manage that counselor. When the principal decides that he or she does not want to do that then you go to the next level. You don't go to the next level before you give a manager the opportunity to solve a problem. And that's how you handle it, you go up the chain, you go up the chain of commands whatever that chain would be. So, I mean, I think this is something that will be solvable.

Daughter: Oh yeah, well that's what I'm saying I mean we can have that conversation quickly.

Father: Exactly. (Father to daughter, age 15, *The Teacher*)

Finally, *Parental Supportiveness* is defined as parents' direct expressions of emotional support of their children's efforts to solve racial dilemmas. Supportiveness can be expressed verbally and nonverbally. Parents' emotional support of their adolescents' problem-solving efforts represents a type of emotional scaffolding (Neitzel & Stright, 2003). Parental emotional supportiveness in the context of adolescents' racial problem-solving fosters their confidence and persistence in addressing racial dilemmas.

"Well, I applaud what you just said, that's great, I never, that thought never occurred to me to just ignore them" (Father to daughter, age 15, *Soccer*)

"I agree with you 100% because there's no need in you trying to talk to this person because obviously, they gave you [the grade] they thought you should have" (Mother to daughter, age 15, *The Teacher*)

The third category, *Adolescent Strategies for Coping with Discrimination*, represents the collection of behaviors adolescents shared in response to various racial dilemmas. As noted, children's coping responses remains one of the least explored aspects of racial socialization research to date (Anderson & Stevenson, 2019; Scott et al., 2020). While the TRSA articulates two constructs to date, there is room for substantial contributions to understanding how adolescents communicate about their racial coping in conversations with parents about racial discrimination. The two strategies are (1) *adolescent solutions for coping* and (2) *adolescent respect for elders*.

Adolescent solutions for coping represent the collection of individual problem-focused and emotion-focused coping strategies (Compas et al., 2017) that adolescents engage in in response to racial dilemmas. Their solutions vary in complexity and viability such that some might be viewed as impractical or unlikely to resolve the problem in the dilemmas. Some examples of *adolescent solutions* are as follows:

"What I do in this situation instead of going in mad and off at the teacher, I would tell the principal and have a time so I can get everything situated. 'Cause when you argue with the teacher, it makes things worse not better'." (Son, age 17, to *The Teacher*)

"I would really tell one of my friends first though. I'd be like, 'y'all guess what this woman said to me!'" (Daughter, age 16, to mother, *The School Counselor*)

"I'd have been real upset, we'd have tried on clothes, just out of spite...We'd be trying on clothes in that store, she'd have had to call security or something." (Daughter, age 16, to mother, *The Mall*)

The examples of solutions range in detail and quality. The RSOTCS assesses the frequency of the solutions not the quality of the solutions that adolescents generate. Recent research has revealed the importance of emotions in Black youth's coping with racial dilemmas (Dunbar et al., 2022b; Lozada et al., 2022). The RSOTCS does not assess emotional responses to the dilemmas in the vignettes that families discuss. However, an array of emotions can be observed in the video of discussions.

Different emotions are displayed as families shared their solutions, ranging from frustration, anger, bemusement, disbelief, and disgust (Dunbar et al., 2022a, b; Nyborg & Curry, 2003). Recent research using observational methods with younger Black youth indicates that the emotions-focused coping strategies Black youth must master incorporates the (Cooke & Halberstadt, 2021; Dunbar et al., 2022a, b). Thus, this is an area of potential expansion of the RSOTCS.

The second construct, *adolescent respect for elders,* is defined as the degree to which youth displayed respectful communication to the adults in their lives and intentions to "be respectful" in the presence of racial discrimination by White adults. This strategy appears to be reflective of norms in African American culture regarding respect for elders (Boykin & Toms, 1985; Johnson & Staples, 2005). Though understood to be a meaningful practice in African American cultural spaces, it has not been connected explicitly to recent influential models of racial socialization in research in the last 25 years. Teaching children respect for elders appears to be a type of preparation for bias (Smith-Bynum et al., 2021). In the discussions in the RSOT, parents often reminded their children about their previous instruction to "be respectful" in the presence of discrimination from White adults. And adolescents displayed deference to adults in their discussions with their parents. They used phrases such as "yes ma'am" or "no ma'am." They also described strategies for communicating with the White adults in the vignettes intended to defuse a potential confrontation. For instance, adolescents stated the following during the discussions:

> "It's not really my position as a child, a student, to approach an adult in a defensive manner." (Daughter, age 16 to Mother, *The School Counselor*)
>
> "Honestly, I would call you. Because I wouldn't want to say something mean to the lady, by the fact that she was an adult, I wouldn't." (Daughter, age 16, to mother, *The School Counselor*)

In reviewing the discussions, it appeared that some parents deliberately cultivated respect for elders in their children. There was variation in the use of the strategy by the youth in those studies. Our understanding of youth's adoption of parental guidance about racial issues is emergent, and thus new behaviors regarding this global category will evolve. For example, researchers have begun to expand the field's focus on racial coping within the family unit as well as by the youth (Anderson & Stevenson, 2019; Scott et al., 2020).

Applications of the TRSA

The Theory of Racial Socialization in Action (TRSA) attempts to facilitate deeper integration between racial socialization research, research on global aspects of parenting, and other areas of developmental science. Its focus on live conversations between family members gives researchers distinct insights into how the socialization process unfolds in real time. It presents a framework that can guide detailed questions regarding the features of discriminatory events, the ecological contexts shaping Black youth development, timing of the delivery of racial socialization strategies, and the role of the parent-adolescent relationship context (Scott et al., 2020). Given the fresh awareness of the harmful impacts of historic and current racist policies and practices—structural racism—on African American youth and other youth of color, a comprehensive and granular assessment of all the ways families navigate race and racism is long overdue.

Applications to families with different racial-ethnic backgrounds The TRSA and the RSOTCS are relevant to ethnic-racial socialization to other families of color and to children of different ages (Osborne et al., 2022; Osborne et al., in press). The theoretical assumptions of the TRSA likely have direct relevance to other families of color as they focus on family communication processes. Variables identified as *Reactions to the Events* likely apply across families from different ethnic-racial backgrounds. Though the spirit of the variables classified as *Parental Strategies for Coping with Discrimination* and *Adolescent Strategies for Coping with Discrimination* apply to communication about ethnic-racial issues in other families of color, there are likely culturally specific variations that need to be accounted for through construct modification. The RSOTCS has been successfully adapted for pre-adolescent aged youth and Latinx families (Osborne et al., 2022; Osborne et al., in press). Key modifications to the RSOT involved modification and creation of the vignettes to fit the developmental stage and lived experience of youth from Mexican-origin families.

Creations of new vignettes that resonate with the target population will likely be necessary to address the ways racism plays out in the treatment of immigrants of color and families of color presumed to be immigrants (e.g., perpetual foreigner stereotype; Lee 2008), and families of color in general. Furthermore, families in which members vary by nativity and length of residence in the United States should be major considerations in any adaptation of the RSOTCS for families of other backgrounds (Patel et al., 2022). Failure to adapt the ROSTCS will reduce its validity and contribute to ongoing lack of clarity about these highly nuanced dynamics occurring within families contending with complex racial and ethnic dynamics in the US context.

Additionally, researchers should be careful not to generalize the entire TRSA and RSOTCS to White families, Multiracial families, or transracial adoptive families. The theory and observational system were created and conceived with the minoritized and cultural experiences of African American families at the center. Observed ethnic-racial socialization in White families should explicitly incorporate family communication that reflect dominant group perspectives and experiences regarding race and ethnicity (e.g., colorblind ideology, silence about race) where appropriate (Hagerman, 2022; Pahlke et al., 2012). Similarly, attention to multiple perspectives from Multiracial families in with parents and caregivers with one White parent/caregiver, families with two parents/caregivers from racial-ethnic minoritized backgrounds (Atkins & Yoo, 2019; Green et al., 2022), and transracial adoptive families (Hrapczynski & Leslie, 2018).

Another key methodological consideration is in the administration and coding with the RSOTCS. Due to the reticence families will likely have talking about sensitive matters in cross-racial interactions (i.e., cultural mistrust; Shelton & Richeson, 2015), staff administering the RSOT should be matched on race and ethnicity with the family members participating in the study. Evidence shows that errors in cross-cultural misunderstandings can occur when people code behaviors and emotions for participants that have different racial backgrounds (Babcock & Banks, 2019; Gonzales et al., 1996). Thus, the coding should also be completed by trained staff who share the racial and ethnic identities with the participating families to enhance the cultural validty of the coding process.

Future Directions for Observational Methods with Families of Color

Even with the contributions of the TRSA, the universe of family-based racial socialization constructs has not been discovered. The TRSA focuses exclusively on racial socialization family dynamics involving discriminatory incidents specifically. Because it focuses on discrimination experiences, the RSOTCS does not reveal extensive information about the ways parents impart knowledge about Black culture.

Measurement of cultural socialization in action In their classic piece, Boykin and Toms (1985) asserted that Black children master the nuances of Black culture through routine exposure to Black cultural spaces and practices:

> "Black cultural motifs are typically what is [transmitted]...Black culture is socialized more so through a tacit cultural conditioning process, a process through which children pick up 'modes, sequences, and styles of behavior' through their day-to-day encounters with parents and other family members; encounters in which such modes, sequences, and styles of behavior are displayed to ...children in a consistent, persistent, and enduring fashion. This is so even if the cultural import of such displays is unarticulated, and even if they are not accompanied by directives or imperatives to learn them." (p. 42).

Self-report measures of cultural socialization are a necessary first line of attack from a measurement perspective, but they do not capture the richness and vitality of the cultural socialization process. It is critical that researchers continue to create robust measurements of how Black families teach children to master the ways of being embedded in Black culture as cultural socialization is a major source of Black youth's private regard and Black joy (Davis Tribble et al., 2017; Lacy, 2007; Neblett Jr. et al., 2013; Wang et al., 2019).

Investment in large-scale, representative studies with Black children and other Youth of Color With the introduction of the TRSA and the RSOTCS, studies that jointly examine parent and youth self-reported and observed racial socialization are needed to uncover the ways content and process affect acquisition of racial coping and youth developmental outcomes. Studies also need to examine how racial socialization processes affect youth growing up in different developmental contexts.

Black youth across the United States negotiate settings like neighborhoods and schools (see Chap. 8 in this book) that vary by racial composition, social class, urbanicity, region, and other key factors. Moreover, the settings can change over time when children transition to new schools and families move to different communities. Additionally, digital spaces and the COVID-19 pandemic present new challenges for development in youth of color. (Cheah et al., 2020; Weinstein et al., 2021).

More large-scale, representative studies that sample Black youth from a variety of community settings (e.g., rural, urban, suburban, regional)[2] are sorely needed. Public use datasets with representative subsamples of Black youth in the United States often fail to grant sufficient longitudinal coverage to race-related constructs or consider the heterogeneity within the Black population in the United States. In my research, samples resided in Mid-Western and Mid-Atlantic metropolitan areas, regions of the country with different race-related histories and racial-ethnic heterogeneity. The data revealed meaningful differences in the pattern of relationships between self-report and observed racial socialization variables without clear explanations as to why (Smith-Bynum, 2008; Smith-Bynum, 2011a). As the health impacts of racism come into clearer view (Braveman et al., 2011; Price et al., 2022), we need more studies that assess these racial realities and complexities to advance our understanding of impacts on the developmental and health outcomes in Black youth.

Conclusions

Black families and other families of color are negotiating extraordinarily tricky terrain in helping their children build capacity for coping with racism and discrimination. Researchers must continue to develop new theories and measurement tools to assess these complex life experiences. The TRSA provides a road map for new avenues to explore the impact of race and racism on Black children and other children of color. It also extends the reach of the groundbreaking work represented in the TPRES (Hughes et al., 2006) by articulating how the developmental processes outlined by Hughes et al. (2006) occur. Even with these advancements, more critical discoveries remain about the socialization processes for Black families and other families of color. As researchers, we must remain open to new discoveries in this arena and continue to press forward with the critical work to identify them. The health and well-being of tomorrow's youth depend on it.

[2] The Center for the Study of Black Youth in Context at the University of Michigan and the Family and Community of Health Study (Simons et al., University of Georgia) and are good examples of longitudinal studies with representative samples of Black children containing meaningful, useful coverage of race-related constructs.

Acknowledgments The National Institute of Mental Health (1R21MH083986-01A1) and the Purdue Research Foundation funded this research. The author thanks Tony Johnson and Indy Parks & Recreation, City of Indianapolis, Indiana, for assistance with data collection in the early stages of this research. The author expresses her gratitude to the families who participated in this research described in this chapter.

References

Ahn, L. H., Dunbar, A. S., Coates, E. E., & Smith-Bynum, M. A. (2021). Cultural and universal parenting, ethnic identity, and internalizing symptoms among African American adolescents. *Journal of Black Psychology, 47*(8), 695–717. https://doi.org/10.1177/00957984211034290

Anderson, R. E., Heard-Garris, N., & DeLapp, R. C. T. (2022). Future directions for vaccinating children against the American endemic: Treating racism as a virus. *Journal of Clinical Child and Adolescent Psychology, 51*(1), 127–142. https://doi.org/10.1080/15374416.2021.1969940

Anderson, R. E., Jones, S. C. T., & Stevenson, H. C. (2020). The initial development and validation of the Racial Socialization Competency Scale: Quality and quantity. *Cultural Diversity and Ethnic Minority Psychology, 26*(4), 426–436. https://doi.org/10.1037/cdp0000316

Anderson, R. E., & Stevenson, H. C. (2019). RECASTing racial stress and trauma: Theorizing the healing potential of racial socialization in families. *American Psychologist, 74*(1), 63–75. https://doi.org/10.1037/amp0000392

Atkin, A. L., & Yoo, H. C. (2019). Familial racial-ethnic socialization of multiracial American youth: A systematic review of the literature with MultiCrit. *Developmental Review*, 53. https://doi.org/10.1016/j.dr.2019.100869

Babcock, J. C., & Banks, J. C. (2019). Interobserver agreement and the effects of ethnicity on observational coding of affect. *Journal of Social and Personal Relationships, 36*(9), 2842–2856. https://doi.org/10.1177/0265407518803474

Bañales, J., Lozada, F. T., Channey, J., & Jagers, R. J. (2021). Relating through oppression: Longitudinal relations between parental racial socialization, school racial climate, oppressed minority ideology, and empathy in black male adolescents' prosocial development. *American Journal of Community Psychology, 68*(1–2), 89–99. https://doi.org/10.1002/ajcp.12496

Belsky, J. (1984). The determinants of parenting: A process model. *Child Development, 55*(1), 83–96. https://doi.org/10.2307/1129836

Bonam, C. M., Bergsieker, H. B., & Eberhardt, J. L. (2016). Polluting Black space. *Journal of Experimental Psychology: General, 145*(11), 1561–1582. https://doi.org/10.1037/xge0000226

Bor, J., Venkataramani, A. S., Williams, D. R., & Tsai, A. C. (2018). Police killings and their spill-over effects on the mental health of Black Americans: A population-based, quasi-experimental study. *The Lancet*. https://doi.org/10.1016/s0140-6736(18)31130-9

Boykin, A. W., & Toms, F. D. (1985). Black child socialization: A conceptual framework. In H. P. McAdoo & J. L. McAdoo (Eds.), *Black children: Social, educational, and parental environments* (pp. 33–51). Sage Publications.

Braveman, P., Egerter, S., & Williams, D. R. (2011). The social determinants of health: Coming of age. *Annual Review of Public Health, 32*(1), 381–398. https://doi.org/10.1146/annurev-publhealth-031210-101218

Brown, T. N., & Lesane-Brown, C. L. (2006). Race socialization messages across historical time. *Social Psychology Quarterly, 69*(2), 201–213. https://doi.org/10.1177/019027250606900205

Buchanan, N. T., Perez, M., Prinstein, M. J., & Thurston, I. B. (2021). Upending racism in psychological science: Strategies to change how science is conducted, reported, reviewed, and disseminated. *American Psychologist, 76*(7), 1097–1112. https://doi.org/10.1037/amp0000905

Camoirano, A. (2017). Mentalizing makes parenting work: A review about parental reflective functioning and clinical interventions to improve it. *Frontiers in Psychology, 8*(14). https://doi.org/10.3389/fpsyg.2017.00014

Carter, R., & Flewellen, A. O. (2022). Ethnic-racial socialization among mothers of Black and Black-White biracial daughters during the pubertal transition. *Journal of Research on Adolescence*. https://doi.org/10.1111/jora.12724

Caughy, M. O. B., Randolph, S. M., & O'Campo, P. J. (2002). The Africentric Home Environment Inventory: An observational measure of the racial socialization features of the home environment for African American preschool children. *Journal of Black Psychology, 28*(1), 37–52.

Chae, Y., Ogle, C. M., & Goodman, G. S. (2009). Remembering negative childhood experiences. In *Emotion and memory in development: Biological, cognitive, and social considerations* (pp. 2–27). Oxford University Press.

Cheah, C. S. L., Wang, C., Ren, H., Zong, X., Cho, H. S., & Xiaofang, X. (2020). COVID-19 Racism and mental health in Chinese American families. *Pediatrics, 146*(5) https://doi.org/10.1542/peds.2020-021816

Cliffordson, C. (2001). Parents' judgments and students' self-judgments of empathy: The structure of empathy and agreement of judgments based on the interpersonal reactivity index (IRI). *European Journal of Psychological Assessment, 17*(1), 36–47. https://doi.org/10.1027//1015-5759.17.1.36

Coard, S. I. (2021). Race, discrimination, and racism as "growing points" for consideration: Attachment theory and research with African American families. *Attachment & Human Development, 24*(3), 373–383. https://doi.org/10.1080/14616734.2021.1976931

Condon, E. M., Barcelona, V., Ibrahim, B. B., Crusto, C. A., & Taylor, J. Y. (2022). Racial discrimination, mental health, and parenting among African American mothers of preschool-aged children. *Journal of the American Academy of Child & Adolescent Psychiatry, 61*(3), 402–412. https://doi.org/10.1016/j.jaac.2021.05.023

Compas, B. E., Jaser, S. S., Bettis, A. H., Watson, K. H., Gruhn, M. A., Dunbar, J. P., Williams, E., & Thigpen, J. C. (2017). Coping, emotion regulation, and psychopathology in childhood and adolescence: A meta-analysis and narrative review. *Psychological Bulletin, 143*(9), 939–991. https://doi.org/http://psycnet.apa.org/doi/10.1037/bul0000110

Cooke, A. N., & Halberstadt, A. G. (2021). Adultification, anger bias, and adults' different perceptions of black and white children. *Cognition and Emotion, 35*(7), 1416–1422. https://doi.org/10.1080/02699931.2021.1950127

Cooper, S. M., & McLoyd, V. C. (2011). Racial barrier socialization and the well-being of African American adolescents: The moderating role of mother–adolescent relationship quality. *Journal of Research on Adolescence, 21*(4), 895–903. https://doi.org/10.1111/j.1532-7795.2011.00749.x

Darling, N., & Steinberg, L. (1993). Parenting style as context: An integrative model. *Psychological Bulletin, 113*(3), 487–496.

Davis Tribble, B. L., Smith-Bynum, M. A., Saleem, F. T., et al. (2017). Racial socialization, private regard, and behavior problems in African American youth: Global self-esteem as a mediator. *Journal of Child and Family Studies, 26*, 709–720. https://doi.org/10.1007/s10826-016-0601-8

Davis Tribble, B. L., Allen, S. H., Hart, J. R., Francois, T. S., & Smith-Bynum, M. A. (2019). 'No [right] way to be a Black woman': Exploring gendered racial socialization among Black women. *Psychology of Women Quarterly, 43*(3), 381–397. https://doi.org/10.1177/0361684318825439

Dovidio, J. F., & Gaertner, S. L. (2004). Aversive racism. In M. P. Zanna (Ed.), *Advances in experimental social psychology* (Vol. 36, pp. 1–51). Academic Press.

Duckitt, J. H. (1992). Psychology and prejudice: A historical analysis and integrative framework. *American Psychologist, 47*(10), 1182–1193. https://doi.org/10.1037/0003-066X.47.10.1182

Dunbar, A. S., Perry, N. B., Cavanaugh, A. M., & Leerkes, E. M. (2015). African American parents' racial and emotion socialization profiles and young adults' emotional adaptation. *Cultural Diversity & Ethnic Minority Psychology, 21*(3), 409–419. https://doi.org/10.1037/a0037546

Dunbar, A. S., Leerkes, E. M., Coard, S. I., Supple, A. J., & Calkins, S. (2017). An integrative conceptual model of parental racial/ethnic and emotion socialization and links to children's social-emotional development among African American families. *Child Development Perspectives, 11*(1), 16–22. https://doi.org/10.1111/cdep.12218

Dunbar, A. S., Lozada, F. T., Ahn, L. H., & Leerkes, E. M. (2021). Mothers' preparation for bias and responses to children's distress predict positive adjustment among Black children: An attachment perspective. *Attachment and Human Development, 24*(3), 287–303. https://doi.org/10.1080/14616734.2021.1976922

Dunbar, A. S., HaRim Ahn, L., Coates, E. E., & Smith-Bynum, M. A. (2022a). Observed dyadic racial socialization disrupts the association between frequent discriminatory experiences and emotional reactivity among Black adolescents. *Child Development, 93*(1), 39–57. https://doi.org/10.1111/cdev.13680

Dunbar, A. S., Zeytinoglu, S., & Leerkes, E. M. (2022b). When is parental suppression of Black children's negative emotions adaptive? The role of preparation for racial bias and children's resting cardiac vagal tone. *Research on Child and Adolescent Pychopathology, 50*, 163–176. https://doi.org/10.1007/s10802-021-00779-z

Eberhardt, J. L., Goff, P. A., Purdie, V. J., & Davies, P. G. (2004). Seeing black: Race, crime, and visual processing. *Journal of Personality and Social Psychology, 87*(6), 876–893. https://doi.org/10.1037/0022-3514.87.6.876

Eberhardt, J. L., Davies, P. G., Purdie-Vaughns, V. J., & Johnson, S. L. (2006). Looking deathworthy: Perceived stereotypicality of Black defendants predicts capital-sentencing outcomes. *Psychological Science, 17*(5), 383–386. https://doi.org/10.1111/j.1467-9280.2006.01716.x

Eisenberg, N., Smith, C. L., & Spinrad, T. L. (2004). Effortful control: Relations with emotion regulation, adjustment, and socialization in childhood. In K. D. Vohs & R. F. Baumeister (Eds.), *Handbook of self-regulation: Research, theory, and applications* (2nd ed., pp. 263–283). Guilford Press.

Fisher, C. B., Wallace, S. A., & Fenton, R. E. (2000). Discrimination distress during adolescence. *Journal of Youth and Adolescence, 29*(6), 679–695. https://doi.org/10.1023/A:1026455906512

Frabutt, J. M., Walker, A. M., & MacKinnon-Lewis, C. (2002). Racial socialization messages and the quality of mother/child interactions in African American families. *The Journal of Early Adolescence, 22*(2), 200–217. https://doi.org/10.1177/0272431602022002004

García Coll, C., Lamberty, G., Jenkins, R., McAdoo, H. P., Crnic, K., Wasik, B. H., & Garcia, H. V. (1996). An integrative model for the study of developmental competencies in minority children. *Child Development, 67*(5), 1891–1914. https://doi.org/10.1111/j.1467-8624.1996.tb01834.x

Goff, P. A., Eberhardt, J. L., Williams, M. J., & Jackson, M. C. (2008). Not yet human: Implicit knowledge, historical dehumanization, and contemporary consequences. *Journal of Personality and Social Psychology, 94*(2), 292–306. https://doi.org/10.1037/0022-3514.94.2.292

Goff, P. A., Jackson, M. C., Di Leone, B. A., Culotta, C. M., & DiTomasso, N. A. (2014). The essence of innocence: Consequences of dehumanizing Black children. *Journal of Personality & Social Psychology, 106*(4), 526–545. https://doi.org/10.1037/a0035663

Golinkoff, R. M., Can, D. D., Soderstrom, M., & Hirsh-Pasek, K. (2015). (Baby)talk to me: The social context of infant-directed speech and its effects on early language acquisition. *Current Directions in Psychological Science, 24*(5), 339–344. https://doi.org/10.1177/0963721415595345

Gonzales, N. A., Cauce, A. M., & Mason, C. A. (1996). Interobserver agreement in the assessment of parental behavior and parent-adolescent conflict: African American mothers, daughters, and independent observers. *Child Development, 67*(1), 1483–1498. https://doi.org/10.1111/j.1467-8624.1996.tb01809.x

Gredler, M. E., & Shields, C. C. (2008). *Vygotsky's legacy: A foundation for research and practice*. Guilford Press.

Green, M. N., Christophe, N. K., & Lozada, F. T. (2022). Concordant and discordant patterns of parental racial socialization among biracial Black-White adolescents: Correlates and consequences. *Journal of Youth and Adolescence, 51*(12), 2340–2354. https://doi.org/10.1007/s10964-022-01667-5

Griffith, A. N., Hurd, N. M., & Hussain, S. B. (2019). 'I didn't come to school for this': A qualitative examination of experiences with race-related stressors and coping responses among Black students attending a predominantly White institution. *Journal of Adolescent Research, 34*(2), 115–139. https://doi.org/10.1177/0743558417742983

Habermas, T., Negele, A., & Mayer, F. B. (2010). 'Honey, you're jumping about'—Mothers' scaffolding of their children's and adolescents' life narration. *Cognitive Development, 25*(4), 339–351. https://doi.org/10.1016/j.cogdev.2010.08.004

Hagerman, M. A. (2018). *White kids: Growing up with privilege in a racially divided America.* NYU Press.

Hagerman, M. A. (2022). Considerations for future research on racial learning processes and White youth. *Journal of Research on Adolescence, 32*(3), 943–948. https://doi.org/10.1111/jora.12791

Harrell, S. P. (2000). A multidimensional conceptualization of racism-related stress: Implications for the well-being of people of color. *American Journal of Orthopsychiatry, 70*(1), 42–57.

Harrison, A. O., Wilson, M. N., Pine, C. J., & Chan, S. Q. (1990). Family ecologies of ethnic minority children. *Child Development, 61*(2), 347–362. https://doi.org/10.1111/j.1467-8624.1990.tb02782.x

Hrapczynski, K. M., & Leslie, L. A. (2018). Engagement in racial socialization among transracial adoptive families with White parents. *Family Relations: An Interdisciplinary Journal of Applied Family Studies, 67*(3), 354–367. https://doi.org/10.1111/fare.12316

Heard-Garris, N. J., Cale, M., Camaj, L., Hamati, M. C., & Dominguez, T. P. (2018). Transmitting trauma: A systematic review of vicarious racism and child health. *Social Science & Medicine, 199*, 230–240. https://doi.org/10.1016/j.socscimed.2017.04.018

Hirvonen, R., Väänänen, J., Aunola, K., Ahonen, T., & Kiuru, N. (2018). Adolescents' and mothers' temperament types and their roles in early adolescents' socioemotional functioning. *International Journal of Behavioral Development, 42*(5), 453–463. https://doi.org/10.1177/0165025417729223

Hoggard, L. S., Jones, S. C. T., & Sellers, R. M. (2016). Racial cues and racial identity: Implications for how African Americans experience and respond to racial discrimination. *Journal of Black Psychology, 43*(4), 409–432. https://doi.org/10.1177/0095798416651033

Holman, A. R. (2012). Gendered racial socialization in black families: Mothers' beliefs, approaches, and advocacy (Publication No. 3555722). [Doctoral Dissertation, University of California-Berkeley]. ProQuest Dissertations & Theses Global.

Hughes, D., & Chen, L. (1999). The nature of parent's race-related communications to children: A developmental perspective. In L. Balter & C. S. Tamis-LeMonda (Eds.), *Child psychology: A handbook of contemporary issues* (pp. 467–490). Psychology Press.

Hughes, D., & Johnson, D. (2001). Correlates in children's experiences of parents' racial socialization behaviors. *Journal of Marriage and Family, 63*(4), 981–995. https://doi.org/10.1111/j.1741-3737.2001.00981.x

Hughes, D., Rodriguez, J., Smith, E. P., Johnson, D. J., Stevenson, H. C., & Spicer, P. (2006). Parents' ethnic-racial socialization practices: A review of research and directions for future study. *Developmental Psychology, 42*(5), 747–770. https://doi.org/10.1037/0012-1649.42.5.747

Huguley, J. P., Wang, M. T., Vasquez, A. C., & Guo, J. (2019). Parental ethnic–racial socialization practices and the construction of children of color's ethnic–racial identity: A research synthesis and meta-analysis. *Psychological Bulletin, 145*(5), 437–458. https://doi.org/10.1037/bul0000187

Iruka, I. U., Lewis, M. L., Lozada, F. T., Bocknek, E. L., & Brophy-Herb, H. E. (2021). Call to action: Centering blackness and disrupting systemic racism in infant mental health research and academic publishing. *Infant Mental Health Journal, 42*(6), 745–748. https://doi.org/10.1002/imhj.21950

Jean, E. A., Neal-Barnett, A., & Stadulis, R. (2022). How we see us: An examination of factors shaping the appraisal of stereotypical media images of black women among black adolescent girls. *Sex Roles: A Journal of Research*. https://doi.org/10.1007/s11199-021-01269-8

Johnson, D. J. (2005). The ecology of children's racial coping: Family, school, and community influences. In T. S. Weisner (Ed.), *Discovering successful pathways in children's development: Mixed methods in the study of childhood and family life* (pp. 87–109). University of Chicago Press.

Johnson, L. B., & Staples, R. (2005). *Black families at the crossroads: Challenges and prospects*. (Rev. Ed). New York: Wiley & Sons.

Jones, C. P. (2000). Levels of racism: A theoretic framework and a gardener's tale. *American Journal of Public Health, 90*(8), 1212–1215. https://doi.org/10.2105/AJPH.90.8.1212

Jones, N. L., Gilman, S. E., Cheng, T. L., Drury, S. S., Hill, C. V., & Geronimus, A. T. (2019). Life course approaches to the causes of health disparities. *American Journal of Public Health, 109*(S1), S48–S55. https://doi.org/10.2105/AJPH.2018.304738

Jones, S. C. T., Anderson, R. E., & Stevenson, H. C. (2021). Differentiating competency from content: Parental racial socialization profiles and their associated factors. *Family Process*. https://doi.org/10.1111/famp.12699

Keating, L., Kaur, A., Mendieta, M., Gleason, C., Basello, G., Roth, A., & Brondolo, E. (2022). Racial discrimination and core executive functions. *Stress and Health, 38*(3), 1–7. https://doi.org/10.1002/smi.3116

Lacy, K. R. (2007). *Blue-chip black: Race, class, and status in the new black middle class*. University of California Press.

Landor, A. M., Simons, L. G., Simons, R. L., Brody, G. H., Bryant, C. M., Gibbons, F. X., Granberg, E. M., & Melby, J. N. (2013). Exploring the impact of skin tone on family dynamics and race-related outcomes. *Journal of Family Psychology, 27*(5), 817–826. https://doi.org/10.1037/a0033883

Lavner, J. A., Hart, A. R., Carter, S. E., & Beach, S. R. H. (2022). Longitudinal effects of racial discrimination on depressive symptoms among Black youth: Between- and within-person effects. *Journal of the American Academy of Child & Adolescent Psychiatry, 61*(1), 56–65. https://doi.org/10.1016/j.jaac.2021.04.020

Lee, S. J. (2008). The model minority and the perpetual foreigner: Stereotypes of Asian Americans. In N. Tewari & A. N. Alvarez (Eds.), *Asian American Psychology: Current perspectives* (pp. 119–134). Psychology Press.

Legewie, J. (2016). Racial profiling and use of force in police stops: How local events trigger periods of increased discrimination. *American Journal of Sociology, 122*(2), 379–424. https://doi.org/10.1086/687518

Lozada, F. T., Riley, T. N., Catherine, E., & Brown, D. W. (2022). Black emotions matter: Understanding the impact of racial oppression on Black youth's emotional development. *Journal of Research on Adolescence, 32*(1), 13–33. https://doi.org/10.1111/jora.12699

Maccoby, E. E., & Martin, J. A. (1983). Socialization in the context of the family: Parent-child interaction. In E. M. Hetherington (Ed.), *Handbook of child psychology* (Vol. 4, pp. 1–101). Routledge.

Miles, M. L., Brockman, A. J., & Naphan-Kingery, D. E. (2020). Invalidated identities: The disconfirming effects of racial microaggressions on Black doctoral students in STEM. *Journal of Research in Science Teaching, 57*(10), 1608–1631. https://doi.org/10.1002/tea.21646

Morales, S., Pérez-Edgar, K., & Buss, K. (2016). Longitudinal relations among exuberance, externalizing behaviors, and attentional bias to reward: The mediating role of effortful control. *Developmental Science, 19*(5), 853–862. https://doi.org/10.1111/desc.12320

Mouzon, D. M., & McLean, J. S. (2017). Internalized racism and mental health among African-Americans, US-born Caribbean Blacks, and foreign-born Caribbean Blacks. *Ethnicity & Health, 22*(1), 36–48. https://doi.org/10.1080/13557858.2016.1196652

Mulvaney, M. K., McCartney, K., Bub, K. L., & Marshall, N. L. (2006). Determinants of dyadic scaffolding and cognitive outcomes in first graders. *Parenting: Science and Practice, 6*(4), 297–320. https://doi.org/10.1207/s15327922par0604_2

Murphy, M. C., Richeson, J. A., Shelton, J. N., Rheinschmidt, M. L., & Bergsieker, H. B. (2012). Cognitive costs of contemporary prejudice. *Group Processes & Intergroup Relations, 16*(5), 560–571. https://doi.org/10.1177/1368430212468170

Murry, V. M., Gonzalez, C. M., Hanebutt, R. A., Bulgin, D., Coates, E. E., Inniss-Thompson, M. N., Debreaux, M. L., Wilson, W. E., Abel, D., & Cortez, M. B. (2021). Longitudinal study of the cascading effects of racial discrimination on parenting and adjustment among African

American youth. *Attachment & Human Development*, 1–17. https://doi.org/10.1080/14616734.2021.1976926

Neblett, E. W., Jr., Rivas-Drake, D., & Umaña-Taylor, A. J. (2012). The promise of racial and ethnic protective factors in promoting ethnic minority youth development. *Child Development Perspectives, 6*(3), 295–303. https://doi.org/10.1111/j.1750-8606.2012.00239.x

Neblett, E. W., Jr., Banks, K. H., Cooper, S. M., & Smalls-Glover, C. (2013). Racial identity mediates the association between ethnic-racial socialization and depressive symptoms. *Cultural Diversity and Ethnic Minority Psychology, 19*(2), 200–207. https://doi.org/10.1037/a0032205

Neitzel, C., & Stright, A. D. (2003). Mothers' scaffolding of children's problem solving: Establishing a foundation of academic self-regulatory competence. *Journal of Family Psychology, 17*(1), 147–159. https://doi.org/10.1037/0893-3200.17.1.147

Nelson, S. C., Syed, M., Tran, A. G. T. T., Hu, A. W., & Lee, R. M. (in press). Pathways to ethnic-racial identity development and psychological adjustment: The differential associations of cultural socialization by parents and peers. https://doi.org/10.31234/osf.io/u6sny

Nyborg, V. N., & Curry, J. F. (2003). The impact of perceived racism: Psychological symptoms among African American boys. *Journal of Clinical Child & Adolescent Psychology, 32*(2), 258–266. https://doi.org/10.1207/S15374424JCCP3202_11

O'Brien Caughy, M., O'Campo, P. J., & Muntaner, C. (2004). Experiences of racism among African American parents and the mental health of their preschool aged children. *American Journal of Public Health, 94*(12), 2118–2124. https://doi.org/10.2105/AJPH.94.12.2118

Oliver, A., Andemeskel, G., King, C. R., Wallace, L., McDougal, S., Monteiro, K. P., & Ben-Zeev, A. (2017). 'I'm Black and I'm Proud': A majority ecological context protects affective aspects of Black identity under stereotype threat. *Race and Social Problems, 9*(4), 313–320. https://doi.org/10.1007/s12552-017-9216-y

Osborne, K. R., Walsdorf, A. A., Smith-Bynum, M. A.., Redig, S. L., Brinkley, D., Owen, M. T., & Caughy, M. O. (2022). Responding to racism at school: Ethnic-racial socialization and the academic engagement of Black and Latinx youth. *Child Development*. https://doi.org/10.1111/cdev.13853

Osborne, K., Smith-Bynum, M. A., Walsdorf, A., Anderson, L., & Caughy, M. O. (in press). Preparing Black and Latinx children for police encounters: Caregiver response profiles and child self-regulation. *Journal of Research on Adolescence*.

Pahlke, E., Bigler, R. S., & Suizzo, M. (2012). Relations between colorblind socialization and children's racial bias: Evidence from European American mothers and their preschool children. *Child Development, 83*(4), 1164–1179. https://doi.org/10.1111/j.1467-8624.2012.01770.x

Patel, P., Martin Romero, M. Y., Stein, G. L. & Raval, V. (2022). Promoting pride but missing the need for preparation for bias: Racial-ethnic socialization among Indian American families living in the southeast U.S. *Asian American Journal of Psychology*. https://doi.org/10.1037/aap0000296

Peters, M. F., & Massey, G. (1983). Mundane extreme environmental stress in family stress theories: The case of black families in white America. *Marriage & Family Review, 6*(1–2), 193–218. https://doi.org/10.1300/J002v06n01_10

Price, M. A., Weisz, J. R., McKetta, S., Hollinsaid, N. L., Lattanner, M. R., Reid, A. E., & Hatzenbuehler, M. L. (2022). Meta-analysis: Are psychotherapies less effective for Black youth in communities with higher levels of anti-Black racism? *Journal of the American Academy of Child & Adolescent Psychiatry, 61*(6), 754–763. https://doi.org/10.1016/j.jaac.2021.07.808

Priest, N., Walton, J., White, F., Kowal, E., Baker, A., & Paradies, Y. (2014). Understanding the complexities of ethnic-racial socialization processes for both minority and majority groups: A 30-year systematic review. *International Journal of Intercultural Relations, 43*, 139–155. https://doi.org/10.1016/j.ijintrel.2014.08.003

Quillian, L., & Pager, D. (2001). Black neighbors, higher crime? The role of racial stereotypes in evaluations of neighborhood crime. *American Journal of Sociology, 107*(3), 717–767. https://doi.org/10.1086/338938

Rasmussen, H. F., Borelli, J. L., Smiley, P. A., Cohen, C., Cheung, R. C. M., Fox, S., Marvin, M., & Blackard, B. (2017). Mother-child language style matching predicts children's and mothers' emotion reactivity. *Behavioural Brain Research, 325*, 203–213. https://doi.org/10.1016/j.bbr.2016.12.036

Reynolds, J. E., & Gonzales-Backen, M. A. (2017). Ethnic-racial socialization and the mental health of African Americans: A critical review. *Journal of Family Theory & Review, 9*(2), 182–200. https://doi.org/10.1111/jftr.12192

Riina, E. M., & McHale, S. M. (2012). Adolescents' experiences of discrimination and parent-adolescent relationship quality: The moderating roles of sociocultural processes. *Journal of Family Issues, 33*(7), 851–873. https://doi.org/10.1177/0192513X11423897

Roberts, S. O., Bareket-Shavit, C., Dollins, F. A., Goldie, P. D., & Mortenson, E. (2020). Racial inequality in psychological research: Trends of the past and recommendations for the future. *Perspectives on Psychological Science, 15*(6), 1295–1309. https://doi.org/10.1177/1745691620927709

Roque, L., Veríssimo, M., Fernandes, M., & Rebelo, A. (2013). Emotion regulation and attachment: Relationships with children's secure base, during different situational and social contexts in naturalistic settings. *Infant Behavior and Development, 36*(3), 298–306.

Rothbart, M. K. (2007). Temperament development and personality. *Current Directions in Psychological Science 16*(4) 207–212. https://doi.org/10.1111/j.1467-8721.2007.00505.x

Rothbart, M. K. (2011). *Becoming who we are: Temperament and personality in development*. Guilford Press.

Salvatore, J., & Shelton, J. N. (2007). Cognitive costs of exposure to racial prejudice. *Psychological Science, 18*(9), 810–815. https://doi.org/10.1111/j.1467-9280.2007.01984.x

Sanders Thompson, V. L. (1994). Socialization to race and its relationship to racial identification among African Americans. *Journal of Black Psychology, 20*(2), 175–188.

Scopesi, A. M., Rosso, A. M., Viterbori, P., & Panchieri, E. (2015). Mentalizing abilities in pre-adolescents' and their mothers' autobiographical narratives. *The Journal of Early Adolescence, 35*(4), 467–483. https://doi.org/10.1177/0272431614535091

Scott, J. C., Pinderhughes, E. E., & Johnson, S. K. (2020). How does racial context matter?: Family preparation-for-bias messages and racial coping reported by Black youth. *Child Development, 91*(5), 1471–1490. https://doi.org/10.1111/cdev.13332

Shai, D., & Belsky, J. (2017). Parental embodied mentalizing: How the nonverbal dance between parents and infants predicts children's socio-emotional functioning. *Attachment & Human Development, 19*(2), 191–219. https://doi.org/10.1080/14616734.2016.1255653

Sharp, C., Fonagy, P., & Goodyer, I. M. (2006). Imagining your child's mind: Psychosocial adjustment and mothers' ability to predict their children's attributional response styles. *British Journal of Developmental Psychology, 24*(1), 197–214. https://doi.org/10.1348/026151005X82569

Shelton, J. N., & Richeson, J. A. (2015). Interacting across racial lines. In M. Mikulincer, P. R. Shaver, J. F. Dovidio, & J. A. Simpson (Eds.), *APA handbook of personality and social psychology, volume 2: Group processes* (pp. 395–422). American Psychological Association. https://doi.org/10.1037/14342-015

Shiner, R. L., Buss, K. A., McClowry, S. G., Putnam, S. P., Saudino, K. J., & Zentner, M. (2012). What is temperament now? Assessing progress in temperament research on the twenty-fifth anniversary of Goldsmith et al. *Child Development Perspectives, 6*(4), 436–444. https://doi.org/10.1111/j.1750-8606.2012.00254.x

Slade, A. (2005). Parental reflective functioning: An introduction. *Attachment & Human Development, 7*(3), 269–281. https://doi.org/10.1080/14616730500245906

Smalls, C. (2010). Effects of mothers' racial socialization and relationship quality on African American youth's school engagement: A profile approach. *Cultural Diversity & Ethnic Minority Psychology, 16*(4), 476–484. https://doi.org/10.1037/a0020653

Smedley, B. D., Stith, A. Y., & Nelson, A. R. (2009). *Unequal treatment: Confronting racial and ethnic disparities in health care (with CD)*. National Academies Press.

Smetana, J. G., & Abernethey, A. (1998). *African American families macro-coding manual.* Unpublished manual, University of Rochester.

Smith, C. D., & Hope, E. C. (2020). 'We just want to break the stereotype': Tensions in Black boys' critical social analysis of their suburban school experiences. *Journal of Educational Psychology, 112*(3), 551–566. https://doi.org/10.1037/edu0000435

Smith-Bynum, M. A. (2006). *Talk to me!: A pilot study of racial socialization.* [Unpublished raw data]. University of Maryland-College Park.

Smith-Bynum, M. A. (2008). Observing racial socialization in action: A construct validity study with African American families. In M. Yasui (Chair), *Examining ethnic and racial socialization influences on adolescent adjustment: Clinical implications and assessment.* Paper presented at the biennial meeting of the Society for Research on Adolescence, Chicago, IL.

Smith-Bynum, M. A. (2011a). *Black Parenting Project 2012.* [Unpublished raw data]. University of Maryland-College Park.

Smith-Bynum, M. A. (2011b). *Racial socialization observational task and coding manual.* [Unpublished Research Manual]. University of Maryland.

Smith-Bynum, M. A. (2022). It's about time Black Lives Matter: The urgency of research on Black youth development. *Journal of Research on Adolescence, 32*(1), 302–306. https://doi.org/10.1111/jora.12735

Smith-Bynum, M., & Usher, K. (2004). "Really, he had a racist teacher": African American parent-adolescent conversations about racial discrimination. In. M. S. Bynum (Chair), *Parenting adolescents in ethnic minority families: Perspectives on warmth, control, and ethnic socialization.* Symposium conducted at the biennial conference of the Society for Research on Adolescence, Baltimore, MD.

Smith-Bynum, M., Usher, K. K., Callands, T. A., & Burton, E. T. (2005). An observational measure of African American parent-adolescent communication about racial discrimination. Paper presented at the biennial conference of the Society for Research in Child Development, Atlanta, GA.

Smith-Bynum, M. A., Anderson, R. E., Davis, B. L., Franco, M. G., & English, D. (2016). Observed racial socialization and maternal positive emotions in African American mother–adolescent discussions about racial discrimination. *Child Development, 87*(6), 1926–1939. https://doi.org/10.1111/cdev.12562

Smith-Bynum, M. A., Dunbar, A. S., Ahn, L. H., Coates, E. E., Franco, M. G., & Small, K. (2019), March. Observed mother-adolescent communication about racial issues and adolescent mental health in African American families. In M. Banarjee (Chair), *Examining the relation of ethnic racial socialization and racial discrimination in ethnic minorities.* Symposium presented at the biennial meeting of the Society for Research in Child Development, Baltimore, MD.

Smith-Bynum, M. A., Coates, E. E., Ahn, L. H., Allen, S. H., & Franco, M. G. (2021). *African American mothers' observed racial socialization and adolescents' responses to racial dilemmas.* Department of Family Science, University of Maryland. [Manuscript submitted for publication].

Spencer, M. B. (1983). Children's cultural values and parental child rearing strategies. *Developmental Review, 3*(4), 351–370. https://doi.org/10.1016/0273-2297(83)90020-5

Steele, C. M. (1997). A threat in the air: How stereotypes shape intellectual identity and performance. *American Psychologist, 52*(6), 613–629. https://doi.org/10.1037/0003-066X.52.6.613

Stein, G. L., Coard, S. I., Kiang, L., Smith, R. K., & Mejia, Y. C. (2018). The intersection of racial–ethnic socialization and adolescence: A closer examination at stage-salient issues. *Journal of Research on Adolescence, 28*(3), 609–621. https://doi.org/10.1111/jora.12380

Stern, J. A., Barbarin, O., & Cassidy, J. (2021). Working toward anti-racist perspectives in attachment theory, research, and practice. *Attachment & Human Development*, 1–31. https://doi.org/10.1080/14616734.2021.1976933

Stevenson, H. C. (1994). Validation of the scale of racial socialization for African American adolescents: Steps toward multidimensionality. *Journal of Black Psychology, 20*(4), 445–468.

Stevenson, H. C., Cameron, R., Herrero-Taylor, T., & Davis, G. Y. (2002). Development of the Teenager Experience of Racial Socialization Scale: Correlates of race-related socialization frequency from the perspective of Black youth. *Journal of Black Psychology, 28*(2), 84–106. https://doi.org/10.1177/0095798402028002002

Sue, D. W. (2010). *Microaggressions in everyday life: Race, gender, and sexual orientation*. John Wiley & Sons.

Syed, M., Santos, C., Yoo, H. C., & Juang, L. P. (2018). Invisibility of racial/ethnic minorities in developmental science: Implications for research and institutional practices. *American Psychologist, 73*(6), 812–826. https://doi.org/10.1037/amp0000294

Telzer, E. H., Qu, Y., Goldenberg, D., Fuligni, A. J., Galván, A., & Lieberman, M. D. (2014). Adolescents' emotional competence is associated with parents' neural sensitivity to emotions. *Frontiers in Human Neuroscience, 8*. https://doi.org/10.3389/fnhum.2014.00558

Thomas, A. J., & Blackmon, S. K. M. (2015). The influence of the Trayvon Martin shooting on racial socialization practices of African American parents. *Journal of Black Psychology, 41*(1), 75–89. https://doi.org/10.1177/0095798414563610

Trent, M., Dooley, D. G., & Dougé, J. (2019). The impact of racism on child and adolescent health. *Pediatrics, 144*(2), e20191765. https://doi.org/10.1542/peds.2019-1765

Turner, J. S. (2014). A longitudinal content analysis of gender and ethnicity portrayals on ESPN's SportsCenter from 1999 to 2009. *Communication & Sport, 2*(4), 303–327. https://doi.org/10.1177/2167479513496222

Umaña-Taylor, A. J., & Hill, N. E. (2020). Ethnic–racial socialization in the family: A decade's advance on precursors and outcomes. *Journal of Marriage and Family, 82*(1), 244–271. https://doi.org/10.1111/jomf.12622

van der Voort, A., Linting, M., Juffer, F., Bakermans-Kranenburg, M. J., Schoenmaker, C., & van Ijzendoorn, M. H. (2014). The development of adolescents' internalizing behavior: Longitudinal effects of maternal sensitivity and child inhibition. *Journal of Youth and Adolescence, 43*(4), 528–540. https://doi.org/10.1007/s10964-013-9976-7

Vespa, J. Medina, L., & Armstrong, D. M. (2020, February). Demographic turning points for the United States: Population projections for 2020 to 2060. *Current Population Reports*, P25–1144. U. S. Census Bureau. https://www.census.gov/content/dam/Census/library/publications/2020/demo/p25-1144.pdf

Vygotsky, L. S. (1978). *Mind in society: The development of higher psychological processes*. Harvard University Press.

Wagers, K. B., & Kiel, E. J. (2019). The influence of parenting and temperament on empathy development in toddlers. *Journal of Family Psychology, 33*(4), 391–400. https://doi.org/10.1037/fam0000505

Wang, Y., Cham, H., Aladin, M., & Yip, T. (2019). Parental cultural socialization and adolescent private regard: Exploring mediating pathways through daily experiences. *Child Development, 90*(1), e19–e36. https://doi.org/10.1111/cdev.12911

Wang, M.-T., Henry, D. A., Smith, L. V., Huguley, J. P., & Guo, J. (2020a). Parental ethnic-racial socialization practices and children of color's psychosocial and behavioral adjustment: A systematic review and meta-analysis. *American Psychologist, 75*(1), 1–22. https://doi.org/10.1037/amp0000464

Wang, M. T., Smith, L. V., Miller-Cotto, D., & Huguley, J. P. (2020b). Parental ethnic-racial socialization and children of color's academic success: A meta-analytic review. *Child Development, 91*(3), e528–e544. https://doi.org/10.1111/cdev.13254

Washington, T., Rose, T., Colombo, G., Hong, J. S., & Coard, S. I. (2015). Family-level factors and African American children's behavioral health outcomes: A systematic review. *Child & Youth Care Forum, 44*(6), 819–834. https://doi.org/10.1007/s10566-015-9308-z

Weinstein, M., Jensen, M. R., & Tynes, B. M. (2021). Victimized in many ways: Online and offline bullying/harassment and perceived racial discrimination in diverse racial–ethnic minority adolescents. *Cultural Diversity and Ethnic Minority Psychology, 27*(3), 397–407. https://doi.org/10.1037/cdp0000436

Wertsch, J. V., & Wertsch, J. V. (2009). *Voices of the mind: Sociocultural approach to mediated action*. Harvard University Press.

Williams, L. R., Degnan, K. A., Perez-Edgar, K. E., et al. (2009). Impact of behavioral inhibition and parenting style on internalizing and externalizing problems from early childhood through adolescence. *Journal of Abnormal Child Psychology, 37*(8), 1063–1075.

Wyatt, S. B., Williams, D. R., Calvin, R., Henderson, F. C., Walker, E. R., & Winters, K. (2003). Racism and cardiovascular disease in African Americans. *American Journal of Medical Sciences, 325*(6), 315–331. https://doi.org/10.1097/00000441-200306000-00003

Zimmer-Gembeck, M. J., & Skinner, E. A. (2016). The development of coping: Implications for psychopathology and resilience. In *Developmental psychopathology* (pp. 1–61). https://doi.org/10.1002/9781119125556.devpsy410

Chapter 5
Acculturation and Enculturation: The Intersection of Representational Ethics, Measurement, and Conceptualization

Elma I. Lorenzo-Blanco, Gabriela Livas Stein, Richard M. Lee, and Gail M. Ferguson

In this chapter, we consider the concept of representational ethics in research on acculturation and enculturation and the implications of representational ethics for best practices in this research. In developmental science, acculturation is typically defined as a dynamic, multidimensional construct that captures the processes through which intercultural contact results in behavioral, cultural value, and identity changes and has, until relatively recently, been most applied to research on immigrant populations (Schwartz et al., 2010). Enculturation refers to the processes through which individuals learn, internalize, and maintain the behavioral customs, cultural values, and identity associated with their culture of origin (Schwartz et al., 2010). Acculturation and enculturation processes are not just evident at the individual level but also influence and interact with the contexts in which individuals reside (e.g., family, neighborhood, schools, communities, and broader political contexts). Representational ethics refers to the ways in which researchers represent, write about, and study the experiences, contexts, and identities of the people and communities they include in their research (e.g., Haarlammert et al., 2017). In the social sciences, representational ethics has been considered more deeply within the qualitative research traditions of anthropology, sociology, and feminist studies (e.g., Dwyer & Buckle, 2009; Hellawell, 2006), where scholars have been concerned about researchers depicting and studying participants in stereotypical, marginalized, and inaccurate ways (e.g., Haarlammert et al., 2017; Hodkinson, 2005). More recently, community psychologists have shed light on these principles in work with

E. I. Lorenzo-Blanco (✉)
University of Texas at Austin, Austin, TX, USA
e-mail: elma@utexas.edu

G. L. Stein
University of North Carolina at Greensboro, Greensboro, NC, USA

R. M. Lee · G. M. Ferguson
University of Minnesota, Minneapolis, MN, USA

© The Author(s), under exclusive license to Springer Nature Switzerland AG 2023
D. P. Witherspoon, G. L. Stein (eds.), *Diversity and Developmental Science*, https://doi.org/10.1007/978-3-031-23163-6_5

immigrant communities (Haarlammert et al., 2017). Building on this work, we propose that the concept of representational ethics is as important in the developmental sciences as it is in other fields.

Developmental scientists have to make decisions about the topics, people, and communities they study; the research tools they use; the conceptualization of constructs and research questions; the ways they interact with study participants; what to include in publications; and whether to report findings that may portray the experiences, people, and communities they study in positive, negative, marginalized, or stereotypical ways (e.g., Bridges, 2001; Haarlammert et al., 2017). Although relevant to all research topics, the concept of representational ethics might be particularly important for research on issues related to race, ethnicity, culture, and migration among racial/ethnic minorities (e.g., Haarlammert et al., 2017), including marginalized groups who are often painted with broad, stereotypical brushstrokes that research should not reify (e.g., Garcia & Birman, 2020; Haarlammert et al., 2017). In this chapter, we specifically apply the concept of representational ethics to developmental research on acculturation and enculturation as it relates to conceptualization and measurement.

Representational Ethics: Insider and Outsider Perspectives

Qualitative researchers have long debated whether insiders (i.e., researchers who identify with or are part of the group/people/communities they study and, thus, have special insider knowledge) or outsiders (i.e., researchers who are outside the group/people/communities they study and, thus, lack special insider knowledge) are best suited to accurately study and represent the experiences, contexts, and identities of the people and communities they study (e.g., Bridges, 2001; Hodkinson, 2005). Recently, developmental scientists (Rivas-Drake et al., 2016) have discussed the need to consider insider/outsider status in research with racial/ethnic minority youth and families in all types of developmental research. It has been proposed that a researcher's insider/outsider status influences the research process, the research findings, the interpretation of findings, and the way researchers depict the people and communities they study (e.g., Bridges, 2001; Hodkinson, 2005); in part, because researchers' cultural background, including their academic/research culture, can influence the research process (see Hughes & Seidman, 2002 for a more in-depth discussion). Some scholars have postulated that insiders (compared to outsiders) have better access to and deeper insights into the linguistic, cultural, social, political, psychological, and historical experiences of the people and communities they study (e.g., Hellawell, 2006; Hodkinson, 2005), permitting them to generate novel and nuanced research questions and interpretations of research findings as well as less stereotypical depictions of the people and communities they study. However, insiders may run the risk of conducting biased research because their insider status may not allow them to see beyond the knowledge, insights, and connections they have with the experiences, contexts, and identities of the people and communities included in their research (e.g., Hodkinson, 2005). Outsiders, by contrast, may be less biased and more objective due to their detachment from the topics/people/communities they study (e.g., Bridges, 2001; Berger, 2015), allowing them to generate objective research questions, interpretations, and depictions of the topics/people/communities they study. However, outsiders may run the risk of not accurately capturing and portraying the experiences, contexts, and identities of the people and communities they study. Outsiders also may perpetuate dominant narratives about minoritized and marginalized populations and present such bias as objective. Researchers have further proposed that a researcher's insider/outsider status can also influence a researcher's ability to successfully recruit and retain racial/ethnic minority participants (see Rivas-Drake et al., 2016 for an in-depth discussion). Insiders (compared to outsiders) may have better access to and knowledge about culturally anchored ways to recruit and retain racial/ethnic minorities into developmental research on issues related to race, ethnicity, and culture, and may be better able to cultivate trust with their participants (e.g., Rivas-Drake et al., 2016).

Scholars have also argued that the categorization of researchers into either insiders or outsiders is a false dichotomy that does not attend to issues of intersectionality (e.g., Bridges, 2001; Hodkinson, 2005). Researchers have many identities: they may be insiders to the topics/people/communities they study based on some identities (e.g., ethnicity, class) and outsiders based on other identities (e.g., gender, immigration status; Bowleg, 2008; Haarlammert et al., 2017). As such, most researchers are likely insiders *and* outsiders to their research topics, people, and communities they study (e.g., Haarlammert et al., 2017); and their insider/outsider status may shift throughout the research process (e.g., Rivas-Drake et al., 2016), requiring them to develop awareness of how their identities may impact the research process. Because insiders *and* outsiders may bring a biased perspective into their research, we propose that it is important for researchers to explore their positionality in relation to their scholarship and be clear about how their insider/outsider status across multiple dimensions may be influencing their scholarship. Thus, it is important for researchers to examine how their identities and insider/outsider status may influence how they conduct research, portray the people/communities they study, and the conclusions they draw (e.g., Hellawell, 2006; Rivas-Drake et al., 2016).

In this chapter, we write as partial insiders to the topics and people/communities included in our scholarship. We write as partial insiders because we identify with the experiences, identities, and communities we study. Yet, we recognize that we are also partial outsiders because there are differences between our experiences and identities and those of the communities we study; and as such, we realize that we may bring blind spots into our work (e.g., Haarlammert et al., 2017; Hodkinson, 2005). Below, we share the process by which we decided to write about representational ethics in research on acculturation and enculturation. Three of us share our personal narratives as they relate to our scholarship and discuss how our lived experiences inform our work and subject us to blind spots. We also discuss some ways that we have or could have strengthened the way we study and represent the topics, experiences, and identities of the people and communities we include in our research. We end this chapter with recommendations for how to include more diverse perspectives into research studies on acculturation and enculturation by including insider and outsider perspectives and attending to methodological issues in acculturation research.

How We Decided to Focus on Representational Ethics in Research on Acculturation and Enculturation

Our decision to write about representational ethics (including insider/outsider perspectives) developed during our first meeting about how this chapter could move the field of acculturation and enculturation research forward. We had originally planned to discuss ways to think beyond acculturation and enculturation being mainly applicable to recent immigrants and to move away from typology perspectives of acculturation using bidimensional models (e.g., Titzmann & Lee, 2018). We envisioned highlighting how acculturation occurs across generations and time within families. Similarly, we envisioned highlighting how enculturation can develop and/or change across generations and time within families (e.g., Cobb et al., 2021; Titzmann & Lee, 2018). Given the authors' expertise with studying remote acculturation (i.e., a modern form of non-immigrant acculturation that occurs across distance and without direct contact, e.g., Ferguson & Bornstein, 2012; Lorenzo-Blanco et al., 2019), we also wanted to consider how acculturation and enculturation unfold across borders, challenging frequent assumptions that acculturation requires first-hand and direct contact with receiving countries' institutions. As we discussed the topics above, we talked about the possibility of making this chapter about best practices for measuring acculturation and enculturation by providing information about existing measures and discussing the kinds of information existing measures can give (or not give) to make recommendations for best practices (see recent work that addresses this question: Doucerain et al., 2017).

Importantly, our conversation about the writing process prompted us to share our own stories with acculturation and enculturation, including the challenges we as researchers have faced when making methodological or reporting decisions. For example, in a large research project involving one of the authors (Stein), participants who self described as "truly bilingual" (speaking both languages with equal fluency) were randomly assigned to complete study measures in only one of the two languages in which they reported being fully fluent. Although randomly assigning participants to complete measures in one language allowed the researchers to test for language measurement invariance, it also created some difficulty because participants were not able to select the language of the measures they completed. The research team realized that the measure did not specify comfort but instead focused solely on fluency. Further, as the research team discussed how best to capture language use, their conversation about the complexity of how people self-define fluency, when people choose to disclose fluency, and how the context can shape comfort and frequency of use (e.g., may have to speak English due to work contexts but not preference) was informed both by past research and by the fact that the research team included seven bilingual researchers (of 10 researchers). Those insider perspectives shaped the ultimate methodological decision to let participants know that by selecting truly bilingual they would be randomly assigned to complete assessments in either language. This led to a discussion of the fact that simply because someone is bilingual or speaks English fluently or primarily (due to living in the United States), it does not necessarily mean this is the language in which they feel they can best express themselves. Thus, insider perspectives were key to this discussion.

The discussion on language in research prompted us to consider how *we* would complete acculturation/enculturation measures as research participants. We discussed that (despite the many strengths and contributions of existing measures) they would likely not fully capture our cultural orientations because of our multiple and intersectional identities (e.g., Bowles, 2008). Through our discussion, we recognized that we all had blind-spots in how we thought about acculturation and enculturation based on the literature we were most familiar with or the literature we had or had not included in our own acculturation research. For example, two of us realized that we were most familiar with the Latinx acculturation research literature but not the acculturation research literature relevant for other populations; and one of us was primarily attuned to the international acculturation literature outside of the United States. We noticed that some of us were familiar with the acculturation literature published in developmental science outlets (e.g., Juang & Syed, 2019; Keles et al., 2018; Schwartz et al., 2015) but not the literature published in cross-cultural psychology outlets (Eales et al., 2020; Gattino et al., 2016; Taušová et al., 2019). Our awareness of how our blind-spots, identities, and experiences inform our work, motivated us to write about our personal narratives (which three of us share below), outlining our experiences with acculturation and enculturation and how these experiences inform our work. We also decided to review our narratives to identify similarities and differences and to use this information to identify a topic for this chapter.

In reviewing our narratives, we noticed that all of us wrote about our immigrant background, our acculturation experiences, and our experiences with having to navigate multiple cultural contexts either as children of immigrants or as adult immigrants. We also shared socialization experiences within our families and how these experiences influenced our development. Finally, all of us wrote about identifying with the communities/topics we study—making us likely insiders (e.g., Haarlammert et al., 2017; Hodkinson, 2005). This led to our decision to write about representational ethics and insider and outsider perspectives in research on acculturation and enculturation. Below, we briefly share some of our personal stories as they relate to our scholarship. We point out possible blind-spots we bring into our work and discuss ways to offset the potential impact of our blind-spots in our scholarship.

Personal Stories

Elma Lorenzo-Blanco

I discovered my passion for research with Latinx[1] communities as an undergraduate student when I was enrolled in a culture and mental health class that introduced me to research on acculturation and enculturation. I was for the first time exposed to readings that reflected my lived experiences as a child of immigrants which helped me make sense of my lived experiences.

Briefly, I was born in Germany and raised by immigrant parents from Spain. My parents met in Germany after they were brought to my small hometown during their adolescent years so that they could work as guest workers in factories and contribute to their families' finances. Although my mother would have loved to complete high school and go to university, my grandparents needed her to work and support their family. My father had lost his father at age four and, as the oldest son, he had to work at an early age to help out his mother. As long as I can remember, while growing up, my parents worked in factories and often worked two or three jobs. It was not until much later that my mother went back to school.

When I was born, my parents did not speak German (as they were able to get by without learning it), and so my first language was Spanish. It was not until my parents had to navigate German institutions with me (e.g., doctor's appointments, schools, and German friends) that they began to learn German. I began to learn German when my parents sent me to preschool. I assume my acculturation journey started during this time when I became aware of being different, when I had to learn German, and when I was exposed to German teachers, traditions, children, and families.

Similar to many immigrant children or children of immigrants in the United States, I had to learn to navigate different cultural expectations, values, and traditions. I experienced cultural, linguistic, and systematic barriers to higher education. I had to explore my ethnic, cultural, and national identities; and learn to navigate anti-immigrant attitudes, prejudice, and discrimination. I know that these experiences have influenced who I am today. For example, although I was born in Germany and consider Germany one of my many homes, I do not identify as German. I think this is in part because I could not become a German citizen (because my parents were from Spain). I think because German authorities never saw me as German and because they expressed hostile attitudes towards me and my family, I always felt like an outsider and someone who does not really belong. I also do not strongly identify with being Spanish despite having a Spanish passport and being a citizen from Spain and having fond memories of my summers in Spain. I think the hostile

[1] I use the terms Latinx, Hispanic, Latinx/Hispanic. I identify as Hispanic and conduct research with Latinx communities. I use Hispanic/Latinx as I recognize similarities and differences in cultures, experiences, and identities between myself and Latinx youth and families in my research. Using one terms would not accurately capture these nuances.

attitudes and challenges I have experienced in navigating institutions from Spain, in which I had (and still have) to show proof that I am Spanish, have also made me feel like an outsider and someone who does not truly belong.

My experiences as an Hispanic[1] immigrant child motivate my passion for research on acculturation and enculturation among Latinx youth and families. Much of my research draws from acculturation theories and examines how acculturation, related experiences (e.g., social stressors such as discrimination and acculturative stress) and familial (e.g., family functioning and cultural values) factors influence the well-being of Latinx youth and families. In most of my work, I have used data collected by colleagues through school-based surveys. As such, I have never interacted with the youth and families who have participated in these surveys. As an insider to Hispanic culture and the immigrant experience, I hope that my work has made meaningful contributions to how we understand acculturation and enculturation among Latinx youth and families. However, I often wonder how my scholarship is informed by and contributes to stereotypical notions of Latinx communities/cultures in the United States. I often wonder about my blind-spots (due to differences in cultural upbringing, experiences, identities, and privileges that come with my identities) and how these impact my research. In some of my current work, I am conducting focus groups with Latinx youth and parents, and I hope that this qualitative work and interaction I have with youth and families will provide insights that I may not have through survey research. I would embrace the opportunity to collaborate with other scholars who have different experiences and training to safeguard against my blind-spots. I would also embrace the opportunity to collaborate with youth and families in designing research projects, collecting and analyzing data, and disseminating research findings as a means to offset my blind-spots.

I moved to the United States in 1999 because I wanted to be immersed in the English language. My plan was to live in the United States for a couple of years and return to Germany to work in international business. Little did I know that I would find a path to higher education and make the United States my second home. I arrived in the United States with a lot of knowledge and perceptions about the United States. Much of my knowledge and perceptions came from books I read; news I was exposed to; music I listened to; and TV shows and movies I watched. I also spoke some English as I learned English in school. I know that my acculturation in regard to my life in the United States started way before I set foot on US soil. These experiences (along with my scholarship on acculturation among Latinx youth) have informed my recent work on remote acculturation among youth in Mexico. Traditionally, acculturation frameworks have assumed that acculturation begins when immigrants, including Latinx youth and families, begin to navigate US American institutions and have direct contact with US institutions. However, as a result of globalization, we know that acculturation can occur remotely through indirect exposure to US media; TV; movies; the internet; merchandise; tourism; and information exchange through other sources. Thus, immigrants such as Latinx youth and families likely experience acculturation before they have direct contact with US American institutions. I think that scholarship on acculturation among

Latinx or other immigrant youth could benefit from examining how acculturation unfolds across borders and before immigrant youth and families set foot on US soil.

I was able to conduct work in Mexico by joining an international team of researchers who had lived in Mexico for many years or who were living in Mexico. Members in the Mexico research team questioned and challenged the way US researchers (including myself) often portray Mexican-descent/Latinx youth. They questioned notions of traditional cultural values constituting "Mexican culture." They also questioned traditional gender role values being part of "Mexican culture," resulting in discussions of diversity regarding gender role values across youth and families in Mexico. Based on these experiences, I think research on acculturation and enculturation (especially research involving immigrant communities or communities whose countries of origin are not the United States) could greatly benefit from studying acculturation and enculturation across borders. For example, work on acculturation and enculturation among Hispanic/Latinx youth could benefit from understanding what life is like for youth and families in their countries of origin and how enculturation evolves or stays the same when families leave their country of origin. I think collaborations across borders on enculturation would provide important insights into acculturation and enculturation that we cannot gain from doing research solely in the United States.

Since moving to the United States, I have developed a strong connection with Latinx youth and families through my personal life. I have become keenly aware of similarities and differences in cultures, upbringing, opportunities, and identities. I have moved across different states, navigated different cultural contexts and systems, and I am raising two children who identify as Latinx/Hispanic[1]. From these experiences, I have learned that context matters in how people adjust to new environments and the degree to which they identify with their receiving country/community. I think we need more research that examines contexts (e.g., schools, work places, grocery stores, and recreational centers) and how these contexts and reactions from receiving communities influence the adaptation process for immigrant youth and families. I also realize that I have and continue to experience privileges that have facilitated my acculturation and enculturation across different contexts. For example, the challenges I experienced as an immigrant child in Germany were amplified for my friends from Turkey, for whom, it was more difficult to blend their Turkish cultural norms and expectations with the norms and expectations of German culture. I also realize that my experiences as a White Hispanic in the United States bring privileges that many Latinx youth and families do not have access to. I am aware of the complexity of acculturation and enculturation, and how these experiences intersect with our multiple identities and the contexts we navigate.

Writing my personal narrative was not easy as I had to make decisions about what to share. It was also challenging to write a cohesive narrative, link it to my work, and process my blind-spots. I assume that many youth and families who participate in research also have to make these decisions which can further impact our scholarship. I know that my experiences inform the research I do, how I do it, and why I do it. I also know that I have blind-spots that I bring into my work. I hope to continue to grow as a scholar and engage in research that helps us understand the complexities and nuances of acculturation and enculturation among Latinx youth and families.

Gaby Livas Stein

I immigrated to the United States at the age of four. I came directly from Costa Rica although indirectly from Mexico. My father grew up in a large city in Northern Mexico (Monterrey) and my mother grew up in a small town 2 hours outside this city. We came because my father's company was starting to expand to the US market, and my parents had not planned to live here.

permanently thinking we would move back to Mexico. My father had lived in the United States during his graduate school years (he has a PhD in economics from UT Austin). Although my mother had wanted to go to college, her father had not let her, and as the oldest of 11 children, she went to work to help her family. My mother spoke no English prior to immigrating. My parents never intended to raise American children.

I remember many things about those first few years living in Pasadena, California. My parents enrolled me in a private school, and I knew little English and there were no other Latinx or Mexican families in this school. I remember needing to learn English quickly, and my mother always shares the story that shortly after "mastering" English, I declared that I was never speaking Spanish ever again. When she said, "How will you talk with your abuelitas (grandmothers)?" I said I would only make an exception for them. This pressure to acculturate to European American cultural and language norms was something I carried with me throughout my childhood. I remember the shame I felt when my mother could not communicate clearly with my teachers. Spanish was something that labeled me as an outsider.

Yet, my mother wanted to support my acculturation, and when I came home from school asking that we have a Thanksgiving dinner, she quickly learned more about it and thought—well this is just like the dinner we make in Mexico for Christmas! She made a turkey, camote (a sweet potato dish), and a meat stuffing that was a recipe from one of my abuelitas—she made her own version of Thanksgiving. I remember the pride I felt of being American and having an American tradition. We celebrated Thanksgiving every year since then even when we moved back to Mexico for a short time when I was in late elementary school.

I remained connected with my abuelitas and my family in Mexico as I was lucky enough to get to fly to Mexico for vacations and for weeks in the summer. I got to celebrate posadas, spend time with my large extended family, eat delicious homemade food that my grandmothers made with such love and dedication, and learn what it meant to be Mexican. As I grew up, I described myself not as a Mexican-American but instead as a Mexican and American—two different sides that did not come together into one identity but instead more like two sides of a coin. These sides did not feel in conflict to me at the time, but instead felt separated. I had a Mexican family and did Mexican things at home, but at school, I was American and did American things. I think this was facilitated by my light-skin tone and light hair. The world accepted me for the most part as an "American." Yet, I knew I was not "American," and I remember when we received our "green cards" when I was younger. I remember being surprised that it was pink (why did they call them green cards I wondered) and being struck by the words RESIDENT ALIEN. So at some level, I knew I was not "American" in the true sense of the word. I also knew I was not Mexican enough for my family—as my sister and I were called "las gringas[2]" and I also did not fully fit in there either.

My acculturation journey was punctuated by a short move back to Mexico (to which I credit my retaining Spanish!), after which in my early adolescence, we settled in another suburb of Los Angeles that had a high number of Asian and Asian American families and few Latinxs or Mexicans. Here I connected with my Asian peers as they had similar experiences of navigating two cultures, and I began to learn about Japanese and Korean cultures through my best friends. I started to embrace my Mexican side a bit more as it was valued by new friends. But for the most part, I still felt these two sides were not connected inside me. Further, I felt alienated by my Mexican culture that I felt supported male dominance, and I hated the gendered rules my parents enforced at home. I saw being Mexican as limiting my cousins, aunts, and mother and eschewed the culture that I viewed as sexist with my adolescent gaze. So I dove deep into "American" culture in terms of music, books, (s)heros...everything.

This all changed in my "encounter" summer (a la Cross theory). The summer between high school and college, my parents staged an enculturation intervention and forced me to come and live with them in Mexico City (much to my resentment). My dad found me an "internship" with Bertha Maldonado—a stereotype-shattering Mexican woman who shaped my cultural journey for the rest of my life. She worked in advertising with my dad's company, and I spent the summer shadowing her. She was fiercely independent, steeped in indigenous culture, and passionate about art. That summer she taught me a different way to be Mexican than what I had been exposed to more narrowly in my family. You could be feminist and Mexican. Mexican culture celebrated women and she showed me the many ways how. She brought me to mercados, art museums, excavation sites, and shared books with me. She ignited in me an identity exploration that I carry to this day.

Armed with this new understanding (and pride!) in my culture, I started college. Here I experienced more microaggressions (was it because my hair got darker? Did I start looking less "American"?), and both of these culminated in my deep dive into being Latinx and I started to integrate these two sides of my identity. I relished my Spanish classes, Latinx lit classes, and Latin American history classes. Living in NYC, I was exposed to other Latinx cultures and embraced learning about them as well. When I finished college, I started working with a Guatemalan psychologist at

[2] "Gringo" is a slang term in Latin America for foreigners—especially U.S. Americans and can be pejorative

Tufts and through my work with her, this cemented my academic identity with my newly integrated Latinx identity.

In her lab, I started doing research with the Latinx community—particularly immigrant—on a project helping immigrant families and schools connect at the start of kindergarten. Fifteen years after I had gone through this process, I was helping families and schools navigate this acculturative dance. I appreciated the many ways in which families navigated acculturation differently from my own, but I was aware of how youth and their families experienced so many of the same things that I had (pressure to speak English, learning American traditions, pressures of familial obligations, parent-child acculturation gaps).

These early experiences have cemented the work I do today. Most of my work has been in emerging immigrant communities where some of the Latinx families (mostly Mexican) live in primarily non-Latinx spaces but where others can experience their acculturative journey with other same-ethnic peers—so I am particularly attuned to that. Similarly, given my experiences in California, I came to question whose culture do you acculturate to? It is not just to White middle-class culture (e.g., mainstream culture), but there are so many other cultures in the United States including other Latinx cultures. So in a study of the acculturative process, we included a measure that was based on the AAMAS[3] that separated out acculturation to coethnics, other Latinx, non-Latinx White, and African American cultures in terms of traditions, foods, friends, etc. My experiences have also informed me that enculturative experiences may not just come for parents—especially as considering identity. For me, peers, other adults, and my siblings have been key—while this has not made it into my research yet, it is something I consider when I look at the patterns in my research findings. Given my experiences acculturating, one thing I am very skeptical of is using language as a proxy for acculturation. This is also backed by research but having the lived experience for me really highlights this fact.

I am also acutely aware of skin tone and how that has informed my acculturative process (as in privileging it). My sister is considerably darker skinned than I am, and it was clear to me in which ways she had less access to "being American" than I did. Our work needs to more explicitly consider race and colorism as they influence acculturation. One of our studies found that White-identified Latinx endorsed fewer familism values than non-White identified Latinx (Stein et al., 2017). Finally, I know that my acculturation journey was informed by my parents' SES and my documentation status, as I was able to travel back and forth and see my grandparents and family very frequently. I am aware of the benefit of that frequent contact to my enculturation journey. My SES also afforded me the opportunity to work with Bertha and see the world through her eyes. We need to understand the acculturative experience across SES.

In terms of blind-spots, these are harder for me to identify and I mostly have learned about these from collaborating with other immigrant and BIPOC friends and scholars in conversations as they expose me to different points of view and

[3] Asian American Multidimensional Acculturation Scale.

challenge my assumptions. For example, the role of religion and religious values and how they are part of the acculturative experience. My family was Catholic by default, but my father and mother were not strongly identified with the Catholic faith. Yet, it was clear to me from my friends how church served an acculturative influence (or enculturative). Similarly, the role of religious identity needs to be considered along with racial/ethnic and national identity—particularly for religious communities that are marginalized and discriminated against. In terms of acculturation gaps, some of these may be due to value differences around religious communities. Another place where I have learned to grow is to consider the cultural variability within groups (e.g., Mexican indigenous populations), and especially around experiences of race and ethnicity and cultural value endorsement.

My acculturation and identity journey has greatly informed the work I do today. First, I see myself as a light-skinned Latina academic such that gender, race/ethnicity, and professional identity are intertwined. My experiences provide me with a guiding principle of how to be an academic in terms of the mentorship I provide my students, the questions I ask in research, the colleagues I collaborate with, and most importantly, the communities I partner with.

Richard Lee

I am the youngest of three boys raised by Korean immigrant parents. My father emigrated from South Korea in 1964 but his journey began earlier during the Korean War. My father's family was internally displaced from what is now considered North Korea. With a country split into two nations, my father, along with his mother and two sisters, rebuilt their lives in Seoul. It was in Seoul that my father met my mother whose own family had moved from the countryside to Seoul after the war. My father arrived in the United States with a tourist visa to visit his sister who had already immigrated to marry her husband (whom only recently I learned was adopted as a Korean teen orphan by a White family—a family secret revealed only after his death). Whether my father overstayed his visa or lucked into a work visa sponsorship by his employer, it is not clear. But somehow he was later able to secure visas for my mother and oldest brother to immigrate in 1965, the same year the US Congress passed the Immigration and Naturalization Act which established a new immigration policy based on family reunification and skilled laborers, no longer relying on a racist quota system designed to restrict immigration from outside of Europe. It was in this context of a country changing in terms of immigration and civil rights for racial minorities and women that my parents and brother settled in Connecticut where my middle brother and I were born. I often reflect on how these same issues of family separation, relocation, documentation, and migration play out over and over again in history.

But experiences of acculturation to the US culture, as well as enculturation to Korean culture, began before immigration for my family. The Korean War was not only a civil war but also a global war between capitalism (supported by the United

States and the United Nation) and socialism (supported by China and the Soviet Union). American capitalism and the accompanying views of race and racism were introduced to Koreans by the US military who occupied South Korea. To Koreans recovering from a war, American culture became a symbol of hope and ambition for hundreds of thousands of Koreans who would eventually immigrate to the United States. Immigration to the United States reinforced these pre-embedded cultural values and belief systems, similar to remote acculturation (Ferguson & Bornstein, 2012). At the same time, Korean immigrants were and remain fiercely proud of their culture and heritage as a result of the Japanese occupation of the Korean peninsula from 1910–1945, preceding the Korean War. The forced assimilation to Japanese culture and cultural genocide of Koreans under the Japanese colonial rule significantly contributed to a strong national pride in being Korean and Korean culture in general that remains today. Korean immigrants who have lived through the Japanese occupation and the Korean War carry with them a legacy of intergenerational trauma that deeply informs their identity as Koreans and their attitudes toward American (White) culture. It is this juxtaposition of historical experiences that contributes to understanding an approach to acculturation known as accommodation without assimilation or adhesive adaptation (Hurh & Kim, 1984). It also has led to a long-standing interest of mine in the psychology of diaspora—a desire to understand how displaced people make sense of their disrupted cultural experiences and reconcile their past and present (Gulamhussein, 2021; Lee et al., 2007).

My upbringing in Connecticut as an American-born Korean occurred within this larger sociopolitical and historical context of my family's immigration history, as well as the stark reality that the Asian population in Connecticut during the 1970s and 1980s when I was growing up was 0.2%–0.6%! My parents used to joke that we along with my aunt's family were among the first dozen Korean families to permanently settle in Connecticut and looking back at these numbers I have to believe them. Despite the small number of Koreans in Connecticut, my parents found and nurtured a small Korean American community that I grew up in. Korean families living throughout the state established Korean churches as not only a place of worship but also as a space to provide support and resources for immigrants. My parents eventually sponsored my mother's sisters to also immigrate to Connecticut and so I grew up having extended family around. But beyond these few Korean contacts, I grew up in a very White context. It was not only White but also racially hostile toward racial minorities, as noted by the visible presence of the Klu Klux Klan in the state during the 1980s.

Before I was 10 years old, I lived my life immersed in everyday American life and in Korean life. The two worlds coexisted but separately. I not only went to Korean church on the weekends, but my parents also owned a Chinese/Korean restaurant followed by what we then called an Oriental (Asian/Korean) grocery store. When not hanging out with my White friends at school, I spent my time behind the cash register or in the backrooms of my parents' stores. This alternation acculturation strategy (LaFromboise et al., 1993) was natural and normal to me during my early childhood. Having lived this way, I never understood nor accepted acculturation as a unidimensional process.

Like many but not all immigrant families, over time, my parents were "moving on up" as they were able to save money working multiple jobs each. Though I rarely saw my parents during the week as they worked before the sun rose until after sunset, it afforded us the chance to move into a more affluent neighborhood in the same town where I had always lived. This moved me into an even Whiter community and a better public school. I also moved further away from one of my aunts whose home served as a second home. Fortunately, my paternal grandmother (halmoni) continued to live with us. She spoke no English so my brothers and I retained just enough of our childhood Korean to communicate with her. My halmoni along with my mother always cooked us Korean food for dinner. In addition to the Korean church, food became a primary ethnic socialization agent. At the same time, it was a source of rich embarrassment as we had to grow and make much of the basic Korean ingredients. As I entered adolescence, I became acutely aware and self-conscious of the stinky smell of Korean food. Feelings of shame over being Korean began to creep into my consciousness just as puberty and identity development began. Not surprisingly, around 10 or 11 years old, I made the decision to stop speaking Korean unless it was with my grandmother and my parents acquiesced over time to this language switch. I began to actively engage in cultural assimilation to White mainstream culture even as my parents continued to directly and indirectly engage in both ethnic and racial socialization practices. This discrepancy between my acculturation orientation and my parents' gradually led to increased acculturation conflicts with my parents who increasingly complained about my Americanization as I complained of their traditional Korean ways. It is not surprising that family acculturation conflicts became one of my first research projects (Lee et al., 2000).

Racism plays a big role in how we acculturate and develop our ethnic and racial identities. For me, big and small incidences with discrimination directed toward me, toward my parents, and toward others made me increasingly aware of the disadvantages and stigma associated with being a racial minority and from an immigrant family. I found myself yearning to be White while resisting relinquishing my Korean identity and heritage. This ambivalence in the context of discrimination fueled self-doubt and confusion over who I was and who I should be. Moreover, it led to a lack of belonging and social connection throughout my adolescence, even though on the surface I had friends and a strong support network. And again, not surprisingly, my first publication was on social connectedness (Lee & Robbins, 1995).

As a child of immigrants who was born in the United States, I am acutely aware of the privileges afforded to me as a native-born English speaker who was raised and educated in this country. I did not carry with me the trauma of war and resettlement that my parents endured. While life was not always easy, my parents protected me from many adverse experiences that displaced children and youth experience daily before and after migration. And though my educational and career path has not been necessarily linear or smooth, it is definitely not the same as an immigrant who moves to a new country and culture as an adult without built and available resources and support. I do not always remember how hard it can be to navigate living in this country without a full command of the English language and without the cultural and social capital gained by merely living in this country. When I was completing my postdoctoral fellowship in California, I was fortunate to meet and be mentored by the godfather of Asian American journalism, KW Lee (himself a Korean immigrant from the 1950s). KW taught me a powerful lesson one day when he mentioned that my parents' generation has quietly suffered the trauma of war and migration, as well as racism and discrimination, to raise children in this country. Now they are waiting for their second-generation children who have the "King's English on their tongues" to stand up and fight against racism and discrimination and to advocate for their parents and communities. I often remind myself of this message as I do my research and teaching.

Over time, I have come to greatly appreciate and understand the life experiences of people from diverse migration backgrounds and find myself increasingly drawn to work that seeks to illuminate the life experiences of people who have experienced displacement and marginalization and who are seeking to empower themselves, their families, and their communities. For the past 20 years, I have studied Korean adoptees raised in White families. This research has required me to think beyond the binary of who is and what it means to be Korean. Growing up, I did not know about the history of Korean adoption (including the adoption backstory to my aunt's husband) and had never met a Korean adoptee until I was an adult. As I became friends with Korean adoptees, I found similarities and differences in our upbringing. I also learned about the stigma of adoption in Korean culture and the challenges adoptees felt trying to find a sense of belonging in the Korean community. Because I study adoption, people often mistake me as a Korean adoptee, so I always make it a point to disclose that I am not adopted. In the spirit of representational ethics, I have learned it is important for me to be clear in my positionality and to acknowledge that I am an outsider to the adoption experience. Yet, in other ways, I am not. Years after I began to conduct research on Korean adoption, I married a Korean American who was adopted transracially and transnationally by a White family. Our children are a unique blend of immigrant and adoptee families. These personal lived experiences surely inform my work on acculturation and enculturation.

Most recently, I have begun to look beyond traditional models and methods of studying acculturation. My collaborator and I (Titzmann & Lee, 2018) are trying to understand acculturation as more than a static experience that can be classified as assimilated or bicultural. We are trying to understand the process of how acculturation (and enculturation) unfold over time—what we are calling acculturation processes such as the timing, tempo, and synchrony of acculturation. Our hope is that looking at the developmental processes of acculturation will better capture the rich complexity of this experience within individuals and families.

Similarities and Differences Across Our Narratives

While the first three authors wrote about identifying with the topics and communities they study, they recognized differences between their experiences and identities and the experiences and identities of the people and communities they study because of the multiple and intersecting identities of their participants versus themselves (e.g., Bowleg, 2008; Haarlammert et al., 2017). For example, although the second author may be from the same country/ies and may have experienced acculturation pressures similar to many of her research participants, there are also significant differences between the author and her research participants (e.g., she attended a private school, has a father with a PhD, and had a stable job when immigrating to the United States), making her both an "insider" and an "outsider" to the communities/topics she studies. Similarly, although the first author strongly identifies with the Latinx community in the United States, speaks Spanish, is the parent of children who identify as Hispanic/Latinx, and has experienced similar acculturation/immigration experiences as the communities/people she studies, the fact that she grew up in Europe and has secure employment in the United States distinguishes her experiences from those of youth and families included in her research, making her an "insider" and "outsider" to her scholarship. The third author is a US-raised Korean American who grew up in a predominantly White US community. While he shares the same ethnicity and ethnic heritage as Korean adoptees whom he studies, as well as similar assimilation pressures growing up, his experiences as nonadopted in a same-race/ethnic family are nevertheless very different from Korean adoptees raised in White families.

Application to Acculturation and Enculturation Research

While our narratives highlight the complexities of the acculturation process, especially how they relate to our current work, it is important to note that we wrote our own narratives in part because we wanted to shed light on how the acculturation measures, we use do not capture important aspects of our acculturation experiences. In particular, psychological and public health researchers—especially those not specifically assessing the acculturative and enculturative experience—tend to rely on language use/preference, generation status, and length of stay in the United States as proxies for the psychological experience of acculturation and enculturation (Cabassa, 2003; Lara et al., 2005; Lopez-Class et al., 2011). Indeed, a recent review found that 46% of studies on health and acculturation used proxy measures (Schumann et al., 2020). This decision is in part driven by older work pointing to language representing the largest portion of variance in acculturation measures (Lara et al., 2005). Yet, although these variables are useful to demographically describe a sample (which should be done when reporting on populations that tend to have higher portions of immigrants), it is imperative that these variables are not used in analyses that endeavor to explicate acculturative processes in development (i.e., identity processes, cultural value endorsement, behavioral acculturation) as these imprecise measures may be leading to more confusion and seemingly contradictory findings in the literature (Alegria, 2009; Doucerain et al., 2017). Further, generation status may not directly relate to exposure to US culture (e.g., remote acculturation) and language use may be a measure of context (e.g., ability to speak with co-ethnics; Lorenzo-Blanco; Lee; Stein). The very notion of what ethnic heritage culture means is not universal either, as Korean adoptee culture is just as Korean as Korean immigrant culture. Thus, as argued by the many cited in this paragraph, these variables function much more as descriptives and should be used cautiously to understand psychological processes like self-concept, mental health, self-esteem, and academic outcomes. For these reasons, more nuanced measures are necessary to capture the complex behavioral, value, and identity processes across enculturation and acculturation as they intersect individual and familial characteristics and contexts (e.g., neighborhood, schools). Lee et al. (2006), for instance, found the bidimensionality of acculturation to host culture and heritage culture varied between study participants from the West Coast and the Southwest region of the United States. Here, insider perspectives may be crucial to understand the reasons for why the bidimensionality of acculturation varied among participants from the West Coast and Southwest regions.

Clearly, our narratives shed light on the complexity of acculturation processes that move beyond language and generation status across development involving identity exploration and resolution, navigating nationality, endorsement of cultural values, and engagement in behaviors and practices of multiple cultural groups within the context of our neighborhood, schools, and families. For all of us, we personally experienced isolation and were forced to navigate cultural spaces foreign to our parents and community, so acculturation was about survival whether at school, in our neighborhoods, or even now in academia. The intersection of race, phenotype, skin tone, and socioeconomic status (SES) also shaped our acculturation experiences significantly as they intersected with inhospitable receiving contexts that may have pressured our acculturation and enculturation journeys. These nuances cannot be captured by acculturation typologies (e.g., integration, assimilated) that fail to consider finer facets of acculturation, and caution is further warranted in using acculturation typologies as there is lack of consensus on the best methodological approaches to capture these typologies (e.g., Doucerain et al., 2017). Furthermore, even dimensional measures of acculturative processes that include enculturation and acculturation need further consideration. Two recent studies examined how two distinct measures of acculturation (ARMSA-II (Cuellar et al., 1995) and BIQ-S (Szapocznik et al., 1980)) functioned with Latinx parents and youth found that even these deeper measures of acculturation (i.e., including cultural practices, language use, identity) differed significantly across youth and parents, in their association with other acculturative processes (e.g., language challenges), how they operated across time, and how they related to outcomes (youth alcohol use, internalizing and externalizing symptoms) (Martinez et al., 2018; Meca et al., 2018). Accordingly, we (along with many others) recommend that researchers fully consider the acculturative processes specifically being theorized and tested as the domain of acculturation (i.e., language use, values, practices, identity) may differentially relate to outcomes. How can a research team accomplish this Herculean task? Here is where it is critical to include a diverse research team with multiple perspectives to offer not only in terms of scientific knowledge but also in terms of their personal experiences as an "insider" and/or "outsider."

We argue that the composition of teams that include insiders cannot only identify aspects of acculturation that need to be included in a particular study (e.g., identity processes, cultural value gaps), but can also consider how past literatures and measures of acculturation have (or have not) captured important processes in the local contexts of reception for specific populations. For example, in a study that involved one of the authors, the senior PIs were cultural outsiders, but the author was a junior member of the team with an insider perspective. The outsider PI wanted to include an acculturation measure, and the insider author recommended that they should choose a measure that included not just language acculturation but other aspects as well (behaviors, values, identity) to capture the variability in acculturation in the specific context which was primarily an immigrant context. However, another more senior investigator with no insider perspective said that a language variable would suffice as that is what they had done in previous work. The PI ignored the more junior insider scholar, and when the study concluded, the language variable showed no variability (as the insider had predicted) which was a major loss to the study as they were not able to understand how acculturation impacted the findings. Thus, as insiders may have insight into the specific populations outside of past research, they are able to best consider what measures would be most scientifically useful.

In the same vein, insider perspectives can help identify aspects of acculturation that have not been considered in past measurement (e.g., the AAMAS scale includes multiple acculturating influences, including pan-ethnic processes), and can help to develop and test measures that capture the nuances of these processes. In our narratives, it was clear that there was an intersection of developmental, social, contextual, and emotional processes at play in our acculturation experiences that are yet to be captured in measures. Insider perspectives are needed to flesh these out.

Summary and Recommendations for Bridging Insider and Outsider Perspectives

We propose that engaging in a reflective process (i.e., the process by which researchers consciously and critically reflect on their beliefs, values, biases, and identity as it may influence their research choices, methodology, and the way in which they interpret the research) has the potential to enhance the rigor of research on and understanding of acculturation processes. Research reflexivity allows researchers to make decisions about research designs and tools to best capture and represent research topics; and the experiences, identities, and contexts of the people and communities included in research. This, in turn, can strengthen the way we study and represent the experiences, identities, and contexts of the people and communities we include in research on acculturation (e.g., Berger, 2015; Dwyer & Buckle, 2009). Researchers have pointed out that having insider *and* outsider perspectives in the design and interpretation of research findings will allow for a more nuanced and ethical representation of the communities included in research (e.g., Dwyer &

Buckle, 2009). As such, we share recommendations of how researchers can bring insider and outsider perspectives into their scholarship, and thereby further strengthen their scholarship and representation of their research participants.

One first step for researchers is to practice reflexivity and/or seek training on how to engage in reflexivity (including offering training to graduate students; e.g., Berger, 2015; Hellawell, 2006). Through this process, researchers will be able to identify their unique insights, privileges, and blind-spots, and make decisions about ways to minimize potential blind-spots (e.g., Berger, 2015) and incorporate diverse perspectives into their scholarship. Questions to consider for practicing or engaging in reflexivity include (Hellawell, 2006): (1). Do I engage in insider or outsider research or both? (2). What makes me an insider or outsider? (3). How may I not be engaged in insider or outsider research? (4). What are the areas in my research in which I am not sure of how I am engaged in insider or outsider research? (5). How may my insider or outsider status inform and influence my work? (6). What kind of privileges and blind-spots may I bring into my research? (7). What perspectives of the topics, experiences, and identities of the people and communities I study may be under- or overemphasized or altogether missing? (8). What are ways that I can offset my blind-spots?

One approach researchers can take to incorporate insider and outsider perspectives into their research and scholarship is the use of insider/outsider collaborative research teams (e.g., Rivas-Drake et al., 2016). Insider/outsider collaborative research teams tend to include at least one or more scholars and/or research team members (e.g., community liaisons, students) who are relative insiders and one or more scholars who are relative outsiders. Such insider/outsider collaborative research teams allow for the integration of diverse perspectives on acculturation processes. For example, Lee and Stein are part of a larger research team studying Multiracial youth (the LOVING Project). This team includes multiple scholars who identify as Multiracial as well as parents raising Multiracial children. This intentionally created collaborative team allowed for the inclusion of questions in the study that spoke to the experiences of Multiracial youth from multiple perspectives, enriching the study (see Chap. 7 of this book). Further, we also balance insider/outsider perspectives in the team as we interpret and write about our results with all of the first authors being Multiracial themselves. We have made decisions about the language we use, the types of questions we ask (e.g., comparing groups of Multiracial individuals), and how we respond to reviewers that espouse monoracist views. As another example, Ferguson leads a large transdisciplinary, international research team (spanning multiple universities in the United States and Jamaica) studying the roles of acculturation and media in adolescent nutrition within Black immigrant and refugee families (the Food, Culture, and Health Study; see Ferguson et al., 2019 for more details). The team also designed and is culturally adapting a global health intervention for acculturating youth and families across the two countries (the JUS Media? Programme). This team ensures that they include insider and outsider perspectives through intentional study team assembly (e.g., coinvestigators and research assistants from inside and outside each cultural group being studied, hiring of community cultural brokers for each cultural group), and they maintain a clear focus on cultural humility and antiracist research practices through mutually agreed upon (then documented) team values undergirded by joint readings and discussions (e.g., in lab meetings, see APA, 2019). Project meetings also dedicate time to sharing both insider and outsider perspectives, integrating cultural expertise with content expertise in transparent and collaborative decision making. The skill of the research team leader in eliciting and integrating insider and outsider perspectives is critical and dovetails with skills needed for transdisciplinary team leadership (see Ferguson et al., 2019 for more details). Finally, an important point to make about research teams with insider/outsider members is to consider how hierarchy and rank play a role in how methodological decisions are made. As with the example described above on the acculturation measure, the junior scholar with the insider perspective was ignored when they had a recommendation. Therefore, paying attention to hierarchies and power dynamics is critical when building teams with insider/outsider members. Ensuring the voices of all are heard and respected benefits the research quality.

We also recognize it is necessary to grow the academic pipeline of scholars of color from diverse backgrounds. While Lee has had an interest in studying Korean adoption as part of a larger Korean diaspora, he is not an insider to the adoption community. As such, he has actively recruited Korean adoptee graduate students and undergraduate students to his research team in order to ensure representation. This type of insider/outsider research collaboration is also an act of resistance and reclamation. Adoption scholarship is often led by scholars without any personal connection to adoption but who may implicitly hold humanitarian views on the subject or scholars who are themselves adoptive parents or affiliated with adoption agencies and thus hold particular biases and privileges on the subject. Lee and his team have worked toward creating a research team led by adoptees to challenge dominant adoption narratives and create new narratives on the complexity of transracial and transnational adoption.

Another way for researchers to ensure that insider and outsider perspectives are represented in their research and scholarship is through the use of community-based participatory research (CBPR) approaches (e.g., Jason et al., 2004; see Chap. 15 in this book). In traditional models of research, researchers (i.e., outsiders or partial insiders) generate research questions; make decisions about which experiences and identities to investigate and report; and how to interpret and make meaning of research findings. In contrast, in CBPR research, participants (i.e., insiders; people and communities included in our research) are active partners in all phases of a research study, including the conceptualization and identification of research questions and hypotheses, the identification and creation of the research design; the selection of survey measures; the creation of interview questions, the design and implementation of data collection procedures, the analysis and interpretation of research findings, and the dissemination of research findings. CBPR approaches have been described as innovative strategies that can address the needs of the people and communities researchers study by bringing their insider knowledge into all aspects of the research process (Torres-Harding et al., 2004), thereby including insider and outsider perspectives in research and reducing blind-spots.

In regard to scholarship on acculturation and enculturation in the developmental sciences, we propose that CBPR approaches have great potential to enhance our understanding of how acculturation and enculturation unfold and develop across contexts and populations. For example, much of what we know about acculturation and enculturation among Latinx youth and families stems from survey research where survey measures are often chosen because they are well known or standard but may not always fully reflect the experiences or processes researchers try to capture. Some of the measures used to capture acculturation processes among Latinx youth have been developed over 20 years ago and as such they may not accurately capture how acculturation and related experiences may have changed as a result of globalization, greater diversity in the United States, historical events, and the current positioning of Latinx in the United States. Some include outdated language; for example, the ARSMA-II uses the word "Anglo" that is foreign to many participants. As such, including Latinx youth and parents as partners in research studies has great potential for developing new or adapting existing measures and a nuanced understanding of survey research findings.

Another way to include the insights of research participants into survey research (questions or interpretation) is the use of mixed methods research approaches in scholarship on acculturation and enculturation. Mixed methods research goes beyond the integration of qualitative and quantitative methodologies and can be considered a systematic program of research that includes multiple methods and analyses used across studies that are synthesized to provide deeper insights into psychological processes—especially those including voices that are not often represented in scientific research (Mistry & Berardi, 2016). Part of the challenge of doing acculturation research is quantifying the complexities of these cultural, contextual, and psychological experiences in scales. While quantitative research allows for generalizable data that can be leveraged to determine the allocation of resources or policies and also provides insights into processes at a population level, it is by definition reductionist, hindering our understanding of these complexities. Qualitative data provide a methodology that allows for this deeper understanding highlighting the voices and lived experiences that not only provide nuance but may not have been originally conceived by theory or developed by the research team, but this approach is less generalizable. Mixed methods are well suited to ask confirmatory and exploratory questions and allow researchers to test broader questions in both methodologies, but the qualitative portion can serve to provide an insider perspective on the processes under study by allowing participants to share their own lens of understanding of how the various factors intersect and influence their lives (Mistry & Berardi, 2016; Yoshikawa et al., 2008). This can inform the quantitative research questions while also providing a contextual and interpretive frame for understanding findings. These approaches can take multiple forms from focus groups, ethnographies, interviews, to photovoice, and can combine with a larger collection of the survey, observational, biological and/or other forms of quantitative data. Importantly, how these forms are integrated can differ such that the sequence, integration, and interpretation of both types of data depend on the methodological tradition and question at hand (see Axinn & Pearce, 2006; Creswell, 2009; Morse & Niehaus,

2009; Teddlie & Tashakkori, 2009 for resources). As an example, the fourth author pioneered the study of remote acculturation using an explanatory sequential mixed methods study across a number of years. This study—the "Culture and Family Life Study"—started with a large sample quantitative phase surveying nearly 400 adolescents and their mothers in Jamaica along with a comparison of Jamaican immigrant and nonimmigrant subsamples in the United States to document the acculturation profiles of remotely integrated individuals in Jamaica (Americanized Jamaicans) relative to culturally traditional peers there and US-based comparison groups. This was followed by a qualitative phase wherein a small subset of these adolescents and mothers in Jamaica participated in focus groups to describe in greater depth their construal's of Jamaican culture, US cultures, and the Americanized Jamaican experience (Ferguson, 2018).

Just as important as the inclusion of community voices to ensure insider perspectives, we recommend approaching the study of acculturation and enculturation from ethnic studies (e.g., Omi & Winant, 2014) and critical race perspective (e.g., Salter & Adams, 2013). Ethnic studies and critical race theory look beyond individuals and groups, taking into account the historical, systemic, and structural ways in which racism and White supremacy influenced immigration policy, race relations, and more broadly the lived experiences of youth, families, and communities. By taking a reflexive, community-first approach that acknowledges White supremacy, we can move beyond dominant narratives of adaptation and adjustment. For example, we posit it is time to push against static typologies of acculturation that depict immigrants and communities as assimilated, integrated, marginalized, or separated. Rather than placing people in one or the other category, we urge researchers to look at the dynamic, agentic ways in which people navigate, negotiate, and maneuver through society (Titzmann & Lee, 2018). Rather than view adoption as a natural experiment or intervention that paints the adoptive family as humanitarians and saviors, we can view adoptees as empowered individuals who experience adoption as a lifelong journey in which they must constantly negotiate the loss and search for family, culture, and identity in the context of capitalism, racial hierarchies, and transnationalism. Similarly, any kind of immigration is not always rosy and better than what was left in the homeland. Rather than just considering acculturation at the family level as full of strife and replete with acculturation conflict that may lead to maladjustment, identify how families are flexible and resilient by negotiating acculturation together pooling resources across and within families, benefiting from language acquisition and cultural brokering, and working together (sometimes from afar and via technology) to thrive as a unit.

Conclusion

As developmental scientists, we aim to conduct the most rigorous science that can be translated to policy and interventions that support the well-being of our communities. While the history of our field has privileged work that is "objective" such that the researchers can be at an arm's length from the processes in question, we argue that this is an impossible task and that instead research teams should be built intentionally to incorporate insider/outsider perspectives for the communities under study. These perspectives should not only be explicitly attended to in the research team assembly, but at every point in research design—from the questions that are being pursued, the methodologies that are used; the surveys and questions that are included, the way participants are involved in the research process; the analyses that are employed, the way the data is interpreted, the way communities are portrayed, and finally the way that information is disseminated. This is the most rigorous science that we can do and will result in the most translatable work as it will attend to the nuances of the communities our science seeks to support. When working with multiethnic samples, assessing demographic information on immigrant/generation status is critical to inform generalizability, especially in Black, Asian, and Latinx samples in the United States (and other groups as well). When thinking about acculturation, it is important to not use proxy variables in data analysis to inform the acculturation literature, and instead specific measures need to be selected and included that attend to issues most relevant to the study. Finally, at the point of dissemination, insider/outsider perspectives need to be considered in writing research findings that are contextualized and nuanced, and do not contribute to negative stereotypes or support harmful narratives of immigrant communities. Writing and dissemination should highlight the narratives that are born from the community themselves as opposed to those that have been applied to them, that focus on resilience instead of deficits, and that further our nuanced understandings of their experiences instead of reducing their experiences to one narrative.

References

Alegria, M. (2009). The challenge of acculturation measures: what are we missing? A commentary on Thomson & Hoffman-Goetz. *Social Science & Medicine (1982), 69*(7), 996–998. https://doi.org/10.1016/j.socscimed.2009.07.006

American Psychological Association, APA Task Force on Race and Ethnicity Guidelines in Psychology. (2019). *Race and ethnicity guidelines in psychology: Promoting responsiveness and equity*. https://www.apa.org/about/policy/guidelines-race-ethnicity.pdf.

Axinn, W. G., & Pearce, L. D. (2006). *Mixed method data collection strategies*. Cambridge University Press.

Berger, R. (2015). Now I see it, now I don't: researcher's position and reflexivity in qualitative research. *Qualitative Research: QR, 15*(2), 219–234. https://doi.org/10.1177/1468794112468475

Bowleg, L. (2008). When Black+ lesbian+ woman≠ Black lesbian woman: The methodological challenges of qualitative and quantitative intersectionality research. *Sex Roles, 59*(5), 312–325. https://doi.org/10.1007/s11199-008-9400-z

Bridges, D. (2001). The ethics of outsider research. *Journal of Philosophy of Education, 35*(3), 371–386. https://doi.org/10.1111/1467-9752.00233

Cabassa, L. J. (2003). Measuring acculturation: Where we are and where we need to go. Hispanic *Journal of Behavioral Sciences, 25*(2), 127–146.

Cobb, C. L., Martínez Jr, C. R., Lee, S., Lee, T. K., & Lorenzo-Blanco, E. I. (2021). Acculturation trajectories differ by youth age at arrival and time in residency among Latino immigrant families in a US emerging immigrant context. *International Journal of Intercultural Relations, 81*, 79–93.

Creswell, J. W. (2009). Mapping the field of mixed methods research. *Journal of Mixed Methods Research, 3*(2), 95–108. https://doi.org/10.1177/1558689808330883

Cuellar, I., Arnold, B., & Maldonado, R. (1995). Acculturation rating scale for Mexican Americans-II: A revision of the original ARSMA scale. *Hispanic Journal of Behavioral Sciences, 17*(3), 275–304.

Doucerain, M. M., Segalowitz, N., & Ryder, A. G. (2017). Acculturation measurement: From simple proxies to sophisticated toolkit. In *The Oxford handbook of acculturation and health* (1st ed.). Oxford University Press. https://doi.org/10.1093/oxfordhb/9780190215217.013.7

Dwyer, S. C., & Buckle, J. L. (2009). The space between: On being an insider-outsider in qualitative research. *International Journal of Qualitative Methods, 8*(1), 54–63. https://doi.org/10.1177/160940690900800105

Eales, L., Gillespie, S., Eckerstorfer, S., Eltag, E., Global Educators Group, & Ferguson, G. M. (2020). Remote acculturation 101: A primer on research, implications, and illustrations from classrooms around the world. *Online Readings in Psychology and Culture, 8*(1). https://doi.org/10.9707/2307-0919.1168

Ferguson, G. M. (2018). Caribbean migration and globalization: Illuminating new global patterns of acculturation and adaptation in the 21st century. *Caribbean Journal of Psychology, 10*, 192–222.

Ferguson, G. M., & Bornstein, M. H. (2012). Remote acculturation: The "Americanization" of Jamaican islanders. *International Journal of Behavioral Development, 36*(3), 167–177. https://doi.org/10.1177/0165025412437066

Ferguson, G. M., Fiese, B. H., Nelson, M. R., & Meeks Gardner, J. M. (2019). Transdisciplinary team science for global health: Case study of the JUS Media? Programme. *American Psychologist, 74*(6), 725–739. https://doi.org/10.1037/amp0000383

Garcia, M., & Birman, D. (2020). Ethical issues in research with late-arriving and unaccompanied immigrant youth. *Translational Issues in Psychological Science, 6*(3), 207–213. https://doi.org/10.1037/tps0000268

Gattino, S., Miglietta, A., Rizzo, M., & Testa, S. (2016). Muslim acculturation in a catholic country: Its associations with religious identity, beliefs, and practices. *Journal of Cross-Cultural Psychology, 47*(9), 1194–1200.

Gulamhussein, S. A. (2021). A Shia Ismaili Muslim's ringette experiences on and off the ice: An autoethnography. In Leisure and Forced Migration (pp. 191–204). Routledge.

Haarlammert, M., Birman, D., Oberoi, A., & Moore, W. (2017). Inside-out: Representational ethics and diverse communities. *American Journal of Community Psychology, 60*(3–4), 414–423. https://doi.org/10.1002/ajcp.12188

Hellawell, D. (2006). Inside–out: Analysis of the insider–outsider concept as a heuristic device to develop reflexivity in students doing qualitative research. *Teaching in Higher Education, 11*(4), 483–494. https://doi.org/10.1080/13562510600874292

Hodkinson, P. (2005). 'Insider research' in the study of youth cultures. *Journal of Youth Studies, 8*(2), 131–149. https://doi.org/10.1080/13676260500149238

Hughes, D. L., & Seidman, E. (2002). In pursuit of a culturally anchored methodology. In T. A. Revenson, A. R. D'Augelli, S. E. French, D. L. Hughes, D. Livert, E. Seidman, M. Shinn, & H. Yoshikawa (Eds.), *Ecological research to promote social change: Methodological advances from community psychology* (pp. 243–255). Kluwer Academic/Plenum Publishers. https://doi.org/10.1007/978-1-4615-0565-5_10

Hurh, W. M., & Kim, K. C. (1984). Adhesive sociocultural adaptation of Korean immigrants in the US: An alternative strategy of minority adaptation. *International Migration Review, 18*(2), 188–216.

Jason, L. A., Keys, C. B., Suarez-Balcazar, Y., Taylor, R. R., & Davis, M. I. (Eds.). (2004). *APA decade of behavior Vols. Participatory community research: Theories and methods in action*. American Psychological Association. https://doi.org/10.1037/10726-000

Juang, L. P., & Syed, M. (2019). The evolution of acculturation and development models for understanding immigrant children and youth adjustment. *Child Development Perspectives, 13*(4), 241–246. https://doi.org/:10.1111/cdep.12346

Keles, S., Friborg, O., Idsoe, T., Sirin, S., & Oppedal, B. (2018). Resilience and acculturation among unaccompanied refugee minors. *International Journal of Behavioral Development, 42*(1), 52–63. https://doi.org/10.1177/0165025416658136

LaFromboise, T., Coleman, H. L., & Gerton, J. (1993). Psychological impact of biculturalism: Evidence and theory. *Psychological Bulletin, 114*(3), 395–412. https://doi.org/10.1037/0033-2909.114.3.395

Lara, M., Gamboa, C., Kahramanian, M. I., Morales, L. S., & Hayes Bautista, D. E. (2005). Acculturation and Latino health in the United States: a review of the literature and its sociopolitical context. *Annual Review of Public Health, 26*, 367–397.

Lee, R. M., & Robbins, S. B. (1995). Measuring belongingness: The social connectedness and the social assurance scales. *Journal of Counseling Psychology, 42*(2), 232–241. https://doi.org/10.1037/0022-0167.42.2.232

Lee, R. M., Choe, J., Kim, G., & Ngo, V. (2000). Construction of the Asian American family conflicts scale. *Journal of Counseling Psychology, 47*(2), 211–222. https://doi.org/10.1037/0022-0167.47.2.211

Lee, R. M., Noh, C., Yoo, H. C., & Doh, H. (2007). The psychology of diaspora experiences: Intergroup contact, perceived discrimination, and the ethnic identity of Koreans in China. *Cultural Diversity and Ethnic Minority Psychology, 13*(2), 115–124. https://doi.org/10.1037/1099-9809.13.2.115

Lee, R. M., Yoon, E., & Liu-Tom, H. T. T. (2006). Structure and measurement of acculturation/enculturation for Asian Americans using the ARSMA-II. *Measurement and Evaluation in Counseling and Development, 39*(1), 42–55. https://doi.org/ghw2zd

Lopez-Class, M., Castro, F. G., & Ramirez, A. G. (2011). Conceptions of acculturation: A review and statement of critical issues. *Social Science & Medicine, 72*(9), 1555–1562.

Lorenzo-Blanco, E., Arillo-Santillán, E., Unger, J., & Thrasher, J. (2019). Remote acculturation and cigarette smoking susceptibility among youth in Mexico. *Journal of Cross-Cultural Psychology, 50*(1), 63–79. https://doi.org/10.1177/0022022118807578

Martinez, C., Schwartz, S., Thier, M., & McClure, H. (2018). A tale of two measures: Concordance between the ARSMA-II and the BIQ acculturation scales among Latino immigrant families. *Psychological Assessment, 30*(4), 459–473. https://doi.org/10.1037/pas0000491

Meca, A., Schwartz, S. J., Martinez, C. R., & McClure, H. H. (2018). Longitudinal effects of acculturation and enculturation on mental health: Does the measure of matter? *Development and Psychopathology, 30*(5), 1849–1866. https://doi.org/10.1017/S0954579419000129

Mistry, J., & Berardi, A. (2016). Bridging indigenous and scientific knowledge. *Science (American Association for the Advancement of Science), 352*(6291), 1274–1275. https://doi.org/10.1126/science.aaf1160

Morse, J. M., & Niehaus, L. (2009). *Mixed methods design: Principles and procedures*. Left Coast Press.

Omi, M., & Winant, H. (2014). *Racial formation in the United States*. Routledge.

Rivas-Drake, D., Camacho, T. C., & Guillaume, C. (2016). Just good developmental science: Trust, identity, and responsibility in ethnic minority recruitment and retention. In S. S. Horn, M. D. Ruck, & L. S. Liben (Eds.), *Equity and justice in developmental science: Theoretical and methodological issues* (pp. 161–188). Elsevier Academic Press. https://doi.org/10.1016/bs.acdb.2015.11.002

Salter, P., & Adams, G. (2013). Toward a critical race psychology. *Social and Personality Psychology Compass, 7*(11), 781–793. https://doi.org/10.1111/spc3.12068

Schumann, M., Bug, M., Kajikhina, K., Koschollek, C., Bartig, S., Lampert, T., & Santos-Hövener, C. (2020). The concept of acculturation in epidemiological research among migrant populations: A systematic review. *SSM-Population Health, 10*(100539), 1–10. https://doi.org/10.1016/j.ssmph.2020.100539

Schwartz, S. J., Unger, J. B., Zamboanga, B. L., & Szapocznik, J. (2010). Rethinking the concept of acculturation: Implications for theory and research. *American Psychologist, 65*(4), 237–251. https://doi.org/10.1037/a0019330

Schwartz, S. J., Unger, J. B., Zamboanga, B. L., Córdova, D., Mason, C. A., Huang, S., Baezconde-Garbanati, L., Lorenzo-Blanco, E. I., Des Rosiers, S. E., Soto, D. W., Villamar, J. A., Pattarroyo, M., Lizzi, K. M., & Szapocznik, J. (2015). Developmental trajectories of acculturation: Links with family functioning and mental health in recent-immigrant Hispanic adolescents. *Child Development, 86*(3), 726–748. https://doi.org/10.1111/cdev.12341

Stein, G. L., Rivas-Drake, D., & Camacho, T. C. (2017). Ethnic identity and familism among Latino college students: A test of prospective associations. *Emerging Adulthood, 5*(2), 106–115. https://doi.org/10.1177/2167696816657234

Szapocznik, J., Kurtines, W. M., & Fernandez, T. (1980). Bicultural involvement and adjustment in Hispanic-American youths. *International Journal of Intercultural Relations, 4*(3–4), 353–365.

Taušová, J., Bender, M., Dimitrova, R., & van de Vijver, F. (2019). The role of perceived cultural distance, personal growth initiative, language proficiencies, and tridimensional acculturation orientations for psychological adjustment among international students. *International Journal of Intercultural Relations, 69*, 11–23. https://doi.org/10.1016/j.ijintrel.2018.11.004

Teddlie, C., & Tashakkori, A. (2009). *Foundations of mixed methods research: Integrating quantitative and qualitative approaches in the social and behavioral sciences*. Sage.

Torres-Harding, S. R., Herrell, R., & Howard, C. (2004). Epidemiological research: Science and community participation. In L. A. Jason, C. B. Keys, Y. Suarez-Balcazar, R. R. Taylor, & M. I. Davis (Eds.), *Participatory community research: Theories and methods in action* (pp. 53–69). American Psychological Association. https://doi-org.ezaccess.libraries.psu.edu/10.1037/10726-003

Titzmann, P. F., & Lee, R. M. (2018). Adaptation of young immigrants: A developmental perspective on acculturation research. *European Psychologist, 23*(1), 72–82. https://doi.org/10.1027/1016-9040/a000313

Yoshikawa, H., Weisner, T. S., Kalil, A., & Way, N. (2008). Mixing qualitative and quantitative research in developmental science: Uses and methodological choices. *Developmental Psychology, 44*(2), 344–354. https://doi.org/10.1037/0012-1649.44.2.344

Chapter 6
Applying Critical Multiracial Theory to Conceptualizing and Measuring Multiracial Experiences and Identity

Abigail K. Gabriel, Clarissa Abidog, Hyung Chol Yoo, Gabriela Livas Stein, N. Keita Christophe, Annabelle Atkin, Christine Wu, and Richard M. Lee

The number of Americans identifying as two or more races on the 2020 Census increased from 9 million to 33.8 million people (276%) since 2010, making multiracials the fastest growing segment of the US population (U.S. Census Bureau, 2021). Yet, it has only been since the 2000 Census that the choice to identify as more than one race has become an option, and researchers who have examined this population have struggled to do so with any shared approach. In the past 20 years, the literature and theories related to this hidden population have undoubtedly advanced (see Harris, 2016; Rockquemore & Brunsma, 2002; Root, 2003 for some examples), but developmental science has room yet to grow. In this chapter, we address a range of approaches and challenges to conducting research with multiracial participants. Our chapter applies critical multiracial theory (MultiCrit; Harris, 2016), which frames how racism uniquely positions multiracials in the US racial hierarchy, to advance the conceptualization and measurement of multiracial experiences and identity in developmental science. Stressing the importance of an ecological approach to theory and research, we introduce a new Model of Multiracial

A. K. Gabriel (✉) · C. Abidog · H. C. Yoo
Arizona State University, Tempe, AZ, USA
e-mail: akgabri1@asu.edu

G. L. Stein
University of North Carolina at Greensboro, Greensboro, NC, USA

N. K. Christophe
McGill University, Montreal, QC, Canada

A. Atkin
Purdue University, West Lafayette, IN, USA

C. Wu
Oberlin College, Oberlin, OH, USA

R. M. Lee
University of Minnesota Twin Cities, Minneapolis, MN, USA

© The Author(s), under exclusive license to Springer Nature Switzerland AG 2023

D. P. Witherspoon, G. L. Stein (eds.), *Diversity and Developmental Science*,
https://doi.org/10.1007/978-3-031-23163-6_6

Racialization that situates multiracial racialization experiences (including racial identity, racial identification, and racial category) within six ecological levels: (1) Individual Characteristics; (2) Interpersonal Experiences; (3) Contextual Factors; (4) Social, Economic, and Political Environments; (5) Systems of Oppression; and (6) Time. Finally, we conclude this chapter by offering questions for researchers to consider when crafting thoughtful and critical programs of study.

One essential foundation in cultivating a body of research with any consistency is defining clear terminology. Within this chapter, we contend that an individual is considered multiracial if their biological parents belong to two or more different racial groups (e.g., a person with one Latinx parent and one Native American parent), or whose biological parents are themselves multiracial. Similarly, yet distinct, an individual is considered multiethnic if their biological parents belong to two or more different ethnic groups. Multiracial people are inherently multiethnic; however, not all multiethnic people are multiracial. For example, someone with one Irish parent and one Nigerian parent would be considered multiethnic and multiracial. However, someone with a Welsh parent and an Italian parent would be considered multiethnic but not multiracial, as their race would still be designated as monoracial White. As developmental scientists, it is important to recognize how racial categories as social constructs require an understanding of how power, privilege, and oppression have been historically and currently enacted to maintain race as a salient social construction. In our example of a multiracial (Black/White) and multiethnic (Nigerian/Irish) individual, it is fundamental to recognize the inextricable link between race and ethnicity in the United States, wherein Irish ethnic membership embodies Whiteness and Nigerian ethnic membership embodies Blackness. Currently, there is a dearth of literature in developmental science, which teases out such nuances between multiethnicity and multiraciality—particularly from a global perspective where race and ethnicity can have different meanings across countries.

In this chapter, the term "multiracial" is not capitalized or conceptualized as a new, distinct racial group. Although some multiracial scholars and activists view the capitalization of "multiracial" to be an affirmation of this group's unique racialization experiences, others argue that capitalizing "multiracial" reinforces existing racist scripts of white supremacy through racial essentialism. Ultimately, we want to emphasize that social scientists are in a precarious position to inform racial formation for future generations. The ways that researchers conceptualize and define multiraciality (in other words, how they delineate racial categories) today will have implications for how society understands labels of race as the US population diversifies. We strongly encourage readers to consider the question: *How is my developmental science research reinforcing and/or disrupting racial formations and racism?* Moreover, the recommendations in this chapter are specific to the sociopolitical context of the United States, and this unique history may not accurately apply to other global contexts with varying racial formations. While our recommendations may have valuable implications when examining multiracial experiences and identities beyond the borders of the United States, international scholars should consider the forthcoming points within their own sociopolitical and historical contexts.

The Exclusion of Multiracials in Developmental Science

The field of developmental science has and continues to refine examinations of normative developmental tasks unique to monoracial populations. Mainstream theories such as the Integrative Model (García Coll et al., 1996) and Phenomenological Variant of Ecological Systems Theory (Spencer, 2007) aid researchers in considering the culturally relevant aspects of development throughout an individual's lifetime such as ethnic-racial identity and social-cognitive development, particularly for historically minoritized US populations. Yet, while developmental science continues to evolve beyond deficit models (e.g., White developmental standards as normative) among monoracial minority populations, the same should be applied when examining multiracial populations. Multiracials in research samples are too often subsumed into researcher-chosen monoracial labels, categorized as "Other," or purposefully excluded, consequently resulting in the continued erasure of multiracial experiences (Nishina & Witkow, 2020). When researchers make the decision to classify these people into another racial group, they impose research bias and remove personal choice. Appropriate multiracial theories have been developed to guide researchers seeking to intentionally examine multiracial participants in their studies (see Harris, 2016; Rockquemore & Brunsma, 2002; Root, 1992, 1996, 2003 for some examples). However, as the multiracial population in the United States continues to grow, intentional and unintentional exclusion of this understudied population hinders ethically responsible research within developmental science.

How Multiracials Challenge Developmental Science's Understandings of Race and Racism

One advantage multiracial research offers developmental science is that of an arena to challenge understandings of race, racial formation, and racism. Research explicitly examining multiracial realities provides precise examples of the malleability of race as a social construct, further rejecting biological assumptions of race. Furthermore, multiracial people can have an ambiguous phenotypic appearance that challenges the discrete categories of monoracial paradigms. Another challenge to racial formation is embedded in multigenerational multiracial families, where racial identity can vary drastically across time, between parents and children, and even between siblings from the same biological parents (Mauer et al., 2020). Such cases disrupt assumptions of stratified homogeneous experiences not only within multiracials as a growing population, but within monoracial realities as well. As Harris (2016) highlights, critical multiracial theory, or "MultiCrit, allows for a critique of the role that white supremacist structures play in the (re)construction of multiraciality, thus uncovering far more profound effects of racism for multiracial and monoracial people of color than 'a lack of place'" (p. 797). A more nuanced analysis of how multiracials fit into the current American racial structure is needed to appropriately critique racial myths. For example, researchers could investigate how the commercialization of "ancestry" genetic testing has influenced who claims a multiracial identity as well as a renewal of understanding race as a biological, versus social, construction. Rather than inattentively subsuming multiracials into monoracial-specific racial identity models, developmental scientists must thoughtfully critique the role that multiracials play in our current racial structure, as well as how multiracial and monoracial experience intersects, to effectively challenge the racial myths upheld by white supremacy.

Limitations of Monoracial Racial Identity Models

Most renowned models of racial identity (e.g., Cross, 1995; Helms, 1995; Phinney, 1996; Sellers et al., 1997) were conceptualized for monoracial populations, and often do not adequately capture the distinct racialization of individuals with multiple racial heritages. Racial identity for monoracial groups implicitly, if not explicitly, assumes that an individual's ascribed racial heritage and claimed racial identity are equivalent (e.g., an individual with Asian American parents will racially identify as Asian American). In reality, even multiracials who share identical racial heritages (e.g., individuals with African American/Black and Latinx ancestry) may choose to racially identify in ways that are different from each other (e.g., exclusively African American/Black and exclusively multiracial), or even in ways that are different from themselves across time (Brunsma, 2005; Mihoko Doyle & Kao, 2007). Monocentric models of racial identity negate how race and racial boundaries change to reflect particular sociopolitical moments in history (Miller, 1992). For example, consider how the willingness to claim a multiracial identity might have changed after the repeal of anti-miscegenation laws which, prior to 1967, explicitly deemed interracial unions as socially deviant and punishable by law. Similarly, the United States Census Bureau did not allow individuals to select more than one race on the decennial Census until 2000—effectively erasing the presence of a multiracial population in the country on an institutional level. A more suitable theory of multiracial identity would need to incorporate these historical and contextual perspectives to account for nuances in how multiracials understand and construct their racial identity relative to their immediate environments.

Critical Multiracial Theory (MultiCrit)

The emerging literature on multiracial identity highlights the need for theoretical pluralism, or the integration of multiple theoretical perspectives in order to account for the dynamic sociopolitical construction of race and racial hierarchies (Jackson & Samuels, 2019). This chapter conceptualizes race and race-related developmental processes according to critical multiracial theory (MultiCrit; Harris, 2016). MultiCrit builds on the foundations of intersectionality (Combahee River Collective, 1977; Crenshaw, 1989), racial formation theory (Omi & Winant, 2014), and critical race theory (CRT; Delgado & Stefancic, 2001) to delineate the distinct, systematic racialization of multiracial individuals who live in the United States (Harris, 2016). Mirroring original tenets from critical race theory, Harris (2016) proposes eight tenets toward a MultiCrit framework: (1) *challenge to ahistoricism*; (2) *interest convergence*; (3) *experiential knowledge*; (4) *challenge to dominant ideology*; (5) *racism, monoracism, and colorism*; (6) *a monoracial paradigm of race*; (7) *differential micro-racialization*; and (8) *intersections of multiple racial identities*. Central to MultiCrit is the belief that social scientists should challenge ahistorical and dominant ideologies that uphold a monoracial paradigm of race, including the endemic effects of racism and monoracism (Harris, 2016). *Racism* is the production and perpetuation of a racial structure through capitalist racial domination projects such as slavery, nationalism, colonialism, imperialism, and neoliberalism (Bonilla-Silva, 2015). A *racial structure* is a hierarchical classification system that systematically privileges some racial groups at the cost of disenfranchising others (Bonilla-Silva, 2015). Moreover, racial structures are created through *racialization*, which is the process by which individuals and institutions "socially construct" race by ascribing meaning to perceive racial differences (Omi & Winant, 2014). The contemporary racial structure in America can be traced back to the cataloging of different racial categories in the eighteenth century (Gossett, 1997). Namely, the nomenclaturist Carolus Linnaeus conceptualized four separate "varieties," or races, of the human species: *Homo Europaeus*, *Homo Asiaticus*, *Homo Afer*, and *Homo Americanus* (Gossett, 1997). Building on Linnaeus' taxonomy, the phrenologist Johann Friedrich Blumenbach identified five, "color-coded" racial varieties in his thesis, "On the Natural Variety of Mankind" (1775): Caucasian (white), Mongolian (yellow), Ethiopian (black), American (red), and Malay (brown) (Gossett, 1997). The myriad racial taxonomies established by these and other Enlightenment-era scholars have been adapted to justify the social, political, and economic domination of white supremacy across time.

A unique manifestation of racism within the current racial structure of the United States is *monoracism*, which refers to the systematic oppression of multiracials through the preservation of discrete monoracial categories (Johnston & Nadal, 2010; Harris, 2016). Monoracism and the monoracial paradigm of race erase the longstanding history of interracial mixing and multiraciality in the United States (Carter, 2013; Harris, 2016). The existence of multiracials blurs and threatens the rigid monoracial boundaries that characterize our society's understanding of race (Root, 1992). Consequently, multiracials "become one of the most important socio-political signifiers of racial production, presentation, and contestation in American society's racial and social organization" (Williams, 1996, p. 194). Throughout history, monoracism has dictated which racial categories are available to multiracials in ways that serve to uphold the purity of discrete races and white supremacy (Carter, 2013). Consider, for example, how the classification of races on the US Census has changed over time to reflect White resentment and fears about growing rates of interracial unions and the immigration of racial minority groups. Race and racial identities, therefore, are markers of politicized histories across specific contexts (Omi & Winant, 2014). By acknowledging the social construction and political-economic function of racial categories in the United States, social scientists can better understand how multiracials are uniquely racialized and construct their racial identities under a monoracial paradigm of race (Harris, 2016).

Past Theories of Multiracial Identity

One of MultiCrit's tenets, *challenging ahistoricism*, emphasizes the importance of understanding the relevant history of race and racial mixture to fully understand the contemporary experiences of multiracial individuals and families. The discourse around multiraciality in the social sciences has changed dramatically over the past century. Multiracial identity models have moved through a series of four paradigm shifts (Kuhn, 1962) characterized by each era's racial climate. These include the *problem approach* (late Jim Crow era, 1900s to 1960s), the *equivalent approach* (Civil Rights Movement era, 1960s to 1970s), the *variant approach* (Post-Loving v. Virginia era, 1970s to 1990s), and the current *ecological approach* (Post-racial colorblind era, 1990s to present) (Thornton & Wason, 1995; Thornton, 1996; Rockquemore et al., 2009).

Problem Approach (1900s–1950s)

The problem approach politically functioned to delineate clear monoracial categories and to dissuade individuals from racially intermixing for fear of the social and psychological consequences for their families (Thornton, 1996). It was a manifestation of the socio-legal racial segregation of the late Jim Crow era (Rockquemore et al., 2009). The theory of the "marginal man" (Park, 1928; Stonequist, 1937) characterizes what is one of the oldest paradigms in the social sciences for understanding multiraciality. White researchers in the early twentieth century theorized that it was not possible for mixed race children to integrate the innately clashing cultures of their parents' different monoracial groups (Thornton & Wason, 1995). Consequently, multiracials at the margins of these monoracial groups were assumed to be at greater risk of developing psychopathology (Thornton, 1996). However, much of this literature comes from studying clinical samples, including multiracials recruited from the child welfare system and social service agencies who often experience adversities that influence their development, including marginalization and pathologizing (Thornton & Wason, 1995). The problem approach, consequently, is not an accurate reflection of the psychological impact of having multiple racial heritages for the general multiracial population.

Equivalent Approach (1960s–1970s)

The equivalent approach politically functioned to promote racial equity and the affirmations of Black identity and lives in response to blatant racism and the dehumanization of Black people (Rockquemore et al., 2009). It was a manifestation of the values and goals of the rising Civil Rights and Black Power Movements (Rockquemore et al., 2009). Much of this literature was rooted in this era's lingering beliefs about hypodescent and the one-drop rule; namely, that Black multiracials should be racially categorized and included in the Black community (Rockquemore et al., 2009). Consequently, racial identity development under the equivalent approach was assumed to be identical for multiracial and monoracial groups (Thornton & Wason, 1995). Building on Erikson's (1968) theory of ego identity formation, researchers posited that multiracials and monoracials underwent similar processes of exploration and commitment that led to a stable monoracial identity (Rockquemore et al., 2009). This is exemplified by Cross' (1971) Nigrescence model for developing a positive monoracial Black identity, which was used to explain racial identity development in both monoracial and multiracial Black individuals (Thornton, 1996).

Variant Approach (1980s–1990s)

The variant approach politically functioned to challenge problematic assumptions about multiraciality evident in the previous approaches, including the beliefs that racial identity development for multiracials was inherently pathological yet proceeded identically to monoracial racial identity development (Rockquemore et al., 2009). It was a manifestation of the rapidly growing multiracial population following the repeal of anti-miscegenation laws by Loving v. Virginia (1967). Unlike other paradigms, the variant approach was led by mixed race scholars who were able to voice the distinct race-related developmental processes that set multiracials apart from monoracial groups (Thornton, 1996). This is exemplified by the anthology *Racially Mixed People in America* (1992) by Maria Root, which features multiracial scholars from a wide range of disciplines who describe the unique racialization and identity choices of mixed race individuals (Rockquemore et al., 2009). Researchers from the variant approach were interested in how multiracials integrated their multiple racial heritages and began to theorize models of identity development unique to mixed race individuals. For example, Poston (1990) extended Cross' (1971) Nigrescence model by applying it to biracial identity development among Black-White adolescents. Rather than developing a positive monoracial Black identity, the end stage of Poston's (1990) biracial identity development model was a healthy biracial identity in which Black-White multiracials valued both of their racial heritages. Similarly, Kich (1992) proposed a biracial identity development model for people with Asian American and White ancestry that defined the final stage of development as accepting and asserting an integrated biracial identity.

An Ecological Framework for Understanding Multiracial Identity (2000s–Present)

The problem, equivalent, and variant approaches for studying multiracials in the social sciences exemplify how the construction of racial boundaries and the meaning of multiraciality fluctuate across time and contexts. Our collective understanding that race is actively constructed to serve social, political, and economic ends has culminated in the ecological framework for understanding multiracial identity (Renn, 2003; Rockquemore et al., 2009; Root, 2003). Arguably, the ecological approach is a manifestation of our current era of post-racial colorblindness, and the delegitimization of race as a tool of social stratification. Consistent with MultiCrit tenets, its political function is to *challenge the dominant ideology and monoracism.* Researchers who employ an ecological approach contend that what distinguishes multiracial from monoracial racial identity development is the process of negotiating the meaning of one's multiple racial heritages in a society defined by distinct monoracial boundaries (Root, 1992, 1996, 2003). The goal of studying multiracial identity is to identify contextual factors (e.g., phenotype, community racial composition) that influence these meaning-making and identity development processes, with the knowledge that which racial identity choices are "healthiest" depend on defining characteristics of a multiracial person's immediate context (Rockquemore, 1999). Therefore, researchers should not privilege one type of racial identity over another (e.g., monoracial or multiracial) because this essentializes arbitrary racial group differences, and ignores how race is socially, politically, and economically constructed through racialization (Omi & Winant, 2014; Rockquemore et al., 2009). For a complete list of all multiracial-specific theories and measures grounded in the ecological approach, see Table 6.1.

Conceptualizing and Measuring Multiracial Racialization

The MultiCrit tenet, *experiential knowledge or counter storytelling*, further advocates for centering the perspectives and voices of mixed race people in research, not requiring their experiences to be built or adapted from currently existing monoracial theories and literature. In the effort to build a more cohesive understanding of multiracial identity and experiences using the ecological approach, Rockquemore et al. (2009) recommend disentangling three key features of the racialization process for multiracials: racial identification, racial category, and racial identity. These three features dynamically influence racial identity development and reflect different

Table 6.1 Multiracial theories and measures based on the ecological approach

Name	Description	Authors
Theory		
Critical Multiracial Theory	Drawing on Critical Race Theory (Delgado & Stefancic, 2001) and its tenets examining how race and racism are experienced by multiracials and monoracism challenged	Harris (2016)
Ecological Framework for Understanding Multiracial Identity	Contextualizing the border crossings with consideration of regional and generational histories of race relations, sexual orientation, gender, class, community attitudes, racial socialization, family functioning, and individual's personality traits and aptitudes.	Root (1996, 2003)
Multidimensional Model of Biracial Identity	Distinguishes racial identity according to four racial typologies including singular identity, border identity, protean identity, and transcendent identity	Rockquemore (1999); Rockquemore and Brunsma (2002)
Multiracial-Heritage Awareness and Personal Affiliation Theory	Distinguishes between three dominant racial identity orientations: *singular identity, integrated identity,* and *marginal identity*	Choi-Misailidis (2004)
Multiracial Identity Theory	Three key features of the racialization process for multiracials include racial identification, racial category, and racial identity	Rockquemore et al. (2009)
Multiracial Microaggressions	Describes daily verbal, behavioral, or environmental indignities, whether intentional or unintentional, enacted by monoracial persons that communicate hostile, derogatory, or negative slights toward multiracial individuals or groups	Johnston and Nadal (2010)
Measure		
Multiracial Challenges and Resilience Scale	Identifies four racial identity-related challenges (Lack of Family Acceptance, Multiracial Discrimination, Challenges with Racial Identity, and Others' Surprise and Disbelief Regarding Racial Heritage) and two racial identity-related resiliencies (Multiracial Pride and Appreciation of Human Differences)	Salahuddin and O'Brien (2011)
Multiracial Experiences Measure	Measures two racial identity-related risk factors (Multiracial Discrimination and Perceived Racial Ambiguity) and three racial identity-related resilience factors (Shifting Expressions, Creating Third Space, and Multicultural Engagement)	Yoo et al. (2016)
Multiracial Identity Integration Measure	Examines the extent to which multiracials perceive conflict or distance between their multiple racial heritages	Cheng and Lee (2009)

(continued)

Table 6.1 (continued)

Name	Description	Authors
Multiracial Youth Socialization (MY-Soc) Scale	Measures distinct racial socialization of multiracials including Navigating Multiple Heritages Socialization, Multiracial Identity Socialization, Preparation for Monoracism Socialization, Negative Socialization, Colorblind Socialization, Diversity Appreciation Socialization, Race-Conscious Socialization, and Silent Socialization	Atkin et al. (2021)
Racial Identity Invalidation Instrument	Measures racial identity invalidation, others' denial of an individual's racial identity, including Behavior Invalidation, Phenotype Invalidation, and Identity Incongruent Discrimination	Franco and O'Brien (2018)

facets of understanding one's multiracial background (Rockquemore et al., 2009). When investigated in tandem, these three features of multiracial racialization can help researchers construct a more nuanced and holistic understanding of multiracial identity and experiences.

Racial Identification

Racial identification refers to how other people racially perceive and categorize someone who is multiracial (Rockquemore et al., 2009). Rockquemore and Brunsma (2002) contend that the racial identity claimed by a multiracial person depends on whether that identity is validated by others. In other words, multiracials construct their racial identities relative to how other people treat and racially identify them. Racial identification as a process of multiracial racialization is perhaps best exemplified by the recent breadth of studies on how individuals racially discern multiracial faces (Chen et al., 2014, 2018; Gaither et al., 2016; Pauker et al., 2018). For example, Chen et al. (2018) suggest that people demonstrate a "minority bias" when racially categorizing Black-White multiracials, such that they were more likely to identify these multiracials as non-White than White. Franco and O'Brien (2018) also developed the Racial Identity Invalidation Instrument, which includes Behavior Invalidation, Phenotype Invalidation, and Identity Incongruent Discrimination.

The prominence of the monoracial paradigm in the United States has resulted in the widespread belief that races are distinct, separate categories (Harris, 2016). Social perception research is an important avenue for combatting the racial essentialism that underlies these assumptions about racial categorization. For instance, several studies suggest that observing multiracial faces aids in a person's understanding that race is flexible across time and contexts (Chen et al., 2014; Pauker et al., 2018). Young et al. (2017) found that monoracial Black and White individuals racially categorized Black-White multiracials differently based on phenotypic cues, including racial ambiguity or the ability to racially "pass" as a monoracial group.

Moreover, studies have demonstrated how racial identification is motivated by the perceivers' own motivations, such as their internal motivation to control prejudice (Chen et al., 2014), and their desire for social belonging (Gaither et al., 2016).

Racial Category

Racial category refers to how a multiracial person chooses to identify based on which racial identities are available within a specific context, such as on college applications, health forms, demographic surveys, or the US Census (Rockquemore et al., 2009). Racial category may be viewed as an expression of racial identity, such that how a multiracial person understands their racial identity influences which racial category they are most likely to endorse. Our understanding of racial category has benefited greatly from the increased accessibility of large datasets with multiracial samples (see Rockquemore et al., 2009).

Racial category as defined by multiracial racialization (Rockquemore et al., 2009) has often been studied in terms of the associations between racial identity choices and adjustment-related outcomes. For example, Burke and Kao (2013) used the National Longitudinal Study of Adolescent Health (Add Health) to assess the association between identifying as White and academic achievement among multiracial adolescents, and the authors found that Black/White and Asian/White adolescents who identified as White reported lower grade point averages. Other studies have used publicly available datasets to examine the associations between racial identity choices and depression, sense of belonging, substance abuse, and academic achievement (Campbell & Eggerling-Boeck, 2006; Chavez & Sanchez, 2010; Choi et al., 2006; Cooney & Radina, 2000). However, a severe limitation of these studies is that they rely on single-item measures of racial category (e.g., "What is your race? You may select more than one if applicable: White, Black/African American, Asian American, Latinx, Pacific Islander, or American Indian"). Consequently, the mechanisms underlying any associations between racial category and adjustment remain unclear. Future studies should examine possible mediators—such as the potential stress of being forced to identify with one racial group—for the link between racial category and adjustment. Furthermore, any conclusions made about disparities between multiracial and monoracial individuals risk essentializing racial group differences and pathologizing multiraciality. In order to avoid perpetuating racial essentialism in developmental science, we encourage researchers to investigate why possible differences in adjustment-related outcomes exist between multiracial and monoracial groups. For instance, researchers could examine how endorsing a monoracial paradigm of race mediates the association between racial category and psychosocial adjustment. It may be the case that, in certain instances, multiracials are more at risk for negative outcomes because of their unique racial stressor of having to negotiate the meaning of their multiple racial heritages in a society that defines race on monoracial terms. Importantly, such studies would be able to attribute possible differences between multiracial and monoracial groups to their unique racialization experiences, rather than to any innate racial group differences.

Racial Identity

Racial identity refers to how a multiracial person attributes meaning and significance to their membership in multiple racial groups (Rockquemore et al., 2009). Much of the literature on multiracials has focused on developing theories, models, and measures of how these individuals understand their racial identity, and often in comparison or contrast to the racial identity development of monoracial groups. Importantly, under the ecological framework, there is no monolithic racial identity shared by all mixed race individuals; even multiracials who share identical racial backgrounds (e.g., Black/African American and Latinx) can vary in how they interpret the meaning of their mixed race status, and can consequently construct different racial identities.

In order to account for this heterogeneity, scholars have formulated different theoretical models of conceptualizing multiracial racial identity. For instance, Rockquemore (1999) introduced the Multidimensional Model of Biracial Identity, which distinguishes racial identity according to four racial typologies: singular identity, border identity, protean identity, and transcendent identity. Multiracials with a singular identity identify exclusively as one monoracial group. Multiracials with a border identity identify exclusively as multiracial, biracial, or mixed race. Multiracials with a protean identity identify with whichever racial identity they perceive as situationally appropriate within a particular context. Multiracials with a transcendent identity identify as "human," or do not subscribe to existing racial categories. Similarly, Choi-Misailidis' (2004) Multiracial-Heritage Awareness and Personal Affiliation Theory distinguishes between three dominant racial identity orientations: singular identity (identifying with only one monoracial group), integrated identity (identifying as multiracial, and/or with multiple monoracial groups), and marginal identity (not identifying with any racial groups). According to Choi-Misailidis (2004), these dominant identity orientations describe which racial identity choice (e.g., monoracial, multiracial) is most frequently claimed across time and contexts. Importantly, this status (versus stage) model of conceptualizing racial identity is able to account for fluidity in the racial identity choices of multiracial individuals.

In addition to developing theories and models of multiracial racial identity, scholars have created measurement instruments to empirically assess common shared experiences and ways of understanding one's multiracial identity (Cheng & Lee, 2009; Salahuddin & O'Brien, 2011; Yoo et al., 2016). These measures were developed and validated using multiracial samples in order to account for the unique racialization of individuals with multiple racial heritages. In particular, the Multiracial Identity Integration Measure (Cheng & Lee, 2009) examines the extent to which multiracials perceive conflict or distance between their multiple racial heritages. Specifically, the authors defined multiracial identity integration as low perceived distance and conflict (Cheng & Lee, 2009). Other measures of multiracial racial identity, such as the Multiracial Challenges and Resilience Scale (Salahuddin & O'Brien, 2011) and Multiracial Experiences Measure (Yoo et al., 2016), propose common experiences that are believed to be unique to individuals with multiple racial heritages. The Multiracial Challenges and Resilience Scale (Salahuddin & O'Brien, 2011) identifies four racial identity-related challenges (Lack of Family Acceptance, Multiracial Discrimination, Challenges with Racial Identity, and Others' Surprise and Disbelief Regarding Racial Heritage) and two racial identity-related resiliencies (Multiracial Pride and Appreciation of Human Differences). The Multiracial Experiences Measure (Yoo et al., 2016) identifies two racial identity-related risk factors (Multiracial Discrimination and Perceived Racial Ambiguity) and three racial identity-related resilience factors (Shifting Expressions, Creating Third Space, and Multicultural Engagement).

Levels of Analysis Contextualizing Multiracial Racialization

In the following sections, we introduce a Model of Multiracial Racialization (see Fig. 6.1) that is contextualized within six levels of analysis: (1) Individual Characteristics; (2) Interpersonal Experiences; (3) Contextual Factors; (4) Social, Economic, and Political Environment; (5) Systems of Oppression; and (6) Time. This model is an adaptation of the racialization model proposed by Rockquemore et al. (2009). However, this new model synthesizes insights from the ecological framework (Renn, 2003; Root, 2003), ecological systems theory (Bronfenbrenner, 1992), and MultiCrit (Harris, 2016) with the aim of illustrating the widespread areas of investigation on multiracial identity and experiences available to researchers.

Level 1: Individual Characteristics

From an ecological perspective, factors that can influence multiracial racialization include individual-level characteristics such as a person's racial backgrounds (e.g., Asian American and Latinx vs. Black/African American and White), age, gender, socioeconomic status, and phenotype (Gaither, 2015; Rockquemore & Brunsma, 2002; Root, 2003). For instance, Townsend et al. (2012) found that Asian-White multiracials were more likely to identify as multiracial than Black-White or Latinx-White multiracials. In terms of other individual-level characteristics, Townsend et al. (2012) also determined that multiracials from middle-class backgrounds were more likely to identify as multiracial than those from working-class backgrounds. Moreover, Rockquemore (2002) conducted in-depth interviews to better understand how gendered racism uniquely affects the lives of Black biracial men and women. However, the majority of research on individual-level characteristics of multiracial

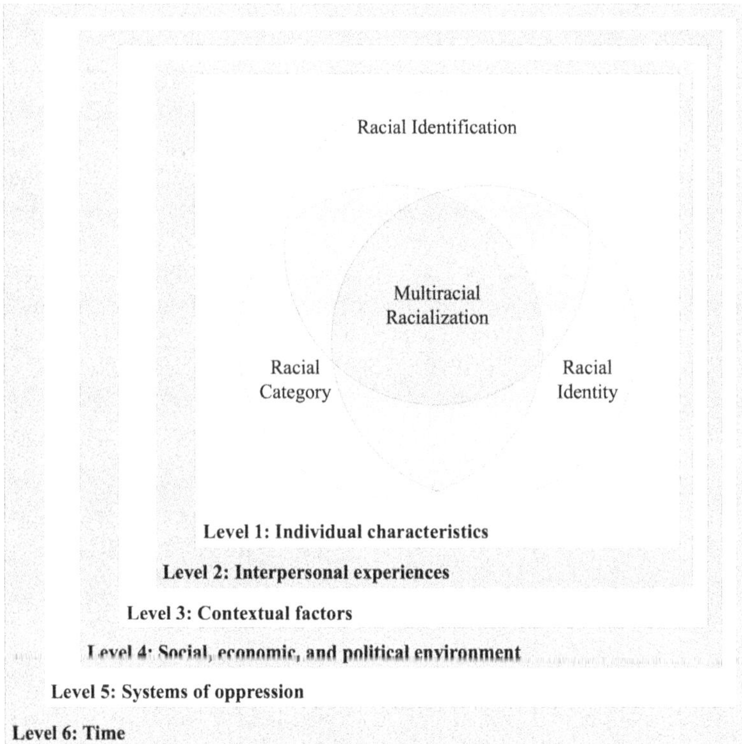

Fig. 6.1 The Model of Multiracial Racialization contextualizes the process of racialization for multiracial individuals across six levels of analysis

racialization has focused on the role of racial phenotype. As an example, Rockquemore and Brunsma (2002) examined whether racial identity varied based on phenotype for Black-White multiracials. Interestingly, phenotype as measured by one's socially perceived appearance significantly predicted racial identity, but phenotype as measured by self-perceived skin color did not. Specifically, multiracial participants identified as either monoracial Black or monoracial White depending on how others racially identified them based on appearances (Rockquemore & Brunsma, 2002). Similarly, AhnAllen et al. (2006) found that, among Asian American-White multiracials, a monoracial appearance predicted perceived racial acceptance and belonging from monoracial peers, as well as self-identification with those monoracial groups. Relatedly, Young et al. (2017) demonstrated how monoracial Black and White individuals racially categorized Black-White multiracials differently based on phenotypic cues, including racial ambiguity or the ability to racially "pass" as a monoracial group.

Level 2: Interpersonal Experiences

Studies on interpersonal-level characteristics that shape multiracial racialization have included multiracial discrimination experiences and racial socialization from family or peers. For example, McDonald et al. (2019) found feeling like an outsider and being ignored by others contributed to lower identity integration with greater racial distance and conflict in multiracial adults. The relationship between discrimination and identity can also vary by the perpetrator's race. Norman and Chen (2020) found perceiving discrimination from ingroup members was associated with stronger multiracial identification, and this was strongest when White ingroup members were the source of discrimination. Also, unique forms of multiracial discrimination including racial identity invalidation (e.g., picked on for not looking or acting like a certain racial group, pressured to pick a race, not accepted by other racial groups) have been linked to higher depressive symptoms, racial homelessness, loneliness and negative affect (Franco, 2019; Franco & O'Brien, 2018; Yoo et al., 2016). Interestingly, Franco et al. (2019) found that incongruent discrimination, or discrimination aimed at a perceived racial identity that does not match the self-identity of a multiracial person, did not relate to mental health outcomes when accounting for race-general discrimination. Other important interpersonal experiences include how multiracial youth are socialized to understand race and racism, often from parents with different racial backgrounds. Using a qualitative approach, Atkin and Jackson (2020) identified three unique themes of multiracial parental support, including connection support, discrimination support, and multiracial identity expression support. In addition, the Multiracial Youth Socialization Scale (Atkin et al., 2021) showed that certain types of socialization messages (e.g., navigating multiple heritages, multiracial identity, race-conscious, and diversity appreciation) were positively associated with racial-ethnic identity, while other messages (e.g., negative, colorblind, silent) were negatively related. Using latent profile analyses, Christophe et al. (2021) identified a four-profile solution in multiracial caregivers' messages with slightly different patterns, with caregiver 1 offering typical (near-average pride and bias messaging with frequent egalitarian messaging), minority (frequent pride and preparation for bias messages centered around the individual's minoritized racial group), high mistrust, and low frequency messages; and caregiver 2 offering typical, negative (high bias and mistrust), promotive (high pride high bias, low mistrust), and low frequency messages. Importantly, these profiles were associated with ethnic-racial identity processes.

Level 3: Contextual Factors

The meaning and developmental implications of multiracial racialization may also vary by contextual factors including but not limited to local, state, country, and international characteristics. Subjective and physical community racial composition are also important in shaping meaning and experiences for multiracial youth and families (see Chap. 9 for more discussion on how the racial compositions of schools and neighborhoods/communities affect development). For instance, Meyers et al. (2020) found that monoracial persons of color and multiracial individuals experienced less discrimination in diverse contexts versus racially homogenous contexts. Using the 1990 US Census data, Holloway et al. (2009) found that context shapes parents' reports of their multiracial children's racial identity. Racial claims made by parents of Latinx-White and Asian-White children significantly varied across metropolitan areas, with increased likelihood of parents reporting their children as White when living in homogeneously White neighborhoods. Black-White parents, however, were more likely to claim a non-White race (e.g., Black) for their children as concentrations of racial/ethnic minority groups increased in their neighborhoods. Supporting this effect, Rockquemore and Brunsma (2002) found that Black-White multiracials were more likely to identify as a monoracial Black when they lived in predominantly Black contexts. Using the 2000 US Census data, Rockquemore (2002) reported mixed race people were more likely to identify as "multiracial" if they lived on either the east or west coast. Using archival sources, oral histories, newspapers, and personal collections and photographs, Guevarra (2012) described the unique racial formation of "becoming Mexipino" for Mexicans and Filipinx living in San Diego, California. Guided by a social-historical interpretation, Guevarra examines the histories of the two ethnic groups tied to Spanish colonialism shaping the discrimination, segregation, and new communities formed in modern San Diego. Using a qualitative approach, Jackson et al. (2013) examined how multiracial Mexican Americans experience unique stress and coping in Arizona. Participants described the importance of external supports (e.g., culturally affirming spaces), interpersonal protective processes (e.g., flexing ethnicity), and internal protective processes (e.g., critiquing race) to navigate monoracism.

Level 4: Social, Economic, and Political Environment

Another important area of investigation is the social, economic, and political environment in which multiracial racialization takes place. In particular, developmental scientists should ask: How can we contextualize racial formation and racism within our current climate or recent events? As an example, the 2008 presidential election of Barack Obama, born to a White mother and Black father, catalyzed a new racial climate in the United States. A few scholars have researched the effects of widespread attention to Obama's racial background on society's understanding of race (Jeffries, 2012; Malahy et al., 2010). For instance, Malahy et al. (2010) found that individuals were more likely to racially identify Obama as monoracial Black than multiracial when they implicitly perceived race as categorical. In addition, Jeffries (2012) interviewed multiracial college students to elucidate the significance of Obama's presidency for American racial history. Future studies should situate multiracial racialization in more recent events, such as the 2021 inauguration of Kamala

Harris, the first Black-Asian American biracial woman to become vice president of the United States. Moreover, multiracial racialization should be investigated in reference to other hallmarks of our current social, economic, and political environment, including the rise of White nationalism and xenophobia against minority racial groups (e.g., the 2019 Muslim Ban), as well as the growth of the Black Lives Matter, Stop Asian Hate, and Free Palestine movements.

Level 5: Systems of Oppression

The macrolevel effects of different systems of oppression (e.g., racism, sexism, classism, and ableism) are severely understudied in the developmental sciences, especially among multiracial youth and families. This absence is understandable given the difficulty in conceptualizing and validating empirical measures of systemic oppression. However, developmental scientists outside of multiracial studies have successfully constructed measures to examine how much individuals internalize and/or challenge different systems of oppression. Neville et al. (2000) developed the Color-Blind Racial Attitudes Scale (CoBRAS), which measures individuals' Unawareness of Racial Privilege, Institutional Discrimination, and Blatant Racial Issues. The Contemporary Critical Consciousness Measure (Shin et al., 2016) assesses levels of critical consciousness across three systems of oppression: Racism, Classism, and Heterosexism. Regarding critical consciousness generally, Diemer et al. (2017) created the Critical Consciousness Scale to measure how individuals perceive and take action against systemic oppression as measured by Perceived Inequality, Egalitarianism, and Sociopolitical Participation. Furthermore, the Anti-Neoliberal Attitudes Scale (ANAS; Grzanka et al., 2020) assesses the intersection of different systems of oppression with neoliberalism through Racism and Sexism Awareness, Communitarian Values, Multicultural Ideology, and Inequality Consciousness. Although the research is limited, multiracial studies scholars have begun to examine how multiracial people internalize and uphold foundational beliefs from different systems of oppression. In particular, Chong and Kuo (2015) found that internalized oppression as measured by the Colonial Mentality Scale (David & Okazaki, 2006) differed based on the racial identity choices of Asian-White multiracials. Specifically, those who identified as monoracial White endorsed the highest levels of internalized oppression, whereas those who identified as multiracial endorsed the lowest levels of internalized oppression. Interestingly, identifying as multiracial had contradictory effects on colorblind racial attitudes. McDonald et al. (2019) found that multiracial identity integration predicted increased colorblind racial attitudes about blatant racial issues and institutional discrimination. We encourage researchers to examine how the internalization of different systems of oppression for multiracial people is affected by other levels of analysis, especially individual characteristics (e.g., racial background, gender, phenotype, class).

Level 6: Time

Finally, time is an important factor for understanding how multiracial racialization changes across lifespans, generations, and histories. Notably, there has been a call for additional research on multiracial developmental processes from a life course perspective (Jackson, 2009; Csizmadia et al., 2012). In one of the first longitudinal studies on social development among multiracial youth, Csizmadia and Ispa (2014) found that parents' racial identification of Black-White biracial children predicted their externalizing behaviors between kindergarten and fifth grade. Regarding how racial categorization changes across time, Hitlin et al. (2006) determined that many multiracial adolescents aged 14–18 changed how they racially identified over a 5-year period. Moreover, Mihoko Doyle and Kao (2007) demonstrate how the fluidity of multiracial self-categorization interacts with individual-level characteristics, including gender, physical appearance, and socioeconomic status.

Using the Model of Multiracial Racialization to Guide Future Research

Integrating aspects of MultiCrit (Harris, 2016), the ecological framework (Root, 2003) and ecological systems theory (Bronfenbrenner, 1992), and Rockquemore et al.'s (2009) model, we have designed the Model of Multiracial Racialization to help further propel the next wave of research on multiracial individuals in the United States and their experiences of racialization. To implement the newly proposed model and work towards addressing these questions, we suggest that future research utilize diverse methods (e.g., qualitative interviews, observation, ecological momentary assessment, daily diary, longitudinal approaches) to understand the various levels and how they interact with one another. For example, a survey might examine how individual characteristics (Level 1) relate to identity choices depending on the racial composition (Level 3) and political views (Level 4) of one's context (e.g., Do gender, phenotype, and age uniquely predict identity choice for Latinx multiracial individuals living on the Mexican border in conservative states in comparison to those living in a racially diverse and politically progressive area such as New York City?). Longitudinal studies could be conducted on a short- or long-term basis (Level 6) to understand how political events (Level 4) influence parents' discussions about race with their multiracial children (Level 2), and how this varies based on the child's phenotype (Level 1) and affects the child's development of identity and critical consciousness (Level 5). By considering the unique impact that each level of the model has on the development of the incredibly diverse population of multiracials in the United States, researchers can contribute to a more nuanced understanding of multiracial developmental processes.

There are many other questions left to be asked and answered within each of the six levels we have presented. To this end, we present several more sample research topics and questions that go within and cut across levels to—hopefully—give readers inspiration as to how this model may be used to advance research on multiracial racialization and the experiences of multiracial individuals in the United States:

- How does fluctuation in socioeconomic status across development and the lifespan influence multiracials' racial identification (Levels 1 & 6)?
- How does the concordance, or discordance of racial-ethnic socialization messages across parents influence multiracial youth's identity development and psychosocial functioning? (Level 2)?
- How do parental racial-ethnic socialization messages intersect with socialization messages received at school and/or socialization from those in the neighborhood to impact multiracial youth's outcomes? (Levels 2 & 3).
- How are patterns of multiracial racialization and identification impacted by the political leanings of multiracial person's state and location (e.g., specific town/city/county) within that state? (Level 4).
- How does a multiracial individual's level of understanding of systemic inequality, particularly racial inequities, influence their racial identity endorsement and the integration of their multiple racial heritages into their sense of self? (Levels 1 & 5).
- Are the mechanisms behind the formation of internalized racism similar for monoracial and multiracial youth? How may internalized oppression function differently due to multiracial individuals' membership in multiple racial categories? (Level 5).
- Do normative changes in appearance across the lifespan influence others racial categorization of multiracial individuals, and does this differ based on the age/generation of the observer? (Levels 6 & 1).

Conclusion

As the fastest-growing segment of the US population, the unique and shared experiences of multiracial youth and families will continue to be an important area of research for developmental science. In this chapter, we described how monoracism—both past and present—impacts social science and developmental science, in particular making invisible or essentializing the experiences of multiracial people. We drew on MultiCrit to illustrate the complexity, multiplicity, and hybridity in the lived experiences of multiracial people and how they navigate, negotiate, and challenge monoracism and white supremacy. To stimulate more research and advance the field in this area, we introduced a new Model of Multiracial Racialization that synthesizes the complexity in multiracial identity and experiences (including racial identification, racial category, and racial identity) with the ecology of the individual (including individual characteristics, interpersonal experiences, contextual factors, social economic, and political environment, systems of oppression, and time).

Grounded in MultiCrit, this model offers developmental science a new vantage point in the study of multiracial youth and families, raising new questions. What are the multiple ways in which research studies capture multiracial racialization, centering mixed race voices, and perspectives? How does the process of multiracial racialization vary by ecological contexts and history? How does the presumed use of monoracial theories, measures, and literature limit the understanding of multiracial youth and families? How do developmental theory and methods account for the diversity of multiracial meaning and lived experiences? How does the study of multiracial youth and families perpetuate (e.g., through essentialism or deficit perspectives) or challenge (e.g., through critical consciousness) monoracism? These questions cannot be answered until developmental scientists intentionally explicate the meaning of multiraciality for our understanding of race and racism.

References

AhnAllen, J. M., Suyemoto, K. L., & Carter, A. S. (2006). Relationship between physical appearance, sense of belonging and exclusion, and racial/ethnic self-identification among multiracial Japanese European Americans. *Cultural Diversity and Ethnic Minority Psychology, 12*(4), 673. https://doi.org/10.1037/1099-9809.12.4.673

Atkin, A. L., & Jackson, K. F. (2020). "Mom, you don't get it": A critical examination of multiracial emerging adults' perceptions of parental support. *Emerging Adulthood, 9*(4), 305–319. https://doi.org/10.1177/2F2167696820914091

Atkin, A. L., Yoo, H. C., White, R. M. B., Tran, A. G. T. T., & Jackson, K. F. (2021). Validation of the Multiracial Youth Socialization (MY-Soc) Scale among racially diverse Multiracial emerging adults. *Journal of Family Psychology, 36*(1), 12–22. https://doi.org/10.1037/fam0000879

Bonilla-Silva, E. (2015). The structure of racism in color-blind, "post-racial" America. *American Behavioral Scientist, 59*(11), 1358–1376. https://doi.org/10.1177/2F0002764215586826

Bronfenbrenner, U. (1992). *Ecological systems theory*. Jessica Kingsley Publishers.

Brunsma, D. L. (2005). Interracial families and the racial identification of mixed-race children: Evidence from the early childhood longitudinal study. *Social Forces, 84*(2), 1131–1157.

Burke, R., & Kao, G. (2013). Bearing the burden of whiteness: The implications of racial self-identification for multiracial adolescents' school belonging and academic achievement. *Ethnic and Racial Studies, 36*(5), 747–773. https://doi.org/bxfw72

Campbell, M. E., & Eggerling-Boeck, J. (2006). "What about the children?" The psychological and social well-being of multiracial adolescents. *The Sociological Quarterly, 47*(1), 147–173. https://doi.org/10.1111/j.1533-8525.2006.00041.x

Carter, G. (2013). *The United States of the united races: A utopian history of racial mixing*. New York University Press.

Chavez, G. F., & Sanchez, D. T. (2010). A clearer picture of multiracial substance use: Rates and correlates of alcohol and tobacco use in multiracial adolescents and adults. *Race and Social Problems, 2*(1), 1–18. https://doi.org/10.1007/s12552-010-9023-1

Chen, J. M., Moons, W. G., Gaither, S. E., Hamilton, D. L., & Sherman, J. W. (2014). Motivation to control prejudice predicts categorization of multiracials. *Personality and Social Psychology Bulletin, 40*(5), 590–603. https://doi.org/f53zg8

Chen, J. M., Pauker, K., Gaither, S. E., Hamilton, D. L., & Sherman, J. W. (2018). Black+ White = Not White: A minority bias in categorizations of Black-White multiracials. *Journal of Experimental Social Psychology, 78*, 43–54. https://doi.org/10.1016/j.jesp.2018.05.002

Cheng, C. Y., & Lee, F. (2009). Multiracial identity integration: Perceptions of conflict and distance among multiracial individuals. *Journal of Social Issues, 65*(1), 51–68. https://doi.org/bwhw84

Choi-Misailidis, S. (2004). *Multiracial-Heritage Awareness and Personal Affiliation: Development and validation of a new measure to assess identity in people of mixed race descent.* [Doctoral Dissertation, Fordham University]. ProQuest Information & Learning.

Choi, Y., Harachi, T. W., Gillmore, M. R., & Catalano, R. F. (2006). Are multiracial adolescents at greater risk? Comparisons of rates, patterns, and correlates of substance use and violence between monoracial and multiracial adolescents. *American Journal of Orthopsychiatry, 76*(1), 86–97. https://dx.doi.org/10.1037%2F0002-9432.76.1.86

Chong, V., & Kuo, B. C. (2015). Racial identity profiles of Asian-White biracial young adults: Testing a theoretical model with cultural and psychological correlates. *Asian American Journal of Psychology, 6*(3), 203–217. https://doi.org/10.1037/aap0000022

Christophe, N. K., Stein, G. L., & The LOVING Study Collaborative. (2021). A person-centered analysis of ethnic–racial socialization patterns and their identity correlates in multiracial college students. *Cultural Diversity and Ethnic Minority Psychology, 27*(3), 332–342. https://doi.org/10.1037/cdp0000438

García Coll, C., Lamberty, G., Jenkins, R., McAdoo, H., Crnic, K., Wasik, B., & García, H. (1996). An integrative model for the study of developmental competencies in minority children. *Child Development, 67*(5), 1891–1914. https://doi.org/10.2307/1131600

Combahee River Collective. (1977). The Combahee River Collective: A Black feminist statement. In *Capitalist patriarchy and the case for socialist feminism* (pp. 362–372).

Cooney, T. M., & Radina, M. E. (2000). Adjustment problems in adolescence: Are multiracial children at risk? *American Journal of Orthopsychiatry, 70*(4), 433–444. https://doi.org/10.1037/h0087744

Crenshaw, K. (1989). *Demarginalizing the intersection of race and sex: A Black feminist critique of antidiscrimination doctrine, feminist theory and antiracist politics*. University of Chicago.

Cross, W. E., Jr. (1971). The Negro-to-Black conversion experience. *Black World, 20*, 13–27.

Cross, W. E., Jr. (1995). The psychology of nigrescence: Revising the Cross model. In J. G. Ponterotto, J. M. Casas, L. A. Suzuki, & C. M. Alexander (Eds.), *Handbook of multicultural counseling* (pp. 93–122). Sage Publications.

Csizmadia, A., & Ispa, J. M. (2014). Black-White biracial children's social development from kindergarten to fifth grade: Links with racial identification, gender, and socioeconomic status. *Social Development, 23*(1), 157–177. https://doi.org/10.1111/sode.12037

Csizmadia, A., Brunsma, D. L., & Cooney, T. M. (2012). Racial identification and developmental outcomes among Black-White multiracial youth: A review from a life course perspective. *Advances in Life Course Research, 17*(1), 34–44. https://doi.org/10.1016/j.alcr.2011.10.003

David, E. J. R., & Okazaki, S. (2006). The Colonial Mentality Scale (CMS) for Filipino Americans: Scale construction and psychological implications. *Journal of Counseling Psychology, 53*(2), 241–252. https://doi.org/10.1037/0022-0167.53.2.241

Delgado, R., & Stefancic, J. (2001). *An introduction to critical race theory*. New York University Press.

Diemer, M. A., Rapa, L. J., Park, C. J., & Perry, J. C. (2017). Development and validation of the critical consciousness scale. *Youth & Society, 49*(4), 461–483. https://doi.org/gf82x9

Erikson, E. H. (1968). *Identity: Youth and crisis*. (No. 7). WW Norton & Company.

Franco, M. (2019). Let the racism tell you who your friends are: The effects of racism on social connections and life-satisfaction for Multiracial people. *International Journal of Intercultural Relations, 69*, 54–65. https://doi.org/10.1016/j.ijintrel.2018.12.005

Franco, M. G., & O'Brien, K. M. (2018). Racial identity invalidation with multiracial individuals: An instrument development study. *Cultural Diversity and Ethnic Minority Psychology, 24*(1), 112–125. https://doi.org/10.1037/cdp0000170

Franco, M., Toomey, T., DeBlaere, C., & Rice, K. (2019). Identity incongruent discrimination, racial identity, and mental health for multiracial individuals. *Counselling Psychology Quarterly*, 1–22. https://doi.org/hp85

Gaither, S. E. (2015). "Mixed" results: Multiracial research and identity explorations. *Current Directions in Psychological Science, 24*(2), 114–119. https://doi.org/f68xr9

Gaither, S. E., Pauker, K., Slepian, M. L., & Sommers, S. R. (2016). Social belonging motivates categorization of racially ambiguous faces. *Social Cognition, 34*(2), 97–118. https://doi.org/f8fwnd

Gossett, T. F. (1997). *Race: The history of an idea in America*. Oxford University Press.

Grzanka, P. R., Miles, J. R., Spengler, E. S., Arnett, J. E., & Pruett, J. (2020). Measuring neoliberalism: Development and initial validation of a scale of anti-neoliberal attitudes. *Social Justice Research, 33*(1), 44–80. https://doi.org/gh3fw3

Guevarra, R. P., Jr. (2012). *Becoming Mexipino: Multiethnic identities and communities in San Diego*. Rutgers University Press.

Harris, J. C. (2016). Toward a critical multiracial theory in education. *International Journal of Qualitative Studies in Education, 29*(6), 795–813. https://doi.org/ghx2r9

Helms, J. E. (1995). An update of Helms's White and people of color racial identity models. In J. G. Ponterotto, J. M. Casas, L. A. Suzuki, & C. M. Alexander (Eds.), *Handbook of multicultural counseling* (pp. 181–198). Sage Publications.

Hitlin, S., Scott Brown, J., & Elder, G. H., Jr. (2006). Racial self-categorization in adolescence: Multiracial development and social pathways. *Child Development, 77*(5), 1298–1308. https://doi.org/drv3vk

Holloway, S. R., Wright, R., Ellis, M., & East, M. (2009). Place, scale and the racial claims made for multiracial children in the 1990 US Census. *Ethnic and Racial Studies, 32*(3), 522–547. https://doi.org/10.1080/01419870802021120

Jackson, K. F. (2009). Beyond race: Examining the facets of multiracial identity through a life-span developmental lens. *Journal of Ethnic & Cultural Diversity in Social Work, 18*(4), 293–310. https://doi.org/10.1080/15313200903310759

Jackson, K. F., & Samuels, G. M. (2019). *Multiracial cultural attunement*. NASW Press.

Jackson, K. F., Wolven, T., & Aguilera, K. (2013). Mixed resilience: A study of multiethnic Mexican American stress and coping in Arizona. *Family Relations, 62*(1), 212–225. https://doi.org/hp87

Jeffries, M. P. (2012). 'Mutts like me': Multiracial students' perceptions of Barack Obama. *Qualitative Sociology, 35*(2), 183–200. https://doi.org/10.1007/s11133-012-9226-4

Johnston, M. P., & Nadal, K. L. (2010). Multiracial microaggressions: Exposing monoracism in everyday life and clinical practice. In D. W. Sue (Ed.), *Microaggressions and marginality: Manifestation, dynamics, and impact* (pp. 123–144). Wiley.

Kich, G. K. (1992). The developmental process of asserting a biracial, bicultural identity. In M. P. Root (Ed.), *Racially mixed people in America* (pp. 304–317). Sage Publications.

Kuhn, T. S. (1962). *The structure of scientific revolutions*. University of Chicago Press.

Malahy, L. W., Sedlins, M., Plaks, J., & Shoda, Y. (2010). Black and White, or shades of gray? Racial labeling of Barack Obama predicts implicit race perception. *Analyses of Social Issues and Public Policy, 10*(1), 207–222. https://doi.org/10.1111/j.1530-2415.2010.01213.x

Mauer, V., Savell, S., Davis, A., Wilson, M. N., Shaw, D. S., & Lemery-Chalfant, K. (2020). Identification of Multiracial adolescents in research samples: An examination and critique of existing practices. *The Journal of Early Adolescence*, 0272431620950471. https://doi.org/10.1177/0272431620950471

McDonald, C. P., Chang, C. Y., Dispenza, F., & O'Hara, C. (2019). Multiracial identity, color-blind racial ideology, and discrimination: Professional counseling implications. *Journal of Counseling & Development, 97*(1), 75–85. https://doi.org/10.1002/jcad.12237

Meyers, C., Aumer, K., Schoniwitz, A., Janicki, C., Pauker, K., Chang, E. C., Gaither, S. E., & Williams, A. (2020). Experiences with microaggressions and discrimination in racially diverse and homogeneously white contexts. *Cultural Diversity and Ethnic Minority Psychology, 26*(2), 250–259. https://doi.org/10.1037/cdp0000293

Mihoko Doyle, J., & Kao, G. (2007). Are racial identities of multiracials stable? Changing self-identification among single and multiple race individuals. *Social Psychology Quarterly, 70*(4), 405–423. https://doi.org/fjzv56

Miller, R. L. (1992). The human ecology of multiracial identity. In M. P. Root (Ed.), *Racially mixed people in America*. Sage Publications.

Neville, H. A., Lilly, R. L., Duran, G., Lee, R. M., & Browne, L. (2000). Construction and initial validation of the color-blind racial attitudes scale (CoBRAS). *Journal of Counseling Psychology, 47*(1), 59. https://doi.org/10.1037/0022-0167.47.1.59

Nishina, A., & Witkow, M. R. (2020). Why developmental researchers should care about biracial, multiracial, and multiethnic youth. *Child Development Perspectives, 14*(1), 21–27. https://doi.org/10.1111/cdep.12350

Norman, J. B., & Chen, J. M. (2020). I am Multiracial: Predictors of Multiracial identification strength among mixed ancestry individuals. *Self and Identity, 19*(5), 501–520. https://doi.org/hp88

Omi, M., & Winant, H. (2014). *Racial formation in the United States* (3rd ed.). Routledge.

Park, R. E. (1928). Human migration and the marginal man. *American Journal of Sociology, 33*(6), 881–893.

Pauker, K., Meyers, C., Sanchez, D. T., Gaither, S. E., & Young, D. M. (2018). A review of multiracial malleability: Identity, categorization, and shifting racial attitudes. *Social and Personality Psychology Compass, 12*(6), 1–15. https://doi.org/10.1111/spc3.12392

Phinney, J. S. (1996). When we talk about American ethnic groups, what do we mean? *American Psychologist, 51*(9), 918–927. https://doi.org/10.1037/0003-066X.51.9.918

Poston, W. C. (1990). The biracial identity development model: A needed addition. *Journal of Counseling & Development, 69*(2), 152–155. https://doi.org/fzrvk7

Renn, K. A. (2003). Understanding the identities of mixed-race college students through a developmental ecology lens. *Journal of College Student Development, 44*(3), 383–403.

Rockquemore, K. A. (1999). Between Black and White: Exploring the "biracial" experience. *Race and Society, 1*, 197–212. https://doi.org/10.1353/csd.2003.0032

Rockquemore, K. A. (2002). Negotiating the color line: The gendered process of racial identity construction among black/white biracial women. *Gender & Society, 16*(4), 485–503. https://doi.org/10.1177/0891243202016004005

Rockquemore, K. A., & Brunsma, D. L. (2002). Socially embedded identities: Theories, typologies, and processes of racial identity among Black/White biracials. *Sociological Quarterly, 43*(3), 335–356. https://doi.org/10.1111/j.1533-8525.2002.tb00052.x

Rockquemore, K. A., Brunsma, D. L., & Delgado, D. J. (2009). Racing to theory or retheorizing race? Understanding the struggle to build a multiracial identity theory. *Journal of Social Issues, 65*(1), 13–34. https://doi.org/10.1111/j.1540-4560.2008.01585.x

Root, M. P. (1992). *Racially mixed people in America*. Sage Publications.

Root, M. P. (1996). The multiracial experience: Racial borders as significant frontier in race relations. In M. P. Root (Ed.), *The multiracial experience: Racial borders as the new frontier* (pp. xii–xxviii). Sage.

Root, M. P. (2003). Racial identity development and persons of mixed race heritage. In M. P. Root & M. Kelley (Eds.), *Multiracial child resource book: Living complex identities*. MAVIN Foundation.

Salahuddin, N. M., & O'Brien, K. M. (2011). Challenges and resilience in the lives of urban, multiracial adults: An instrument development study. *Journal of Counseling Psychology, 58*(4), 494–507. https://doi.org/10.1037/a0024633

Sellers, R. M., Rowley, S. A., Chavous, T. M., Shelton, J. N., & Smith, M. A. (1997). Multidimensional Inventory of Black Identity: A preliminary investigation of reliability and construct validity. *Journal of Personality and Social Psychology, 73*(4), 805–815. https://doi.org/d9fpmg

Shin, R. Q., Ezeofor, I., Smith, L. C., Welch, J. C., & Goodrich, K. M. (2016). The development and validation of the Contemporary Critical Consciousness Measure. *Journal of Counseling Psychology, 63*(2), 210–223. https://doi.org/10.1037/cou0000137

Spencer, M. B. (2007). Phenomenology and ecological systems theory: Development of diverse groups. In R. M. Lerner & W. Damon (Eds.), *Handbook of child psychology: Theoretical models of human development* (6th ed., pp. 829–893). Wiley. https://doi.org/10.1002/9780470147658.chpsy0115

Stonequist, E. V. (1937). *The marginal man: A study in personality and culture conflict.* Scribner Books.

Thornton, M. C., & Wason, S. (1995). Intermarriage. In D. Levinson (Ed.), *Encyclopedia of marriage and the family* (pp. 396–402). Macmillan.

Thornton, M. C. (1996). Hidden agendas, identity theories, and multiracial people. In M. P. Root (Ed.), *The Multiracial experience: Racial Borders as the New Frontier*. Sage Publications.

Townsend, S. S. M., Fryberg, S. A., Wilkins, C. L., & Markus, H. R. (2012). Being mixed: Who claims a biracial identity? *Cultural Diversity & Ethnic Minority Psychology, 18*, 91–96. https://doi.org/10.1037/a0026845

United States Census Bureau. (2021, Aug 12). *2020 Census illuminates racial and ethnic composition of the country*. https://www.census.gov/library/stories/2021/08/improved-race-ethnicity-measures-reveal-united-states-population-much-more-multiracial.html.

Williams, T. K. (1996). Race as process: Reassessing the "What are you?" encounters of biracial individuals. In M. P. Root (Ed.), *The Multiracial experience: Racial Borders as the New Frontier*. Sage Publications.

Yoo, H. C., Jackson, K. F., Guevarra, R. P., Jr., Miller, M. J., & Harrington, B. (2016). Construction and initial validation of the Multiracial Experiences Measure (MEM). *Journal of Counseling Psychology, 63*(2), 198–215. https://doi.org/10.1037/cou0000117

Young, D. M., Sanchez, D. T., & Wilton, L. S. (2017). Biracial perception in black and white: How Black and White perceivers respond to phenotype and racial identity cues. *Cultural Diversity and Ethnic Minority Psychology, 23*(1), 154–169. https://doi.org/10.1037/cdp0000103

Chapter 7
Ecologically Strong: Toward a Strengths-Based and Ecologically Valid Developmental Science

Andrew D. Coppens and Emilie Coppinger

The resources available to psychological and developmental science for advancing strengths-based evidence in our scientific practices, including the social interventions they inform, are now well-developed and characterize the leading edge of scholarship in fields such as critical learning sciences and cultural psychology (e.g., Nasir et al., 2020). In addition, the important contributions of, for example, Indigenous, decolonial, anti-racist, and feminist approaches challenge core traditions of developmental science in terms of relevance to a world increasingly committed to issues of equity, shared social power, and diverse representation.

This volume is a positive step toward developmental science making progress on these issues, particularly in its inward gaze at issues of measurement and conceptualization. The focus of this chapter's reflexive gaze is how developmental science can recalibrate in ways that advance a larger and more robust body of strengths-based research with minoritized (but, indeed, all) communities. Specifically, we revive the long-standing issue of ecological validity as fundamental to advancing strengths-based developmental science. We discuss several examples from developmental literatures and give research recommendations toward advancing an ecologically valid, strengths-based, and ultimately equity-oriented developmental science.

Our view is that a strident methodological commitment to ecological validity is crucial for developmental science advancing its scientific and political relevance. This ecological commitment means deepening connections between, on one hand, a desire to understand and generate valid evidence on the everyday learning and developmental processes common to individuals and cultural communities and, on the other hand, the social interactions and methods in which we ask participants to enact those processes. Ecologically committed developmental science departs from the study of racial and ethnic diversity as an "object" of inquiry and takes up critical

A. D. Coppens (✉) · E. Coppinger
University of New Hampshire, Durham, NH, USA
e-mail: andrew.coppens@unh.edu

© The Author(s), under exclusive license to Springer Nature Switzerland AG 2023

D. P. Witherspoon, G. L. Stein (eds.), *Diversity and Developmental Science*,
https://doi.org/10.1007/978-3-031-23163-6_7

questions about how developmental research is conducted and its role in challenging the marginalization or misunderstanding of underrepresented social and cultural groups.

This chapter aims to be useful to a broad audience of developmental researchers and practitioners with interest in exploring the strengths and contributions of ecologically committed research. We draw on ethnographic and cultural-historical perspectives but do not suggest that those working in other traditions need to "abandon ship" and take up entirely new approaches in their work—a non-starter for many established scientists. Strengths-based research is not partial to method or discipline. Prominent strengths-based research stems from work in fields such as sociolinguistics, anthropology, and cognitive psychology as does research advancing deficit-based findings and frameworks. Any approach to research can benefit from deepening ecological parallels between methods and the activities of participants' everyday lives (Rogoff et al., 2018).

Also, this chapter does not revive oversimplified and now-outdated debates that pit qualitative research against quantitative research; such dichotomous categories are neither helpful nor descriptively accurate. Qualitative research is not necessarily a refuge from the scientific and political problems of deficit frameworks, as we demonstrate in cases below. Similarly, controlled experiments or large-scale numerically or statistically driven approaches do not necessarily generate evidence that perpetuates oppressive interventions and policy (e.g., consider W.E.B. Du Bois' work using large-scale demographic data visual analysis techniques to communicate inequities among southern Black communities in the United States (Du Bois et al., 2018)).

Said differently, our focus is not on the superficial characteristics of data—after all, words can be counted or assumed to carry objective meaning (Packer, 2018) and numerical counts and scores always derive from lived, interactive research moments that require interpretation. Although we firmly support calls for the integration of qualitative and quantitative methods (see Dixon-Román et al., 2020; Hammersley, 1992), even successful mixed-method approaches do not necessarily resolve issues related to contextualization and ecological validity, or ultimately advance strengths-based developmental evidence.

Our chapter centers two overarching considerations for building strengths-based and equity-promoting developmental science findings and interventions: *contextualization* and *politicization*. Key to contextualization, we discuss the long-standing issue of ecological validity in the developmental research designs. As the primary focus of this chapter, we present prominent cases of research in developmental science to explore the pervasive issue of poor ecological validity and its potential role in undermining strengths-based research in the field. Throughout, we argue for the explicit politicization of developmental research as a counter move to the political assumptions of deficit-model research. We then discuss recent methodological advances in research on learning and development which carry with them implications for how the concept of ecological validity might evolve. This chapter concludes with recommendations for strengths-based developmental research across methods and areas of inquiry.

What Is "Strengths-Based" Developmental Research?

A helpful starting place for conceptualizing "strengths-based" developmental research is a decades-long conversation in psychological literatures regarding how to understand the meaning of group-level patterns of variability in individual performance on a wide range of social and psychological tasks—the "difference versus deficit" debate:

> The academic arguments about diversity and deficit at the end of the 1960s when Bruner and I undertook to write on the topic were polarized around explanations for the continuing achievement gap between middle class and working class children, with particular concern focused on African Americans. Most of those who adopted a deficit view assumed that growing up in conditions of poverty deprives children of essential cultural conditions for normal intellectual development. In the parlance of the day, the "*cultural* deprivation" that accompanied poverty and ethnic marginalization produced a "deficit in *psychological* functioning." (Cole, 2013, p. 85, italics in original)

The debate has been a near constant in psychological and developmental research with antecedents prior to the 1960s found in racialized research programs in intelligence and recapitulation theory (Danziger, 1997; Koops & Kessel, 2015), and more recently taking up residence behind terms such as "at risk" and "school readiness." Importantly, opposition to deficit perspectives has been active in educational and psychological fields for at least as long. (For descriptions of the history and mechanisms of deficit-based research see Avineri et al., 2015; Dinishak, 2016; Valencia, 2010; Valencia & Solórzano, 1997.)

The duration and intensity of the difference-vs-deficit debate in psychological and developmental research makes it reasonable to characterize a strengths-based perspective as "critical." Packer and Tappan (2001, p. 14) describe critical approaches in two ways, as involving: (1) *contextualization* or "examination of the conditions of the possibility of some phenomenon" and (2) *politicization* or "identification of forms of exploitation and the criticism of those considered responsible."

Much of the heterodoxical progress that can be attributed to those engaged in the difference-vs-deficit debate has focused on the area of *contextualization*, demonstrating the frequent inadequacy of educational and psychological testing for bringing about or describing the sophisticated skills and competencies of children from marginalized social and cultural groups. Such contextualization efforts have often included keen awareness of the political implications of deficit-model science. Again from Cole (2013), quoting Cole and Bruner (1971): "In the present social context of the United States, the great power of the middle class has rendered differences into deficits because middle-class behavior is the yardstick of success" (p. 87).

Putting in relief the political undercurrents of this work, our view is that a strengths-based approach to developmental science cannot challenge deficit perspectives without an equally explicit political position, which a "difference" perspective may or may not make obvious. We suggest there is an important difference between developmental research on variability per se and research that produces evidence of strengths-based psychological and developmental practices, which can benefit from but does not require evidence of variability.

Our use of the term "strengths-based" as an aspirational descriptor for a contextualized and often ecologically valid developmental science draws on cultural-historical or sociocultural traditions (see also Plaut, 2010). Although developmental and psychological researchers often operationalize context as separate from and a force "acting on" individuals, we oppose this view (see also Cole, 1996; Rogoff, 2003). We argue that the unit of analysis (i.e., individuals, often children or caretakers) assumed by many theories and forms of measurement in developmental research (e.g., academic testing and observational paradigms focused on children's behavioral reactions to imposed stimuli) may predispose researchers to explaining phenomena in terms of individual and often cognitive processes, opening vulnerabilities to findings being linked to deficit explanations.

Contextualized forms of developmental research are, first, sensitive to the mutually constituting relation between individual and cultural/contextual processes of learning and development (Rogoff, 2003). That is, the objects of psychological or developmental inquiry are fundamentally emergent phenomena, rooted in the mutual constitution of individual *psychological* processes and *social, cultural, political, and historical* processes. In this perspective, efforts to explain differences among social groups in terms of some phenomenon of interest must transcend the "structure versus agency" dualism—that *either* individuals *or* the contexts of their learning and development are causally prominent—and reconceptualize the phenomenon as distributed across these individual and supra-individual features of activity.

Second, a contextualized developmental science responds explicitly to the fact that efforts to conduct research—whether via surveys, laboratory tasks, interviews, or even many naturalistic observations—are *context creating*. This notion requires that researchers specifically theorize these created contexts as full social, cultural, and political encounters, which is rare in developmental psychology but has been undertaken admirably in other fields (e.g., Briggs, 1986; Wortham et al., 2011; Vossoughi & Zavala, 2020). Even in research interviews, when ostensibly the entire conversation is oriented toward *denotational* facts such as past events, absent people, and views and values grounded in experiences dissimilar to the interview itself (see Packer, 2018; Wortham et al., 2011), all parties simultaneously undertake efforts to coordinate self-performances with their approximations of what appear to be the purposes and evaluative assumptions of the interview (e.g., reporting views and events that present one's self as a "good parent" in an interview conversation about parenting in general). Other instances of context creation in interviews are more obvious, such as Susan Harding's (1992) brilliant analysis of an attempt to catalyze Harding's own religious conversion amidst a more general interview about a Christian reverend's life and beliefs. These *interactional* moments in created contexts are not inherently "bias" or "noise" but, instead, often can be analyzed as sources of evidence directly relevant to the research questions at hand (e.g., Coppens et al., 2020).

The creation of discrete research activity contexts, which is unavoidable in nearly all methodological designs, also requires that researchers understand and specifically theorize relations between individuals' performance in those contexts and how individuals learn and develop in the activities common to their everyday lives—"ecological validity." Later in this chapter, we present an in-depth discussion of both exemplary cases of ecologically valid empirical research—such as Rheingold's (1982) laboratory-based, developmental research on prosocial helping among young children—as well as cases of developmental research that we believe miss the mark on ecological validity.

Third, a contextualized developmental science that leads to strengths-based insights is often primarily grounded in *actor perspectives*, as opposed to *observer perspectives*.

> As observers... we're looking at the actor, the person doing the behavior we're trying to explain. Thus the actor dominates our literal and mental field... and the circumstances to which he is responding are obscured from view. ... this picture causes a bias when we try to explain the actor's behavior. We emphasize the things we can see. We emphasize things about the actor – characteristics, traits, and so on – that seem like plausible explanations for her behavior. And we deemphasize, as causes of her behavior, the things we can't see very well, namely, the circumstances to which she is adapting. (Steele, 2010, p. 18)

Gharabaghi and Anderson-Nathe (2017) make clear the methodological implications of this observer–actor distinction in perspectives:

> strengths cannot be identified through the objectification of young people as phenomena of interest. They can only be identified through the immersion of the researcher into the everyday life strategies employed by young people in navigating their social contexts and a curiosity on the part of the researcher not about outcomes but about human relationships. (p. 178)

We argue that it may be possible to develop *non-deficit* findings and interpretations while also coming up short of articulating a robust strengths-based view if research findings are not sensitized to actor perspectives and built on politically reciprocal research relationships. As an example of the distinction between *non-deficit* and *strengths-based* insights, the well-known non-deficit "funds of knowledge" framework for recognizing the cultural capital of minoritized communities has made invaluable contributions toward understanding and recognizing the learning and developmental resources available to participants in politically marginalized communities (González et al., 2005). Nonetheless, Esteban-Guitart and Moll (2014) suggest that those resources can only be understood as *strengths* when one understands how such resources are agentically taken up and put to use by families and individuals in self-determined and culturally sustaining ways.

Strengths-based research, in many of its forms, combines these insights when designing studies, posing questions, and interpretating evidence—such as adopting a critical (i.e., contextualized and politicized) stance regarding what it means to have created evidence of cultural or social group differences via measures of some phenomenon of interest. In strengths-based perspectives, individuals' performance in research contexts (or any context of intervention, such as schools) are not necessarily taken to accurately represent their abilities generally. For this reason, strengths-based designs and insights are often cross-contextual or multi-sited (Marcus, 1995; Gutierrez & Larson, 2007; Vossoughi & Gutiérrez, 2014), which facilitates insights into how research contexts and everyday settings both afford and constrain forms of performance deemed as competent or indicative of learning or development.

The Elephant in the Room: Ecological Validity as Foundational to "Calls for Change" in Developmental Research

In the social sciences generally, as well as in developmental psychology, there are several long-standing calls to change the ways that research is conducted to make progress on issues of representation and equity. The most straightforward of these calls is to increase the cultural, racial, ethnic, political, gender, class, ability, nationality, and other diversities of participants who take part in developmental research. Henrich et al. (2010) made a version of this recommendation with widespread citation; however, similar recommendations had existed for years prior and have been made since (e.g., Arnett, 2008; Graham, 1992; Jahoda, 1986; Nielsen et al., 2017). This is a crucial agenda for developmental science.

Many other calls for change in developmental science are deeper than diversifying who takes part in its studies and relate to epistemological and methodological issues in the studies themselves. For example, dominant assumptions of developmental science, for much of the history of the discipline, have assumed that "the child" can be studied as an integrated and bounded whole and that the psychological determinants of developmental outcomes travel with children relatively unfettered from one context (e.g., children's everyday lives in families and communities) to others (e.g., laboratory settings; schools). Kessen (1979) outlines the problem:

> The child – like the Pilgrim, the cowboy, and the detective on television – is invariably seen as a free-standing isolable being who moves through development as a self-contained and complete individual. Other similarly self-contained people – parents and teachers – may influence the development of children, to be sure, but the proper unit of cultural analysis and the proper unit of developmental study is the child alone. The ubiquity of such radical individualism in our lives makes the consideration of alternative images of childhood extraordinarily difficult. … The seminal thinkers about children over the past century have, in fact, been almost undeviating in their postulation of the child as container of self and of psychology. Impulses are in the child, traits are in the child; thoughts are in the child; attachments are in the child. In short, almost every major theory of development accepts the premises of individualism and takes the child as the basic unit of study, with all consequences the choice has for decisions that range from selecting a method of research to selecting a therapeutic maneuver. Uniform agreement on the isolable child as the proper measure of development led to the research paradigms that have dominated child psychology during most of its history; basically, we have observed those parts of development that the child could readily transport to our laboratories or to our testing sites. (p. 819)

In essence, this is the problem of *contextualization*. It seems the issues Kessen (1979) summarized have not been addressed, at least not comprehensively, in the more than 40 years since the critique's publication. Rogoff et al. (2018) contend that "the portfolio of our field needs a course-correction" (p. 5); namely that, "if psychology focuses too narrowly on children's reactions to situations controlled by researchers, without much attention to children's lives outside of this unusual setting, we are in danger of losing track of the phenomena that we are trying to understand" (p. 9). Similarly, Packer and Moreno-Dulcey (2020; see also Packer & Moreno-Dulcey, 2022) cite a "deep issue" in laboratory-based developmental research that "assumes that the psychological abilities or functions under investigation are [solely] characteristics of the individual child" (p. 12). We agree with both Packer and Moreno-Dulcey's (2020) concerns as well as Rogoff et al.'s (2018) urging to ground developmental research in the settings and activities of participants' everyday lives—this would be a productive rebalance for the field of developmental psychology (which would not preclude laboratory- or survey-based research).

Yet, as we previously discussed, strengths-based developmental research may be as much an ethical and political commitment as it is an epistemological and methodological commitment to research in naturalistic settings. Locating the child as inseparable from family and community contexts, when it has happened in developmental theory and research, has not necessarily precluded deficit thinking regarding nondominant families and communities. For example, "culture of poverty" research rooted in the 1960s and continuing today can be read as indeed locating children's development as inseparable from family and community social and cultural processes. However, the *pathologizing* of those processes has led to blaming marginalized communities for their supposed shortcomings in supporting developmental outcomes benchmarked on the values and practices of dominant communities, leaving the cultural resources and strengths of marginalized families and communities either undervalued, not recognized, or erased (Rogoff et al., 2017).

Strident Commitment to Ecological Validity as Key to Strengths-Based Developmental Research

We argue that a transformational reorientation and commitment to ecological validity is needed in developmental research, as a means to better and more politically equitable science—a view with decades of precedent. Much of developmental science is guided by strong and often implicit "ecological commitments" (Dahl, 2017; see also Rogoff et al., 2018), a concern for and interest in generalizing to the everyday developmental contexts and processes shaping children's lives. Yet, ecologically valid developmental research is uncommon.

The confusions that Urie Bronfenbrenner (1977) articulated decades ago about what ecological validity is and what it is not remain relevant to developmental research. The main confusion involves a conflation between superficially *naturalistic* research and *ecologically valid* research. Naturalistic research methods often use words, phrases, objects, tasks, settings, or even people who are familiar to study participants and common to their everyday lives—potentially beneficial methodological choices (Dahl, 2017; see also Valsiner & Benigni, 1986). However, superficially naturalistic research settings and materials do not necessarily align, from study participants' perspectives, the meaning of the research procedures with the meaning of the everyday settings and activities from which their skills and strengths often derive.

Concern for cross-contextual meaning is essential to ecologically valid research, as compatibility in meaning from one situation to the next is likely to evoke in participants similar cognitive, social, or other processes of interest to developmental researchers. To see ecological validity as fundamental to psychological research requires a notion of mind and development that is much more highly contextualized than is the norm in American psychology. Although several definitions of ecological validity exist (e.g., Bronfenbrenner, 1977; Cole et al., 1997; Schmuckler, 2001), we center our discussion on the following:

> Ecological validity refers to the extent to which the psychological processes observed and recorded in a particular psychological experiment reflect the processes that actually occur in other, naturally occurring settings. Ecological validity is also used to address the problem of generalizability, which is the extent to which it is legitimate to apply the lessons learned in an experimental setting to the "real world" of events not deliberately set up for purposes of psychological testing. (*The Story of LCHC: A Polyphonic Autobiography*, http://lchcautobio.ucsd.edu/polyphonic-autobiography/section-2/chapter-4/)

Ecological Validity in Action: Clear and Counter Cases

Like any definition, the concept of ecological validity is probably most helpful put to use. In what follows, we consider several cases in which ecological validity plays a prominent role in their successes or shortcomings as strengths-based research.

Case 1: Studying Children Via Theatrical Tasks: The "Theory of Puppets" Debate

In summer 2020, an interesting and lively debate surfaced on the listserv of the Cognitive Development Society (i.e., COGDEVSOC) about what issues and assumptions are at work when researchers animate otherwise inanimate objects such as dolls and wooden blocks in games or theatrical pretense, with the aim of demonstrating social phenomena to infants and young children and measuring their behavioral responses. The discussion was rich, with many challenges, variations, connections to related topics, and complex technical justifications raised. The listserv conversation resulted in a handful of published papers which source aspects of the debate we take up.

Packer and Moreno-Dulcey (2020) coined the phrase "theory of puppets" to raise objections to the ecological validity of a particular kind of laboratory-based developmental research. They argue that several kinds of normativity—broadly, cultural expectations regarding what children ought to be doing and agreeing to in laboratory tasks—function as unexamined assumptions in pretense- or game-based developmental methodologies often used to study young children's social-cognitive understanding. Although these authors critique a particular type of design known as the protesting paradigm, their points are relevant to many other experimental or otherwise laboratory-based tasks that embed representational stimuli in games or other theatrical forms of pretense for the purpose of provoking behavioral reactions in children.

> the experimental procedure known as the protesting paradigm presumes, but does not study, sophisticated normative competences on the part of its child participants. The research simply could not take place if a child did not respect adult authority, did not understand how to take turns and make moves in a game, and did not understand that puppets can be assigned the status of players in the game. Arguably, these taken-for-granted competences are more advanced than the ability that the research is investigating. (Packer & Moreno-Dulcey, 2020, pp. 10–11).

In fact, we take even this as a relatively generous stance—the concerns may be greater. For example, the issue is not only *whether* children respect authority or *whether* they understand how to take turns. Also at issue is *how* children approach such adult-managed games and whether the culturally normative assumptions built into such highly constrained laboratory scripts are consistent with the cultural backgrounds of children taking part in the study. A young child appearing to not respect adult authority may in fact be approaching adult interactions with a cultural orientation to the interaction that is *different*—not deficient—than the understanding of expected adult–child relations implicit in the design of the laboratory task. And this different approach, in other contexts, may be a cultural and developmental strength.

Notwithstanding, justification for theatrical or game-based pretense was proposed on the basis of two fundamental assumptions (Kominsky et al., 2020). The first assumption is that the presentation of objects, puppets, and plain visual displays to infants and young children "simplifies" the input or stimulus and thus clarifies for children the researcher's intended point of focus. Kominsky et al. argue that this supposed simplification is not only necessary for studying infant cognition but also increases measurement validity, the likelihood that children's behaviors are responding to the focal stimulus and not to other "distracting" or "overwhelming" aspects of the situation. The second assumption is that such simplifications increase the likelihood for researchers to evoke children's specific "cognitive subsystems" of interest (Revencu & Csibra, 2020) such as their understanding of physics, causality, or prosociality.

The seemingly abstract theoretical assumptions undergirding the justifications proposed by Kominsky et al. (2020) are concerning for a developmental science seeking to understand cultural strengths of children across a range of communities, including when researchers and children share cultural backgrounds. As we argued above, all research methods including laboratory-based developmental research tasks are *context creating* rather than context reducing. In the laboratory research methods that are the focus of the "theory of puppets" debate, efforts to simplify laboratory stimuli may instead *replace* or even *add* complexity via the introduction of implicit cultural expectations which may not be shared across cultural communities in the same way as *might* be the case when researchers study children of their same backgrounds.

Kominsky et al. (2020) focus a portion of their discussion on cross-cultural issues but appear not to appreciate the cultural aspects of the research interactions in which children are directed to participate. Instead, their discussion of issues related to cross-cultural generalizability focus on the familiarity of stimuli arguing that "there are some types of stimuli that we would expect to work perfectly well in any human culture" (p. 24). We take this as related to the confusion Bronfenbrenner articulates between superficially *naturalistic* and *ecologically valid* modes of research. What makes any stimulus "work" has less to do with whether a child has seen a puppet before and much more to do with the degree of intersubjective agreement between children and researchers regarding the cultural norms, values, and practices delimiting what's expected and what's going on with the real people orchestrating or taking part in the research.

We raise a second objection to the assumption that "simplification" is possible in the contexts created for the purposes of developmental research, regardless of the method. From a cultural psychological perspective, children's social-cognitive systems are not "subsystems" dressed up superficially and made opaque by the cultural values, practices, and normative expectations of children's families and communities. On the contrary, individual social-cognitive processes are *made possible by* and exist in *mutually constituting relation with* social and cultural processes. The overarching proposal by Kominsky et al. (2020) that developmental researchers can, or even must, "trade" concern with external validity (i.e., ecological validity) for concern with internal validity (i.e., measurement validity) reflects a deep epistemological and methodological fallacy that laboratory experiments can be thought of as closed systems that permit isolation from "external" considerations such as cultural values and practices. Building on decades of experience wrestling with issues of ecological invalidity in cross-cultural research, members of the *Laboratory of Comparative Human Cognition*, with us, instead argue:

> our work, as well as studies by others, showed that even psychological tests and other presumably "closed system" cognitive tasks were permeable and negotiable. Insofar as the psychologist's closed system does not capture the elements of the open system it is presumed to model, experimental results systematically misrepresent the life process from which they are derived. (http://lchcautobio.ucsd.edu/polyphonic-autobiography/section-2/chapter-4/)

Again, that these concerns have decades of precedent (e.g., Cole et al., 1978) raises urgent questions about how the current balance of methods in developmental science is rationalized.

Cases 2 and 3: Exemplary and Ecologically Valid Laboratory-Based Developmental Studies

The foregoing discussion is not a dismissive attack on laboratory-based research. Below, we present two examples of studies using standardized laboratory methods with robust ecological validity. Given our overarching argument that there are numerous parallels between the characteristics of strengths-based developmental research and the criteria for ecologically valid research, we believe that highlighting exemplary cases of ecologically valid developmental studies is especially valuable.

Rheingold's (1982) Study of Children's Early Prosocial Helping

The first case of ecologically valid laboratory-based developmental research is Rheingold's (1982) study on young children's prosocial helping, classically defined as "voluntary behavior intended to benefit another" (Eisenberg et al., 2015, p. 610). Everyday household work is a central context of many young children's early prosocial development, providing regular and accessible opportunities for children to help, collaborate, and contribute to others' efforts. Thus, many contemporary studies of young children's prosociality focus on children's engagement in these tasks and activities. Rheingold's study is frequently cited as foundational to a recent wave of developmental research on young children's prosocial helping, often in reference to a concluding speculation that parents' early exclusion of their children from everyday opportunities to help diminishes children's prosociality as they grow older. Recent research supports this keen speculation (Coppens & Rogoff, 2021; Rogoff & Coppens, in press), and this study continues to be relevant for its accomplishments in ecological validity.

Briefly, Rheingold's (1982) team invited 18-, 24-, and 30-month-old children, their mothers, their fathers, and a couple of unfamiliar adults to complete nine chores set up in a research laboratory space furnished to resemble the rooms of a home. Parents were introduced to the tasks (e.g., chores such as sweeping, setting a table, folding laundry, picking up strewn objects) and asked to "talk about what they would be doing… to maintain a natural milieu" and to "perform the tasks slowly so the children might participate." They were told that reading, playing, or spending time doing things other than tidying up was acceptable, and that all tasks did not need to be completed. In line with research questions focused on children's *prosocial* (i.e., voluntary) helping, parents were instructed to "refrain from telling the children what to do." Children were given no instructions and were not present for parent briefings. Parents and children were then shown the rooms and given 25 minutes to take part.

We focus on a few design features of the study that were crucial to the study's high ecological validity and cross-cultural compatibility. First, Rheingold's (1982) design was not an exact replica of the social situation of participants' households, which would be impossible across the 80 families they studied. Instead, the designed social setting (i.e., the spaces, materials, and instructions for engagement) was sufficiently open-ended to allow for children and adults to fit their own capacities, interests, family and cultural practices, and on-the-spot interpretations into the affordances provided. In this sense, the study is an exemplary model of *ecological validity as a joint accomplishment* between researchers and participants.

Second, the laboratory activity was designed in ways that permitted the research questions to be addressed by delimiting the construct of interest—children's helping—to the level of *shared family activity* rather than delimiting to the more constrained level of stimulus-and-behavioral-response that pervades contemporary laboratory studies of toddlers' prosocial helping. The design was focused, not a simplification; it retained numerous key features of the everyday contextualization of children's helping. By contrast, narrowly constrained stimulus-response designs both present researchers with an enormous burden to "get it right" in an ecological sense and introduce an array of new and potentially confounding contextual features.

Third, social aspects of Rheingold's (1982) design implicitly invited both parents and children to engage *as potential collaborators* by permitting children and parents to share overarching goals of the activity (e.g., spending time together or cleaning up the rooms) in addition to more instrumental goals such as sweeping or picking up pieces of paper. In most contemporary developmental studies of toddlers' helping, children are only able to help an unfamiliar adult and are engaged with adults at only the task, not activity, level.

Together, these ecologically valid design features permitted Rheingold (1982) to creating space for children's voluntary and autonomous helping and subsequently permitted the identification of a type of helping similar to one that is widespread among Indigenous American and other under-represented and politically non-dominant communities: children's *initiative-based helping* (Coppens et al., 2016; Coppens & Rogoff, 2021; Correa-Chávez et al., 2015; de Haan, 1999; Gaskins, 1999; López et al., 2012; Paradise & Rogoff, 2009). To be sure, the cross-cultural compatibility of the design may have been accidental as this type of helping had not yet been formally observed by developmental psychologists, though anthropologists had described young children's autonomous helping among many Indigenous American communities (Lancy, 2020; Modiano, 1973; Whiting & Whiting, 1975). The possible happenstance of the design's cross-cultural compatibility further illuminates the value of conceptualizing strengths-based ecologically valid research as a joint accomplishment among research participant and researchers. Flexible and somewhat open-ended social designs in developmental research permit participants to "bring" or enact cultural strengths in ways that researchers may not anticipate but can observe and measure with precision.

If the families in Rheingold's (1982) study were of middle-class European-heritage backgrounds (likely, but family backgrounds were not reported), it is possible that parents in fact often tell children what to do at home (see Alcalá et al., 2014; Coppens et al., 2016; Coppens & Rogoff, 2021)—an apparent strike against the ecological validity of the study. However, this social design feature may have been justified by research questions that were focused on children's capacity for *voluntary* prosocial helping. We argue that a concept of ecological validity in which

(1) developmental research tasks are required to exactly mirror the social and behavioral processes of participants' everyday activities and (2) that place the burden of such social and cultural isomorphism entirely on researchers is unnecessary, epistemologically under-leveraged, and ultimately doomed to fail (Lave, 1997).

Legare and Souza's (2012) Study of the Causal Efficacy of Ritual-Based Problem Solving

The second case of ecologically valid laboratory-based research is Legare and Souza's (2012) series of studies examining the implicit criteria people use to judge the causal efficacy of rituals used for problem-solving purposes. Although the study is more cognitive than developmental psychology in discipline, the ecologically valid experimental design is applicable across the psychological sciences as an exemplary model of research.

Legare and Souza (2012) sought to examine the implicit causal "logic" embodied in the prescribed behavioral patterns of ritual-based cultural artifacts—that is, the properties of a ritual (e.g., the number of steps it involves, or the level of specificity of its prescriptions) that create the perception that it will be causally effective or "work" when enacted to solve a problem in everyday life. The authors chose a metropolitan area of southeastern Brazil in which curative or remedial behavioral prescriptions, locally called *simpatias* (which are purchased ritual "recipes" for a set of behaviors that, if enacted with fidelity, are believed to intervene on problems of health or interpersonal relations). *Simpatias* were widely "available, endorsed, and used for everyday problem-solving purposes" among the general public in this area of Brazil.

The study began with a content analysis of 50 extant and commonly purchased *simpatias*, from which Legare and Souza (2012) identified shared criteria such as specificity of time or place, repetition of procedures, and number of procedural steps. The researchers then used these derived criteria to design 18 new experimental stimuli (designed as *simpatias*) which systematically varied nine criteria across the later three studies. For example, two paired and otherwise identical experimental *simpatias* varied "specificity of time" by appending, in one, the phrase "in the first day of last quarter phase of the moon" to the behavioral prescription shared by both, "take the milk from a coconut and give it to the affected person to drink..." Differences in participants' estimations of the efficacy of the two *simpatias* were taken as evidence that "specificity of time" was or was not a causally important criterion of everyday ritual.

The authors *specifically tested* the ecological validity of the experimental stimuli with 60 Brazilian adults to ensure they felt familiar but also to ensure they were not conventionally associated with particular everyday problems (e.g., illness, infidelity), which could confound criterion-based judgments of efficacy. The authors reported that, "many of the participants spontaneously answered that they had used several of these particular *simpatias* themselves in the past" (p. 5), which of course they had not, as Legare and Souza had created them. The participants' apparent déjà vu provided excellent support for the ecological validity of the newly created experimental stimuli. Although the study's findings are not central to the present discussion, three experiments found that the presence of a narrow set of "intuitive causal principles" in the experimental *simpatias* such as repetition of procedures, number of procedural steps, and specificity of time increased estimations of ritual efficacy in both Brazil and, subsequently, the United States.

This ecologically valid approach to the standardization of stimuli deserves special focus. Legare and Souza (2012) argue that their "ecologically valid paradigm maximizes the best possible trade-off between internal and external validities." What permitted the success of this trade-off in our view was the high fidelity of the experimental stimuli to both the material "look and feel" as well as the underlying principles of everyday Brazilian *simpatias*, a degree of ecological fidelity made possible by the authors' ethnographic and content analysis work to understand and formally validate their experimental *simpatias*. As a result, their study is an exemplary case of ecological validity in the sense that instead of, "carry[ing] around our laboratory task and mak[ing] it happen in a lot of settings" they first "*discover[ed]* the way it occurs (or doesn't occur) in non-laboratory settings" (Cole et al., 1997, p. 54) and used this information as strict criteria for the design of experimental stimuli.

It is notable that a similar "trade-off" argument is made by Kominsky et al. (2020), however, measurement (i.e., internal) validity is accomplished in the infant research they reference by making use of theatrical pretense, inanimate objects, or "highly simplified stimuli" that are "potentially far less realistic" (p. 6) than how infants encounter the phenomena under investigation in their everyday lives. To be sure, the demands of studying the cognition of pre-verbal infants presents researchers with significant interpretative burdens. Nonetheless, our view is that the trade-off as operationalized by Legare and Souza (2012)—which required neither the *additional* contextual complexities of theatrical pretense (e.g., "imagine this was a *simpatia* that a neighbor had followed in order to cure a cold…") nor the addition of symbolically abstract stimuli—reflects a fundamental commitment to ecological validity that cannot be reduced to the relative ease of studying adults instead of children. Legare and Souza's (2012) deep understanding of the ecological realities of ritual and *simpatias* in Brazil facilitated a high degree of experimental control without the need for dissection or artificiality.

Connections to Other Prominent Cases of Deficit-Oriented Developmental Research

Our hope is that the foregoing discussion articulates and puts to use several general criteria of ecological validity as lenses into the extent to which research is designed to observe and describe the cultural strengths of communities involved in the research. The recommendations we provide in the next section intend to help researchers carry this critical analysis into a variety of methodological approaches. To that end, here we very briefly mention two prominent programs of research in developmental psychology which have been robustly critiqued for having poor ecological validity and, in part as a result, for advancing deficit frameworks regarding the skills and strengths of non-dominant communities. Both literatures explore the limits of the notion that research outside of the lab, in observational contexts, necessarily makes progress on the issue of ecological validity and, by extension, strengths-based developmental science.

First, critiques of developmental research on caretaker–child attachment on conceptual, cross-cultural, and methodological grounds are robust and long-standing (see Burman, 2008; Keller & Bard, 2017; Otto & Keller, 2014; Quinn & Mageo, 2013; Vicedo, 2014). Because such well-developed critiques exist, we do not summarize them in this chapter. Nonetheless, there is much to be learned by studying how large portions of research on attachment—especially those rooted in Bowlby's ethological and psychoanalytical approach—have emerged as a paradigmatic case of deficit-oriented, ethnocentric research characterized by systematic shortcomings with regard to ecological validity. For example, this research has often assumed that biological mothers are young children's primary and most important attachment figure and that positively valenced "secure attachment" consists of universal behavioral characteristics on the part of caregivers and children (i.e., behaviors common to American middle-class mothers and children)—both assumptions have been robustly challenged with cross-cultural evidence. This body of research is a key example of how the ecological invalidity of rigidly standardized laboratory observation methods and ethnocentric analytic approaches, even with cross-cultural samples, can result in empirically questionable and politically/ethically oppressive developmental research findings and interventions (see Morelli et al., 2018b).

Second, developmental research on the supposed "language gap" and the challenges from researchers in linguistic anthropology and from ethnographically sensitive language acquisition researchers in developmental psychology together amount to a master class in the nuances and pitfalls of advancing strengths-based and ecologically valid developmental science. Much of the critique centers on the controversial empirical and policy-oriented legacy of Hart and Risley's (1995) study claiming a "30 million word gap" in the quality of the linguistic environments of children from various racial and socioeconomic backgrounds, with affluent White children argued to be most advantaged. Critique was both swift and robust; however, unfortunately, so were large-scale family interventions and social policy that assumed the veracity of Hart and Risley's findings. Hart and Risley's research used observational methods in families' homes which might tempt an assumption of good ecological validity. However, despite extensive in-home recordings and observations, a key omission in Hart & Risley's research was an appreciation of the importance and developmental significance of overhearing and multi-party storytelling as robust linguistic practices supporting children's language development in many minoritized cultural communities. We recommend the following for more through, nuanced heterodoxical stances against this deficit-oriented research paradigm: Adair et al. (2017); Avineri et al. (2015); Miller and Sperry (2012); Morelli et al. (2018a); Sperry et al. (2019); and a recent special issue of *International Multilingual Research Journal* devoted to the debate (Johnson & Zentella, 2017).

Recommendations for Building Ecologically Valid, Strengths-Based Developmental Research

In this chapter, we have integrated criteria for both strengths-based and ecologically valid developmental research. We have also reviewed the details of several studies that serve as clear and counter cases of ecologically valid research, building on our argument that ecological validity is a key aspect of strengths-based research. In this section, we extend these discussions into a set of recommendations, in no particular order and not a comprehensive "checklist," that invite researchers working from a wide range of methodological approaches and areas of inquiry to enhance the potential of their research to advance strengths-based developmental findings.

Recommendation: Critically Examine Unit of Analysis

It is crucial for strengths-based developmental researchers to understand the vulnerabilities of deficit-thinking that arise when conceptualizing developmental phenomena as delimited or located in individuals and merely "impacted" or "effected" by social/interactional and cultural/institutional processes (see Rogoff, 1995, 2003; Gutiérrez & Rogoff, 2003; and the extensive theoretical tradition of cultural psychology for alternatives to the conventional views of developmental psychology). For example, consider motivation. Classical theories of motivation locate the phenomenon as a property of individuals, creating a binary scale ranging from "intrinsically" (i.e., internally) motivated individuals to "extrinsically" (i.e., externally, or contingently) motivated individuals (e.g., Ryan & Deci, 2000). Observations of apparently low motivation are easily attributed to something—that is, motivation—lacking in individuals, a deficit perspective. However, cultural-historical perspectives locate motivation as a property of cultural activity in which individuals engage at varying and developmentally shifting levels of proximity (Roth, 2011). In this latter view, motivational shortcomings are conceptualized as participatory constraints such as barriers to access rather than something that individuals lack.

Recommendation: Understand All Research as Context Creating

Strengths-based developmental researchers must account for the "research moment"—e.g., taking a survey, completing an educational assessment, performing a laboratory task, or answering a set of interview questions—as a full social, cultural, and political context of activity which may or may not be experienced as representative of seemingly similar activities in participants' everyday lives. In this chapter, we focused on several ways that laboratory-based developmental research is context-creating (see Lave, 1997 for an excellent discussion of how created experimental contexts can be leveraged for comparative, strengths-oriented insights). Semi-structured qualitative research interviews are also a discrete conversational genre and context that is different than "normal" everyday conversation in important ways (see Packer, 2018 for an excellent discussion). Developmental researchers rarely theorize research interviews as unique interactional contexts; however, insights from sociolinguistics suggest that, for this reason, interview data in developmental research are severely under-leveraged for recognizing the skills, strengths, implicit theories, and identity commitments of participants. For an example of developmental research with mothers of two cultural groups that leverages the "interactional text" of interviews (Wortham et al., 2011) as a focal source of strengths-based developmental evidence, see Coppens et al. (2020).

Recommendation: Employ Diverse Samples, with Nonethnocentric Group Comparisons

The problem of lacking diversity across the samples used in developmental research is widespread and widely recognized. However, when developmental studies include samples from more than one social or cultural group new potential problems can arise vis-à-vis comparative methodological designs. We recommend that researchers reflect on whether they are conflating *norm-referenced* comparisons with *criterion-referenced* comparisons. Given the extreme lack of diversity of sampling in developmental research (Nielsen et al., 2017), norm-referenced comparisons are the most vulnerable to deficit-oriented and ethnocentric conclusions as they use the developmental processes or outcomes of an over-sampled and politically dominant social group (often, middle-class European-heritage families and communities) as a normative point of reference for making judgements about the differing processes and outcomes of comparison groups. By contrast, *criterion-referenced* comparisons benchmark evaluations of developmental processes and outcomes on ethnographically grounded and culturally valued points of reference, such as being *acomedido* (López et al., 2012; Rogoff & Coppens, in press) or the amounts and kinds of talk in which children are socialized to be skillful in their own communities (see above citations regarding critical perspectives on "language gap" research). An emphasis on criterion-referenced comparisons prevents consideration of diverse cultural and social group perspectives in developmental science from being reduced to relativism (Rogoff, 2003), whereby judgments about the value and efficacy of behavioral patterns or developmental outcomes are made impossible. Instead, the judgements of developmental scientists are scrutinized according to the values and socialization goals of the same social or cultural groups under investigation.

Recommendation: Prioritize Actor- Over Observer-Perspectives in Developmental Science

Interpretative qualitative researchers often stress the need to address problematic politics in conventional relationships among researchers and study participants. We argued above for prioritizing *actor* over *observer perspectives*, which includes but also has deeper significance than knowing what participants think about the tasks they are prescribed or the questions they are asked in developmental studies. Centering actor perspectives reconsiders relational-epistemological orientations conventional to developmental science. An actor-oriented developmental science becomes attuned to the goals, interests, relationships, and aspirations that children, youth, and adults are pursuing in the social and cultural activities that are the focus of developmental research, *in addition to* understanding the extent to which such aims are consequential for the developmental outcomes that concern researchers. Bang and Vossoughi (2016) argue against just this tendency

> to [only] highlight relationships between social actors *in so far as* they are consequential to subject–object relations, such as the ways particular forms of classroom discourse or collaboration improve normatively defined academic outcomes. These theoretical and methodological decisions risk constraining the forms of knowledge and practice we develop, enact, and share, and thereby narrow the valued forms of learning and development we study and theorize. (p. 179)

Future Considerations for Ecological Validity and Strengths-Based Developmental Science

We have characterized mainstream developmental research in a somewhat monolithic way; however, there are many notable exceptions in the field to the critiques that we have articulated as well as recent efforts underway to address issues that undermine an equity-oriented developmental science. For example, a new task force is being formed through the American Psychological Association focused on promoting anti-racist science and calls for an explicit "diversity science" in psychological research are increasing (e.g., Broesch et al., 2021; Miller et al., 2019; Plaut, 2010).

The concept of ecological validity itself requires scrutiny and theoretical attention to keep pace with innovations underway in research on learning and development. One read of existing concepts of ecological validity is they incline toward a conservative view of the existing social and cultural ecologies to which they allude. What happens when the concept of ecological validity is reconstituted as a forward-looking or imaginative criterion, holding research designs accountable to the family and community aspirations and children and youths' future selves? What new methodological demands might this reconstitution introduce?

Critical learning scientists and developmental researchers working in emancipatory methodologies are swiftly advancing this agenda. For example, the syncretic research-intervention hybridities characteristic of social design-based experiments (e.g., Gutiérrez & Jurow, 2016), formative interventions (e.g., Sannino et al., 2016), participatory design research (see Bang & Vossoughi, 2016), as well as work aligned with politically imaginative frameworks such as culturally sustaining pedagogies (Paris & Alim, 2017) raise important new questions for how commitments to new concepts of ecological validity can advance both strengths-based and strengths-catalyzing forms of developmental research and intervention (see Chap. 9, Banales et al. for a discussion of strengths-based research and YPAR). In the foregoing discussion we suggested that the idea of *ecological validity as a joint accomplishment between researchers and participants* may be a productive avenue for resolving the pervasive but fallacious stalemate between standardization and ecological validity in developmental research designs, highlighting the relational politics of researcher-participant interactions as a key issue. We highlight these emancipatory methodologies and developmental research programs to point out that this work is already underway.

The remaining question, which we pose as closing to the chapter, relates to the new demands of contextualization made visible by learning and developmental research whose explicit aim is to intervene on oppressive ecological realities and to both imagine and organize new future possibilities. To what ecology ought strengths-based developmental research or interventions be valid when the intervention's aim is the creation of new social ecologies (see Chap. 11, Tynes et al. for a discussion of Afrofuturism and how it was used for the development of an intervention)?

References

Adair, J. K., Colegrove, K. S.-S., & McManus, M. E. (2017). How the word gap argument negatively impacts young children of Latinx immigrants' conceptualizations of learning. *Harvard Educational Review, 87*(3), 309–334.

Alcalá, L., Rogoff, B., Mejía-Arauz, R., Coppens, A. D., & Dexter, A. L. (2014). Children's initiative in contributions to family work in Indigenous-heritage and cosmopolitan communities in Mexico. *Human Development, 57*(2–3), 96–115. https://doi.org/10.1159/000356763

Arnett, J. J. (2008). The neglected 95%: Why American psychology needs to become less American. *American Psychologist, 63*, 602–614. https://doi.org/10.1037/0003-066X.63.7.602

Avineri, N., et al. (2015). Invited forum: Bridging the "language gap". *Journal of Linguistic Anthropology, 25*(1), 66–86. https://doi.org/10.1111/jola.12071

Bang, M., & Vossoughi, S. (2016). Participatory design research and educational justice: Studying learning and relations within social change making. *Cognition and Instruction, 34*(3), 173–193. https://doi.org/10.1080/07370008.2016.1181879

Briggs, C. L. (1986). *Learning how to ask: A sociolinguistic appraisal of the role of the interview in social science research*. Cambridge University Press.

Broesch, T., Lew-Levy, S., Kärtner, J., Kanngiesser, P., & Kline, M. (2021). A roadmap to doing culturally grounded developmental science. *Review of Philosophy and Psychology*. https://doi.org/10.1007/s13164-022-00636-y

Bronfenbrenner, U. (1977). Toward an experimental ecology of human development. *American Psychologist, 32*(7), 513–531.

Burman, E. (2008). *Deconstructing developmental psychology* (2nd ed.). Routledge.

Cole, M. (1996). *Cultural psychology: A once and future discipline*. Belknap Press.

Cole, M. (2013). Differences and deficits in psychological research in historical perspective. *Developmental Psychology, 49*(1), 84–91. https://doi.org/10.1037/a0029623

Cole, M., & Bruner, J. S. (1971). Cultural differences and inferences about psychological processes. *American Psychologist, 26*, 867–876. https://doi.org/d5m6f7

Cole, M., Hood, L., & Holzman, L. (1978). *Ecological niche-picking: Ecological invalidity as an axiom of experimental cognitive psychology*. Laboratory of Comparative Human Cognition.

Cole, M., Hood, L., & McDermott, R. P. (1997). Concepts of ecological validity. In M. Cole, Y. Engeström, & O. Vasquez (Eds.), *Mind, culture, and activity: Seminal papers from the Laboratory of Comparative Human Cognition*. Cambridge University Press.

Coppens, A. D., & Rogoff, B. (2021). Cultural variation in the early development of initiative in children's prosocial helping. *Social Development, 31*, 1–22. https://doi.org/10.1111/sode.12566

Coppens, A. D., Alcalá, L., Rogoff, B., & Mejía-Arauz, R. (2016). Children's contributions in family work: Two cultural paradigms. In S. Punch & R. M. Vanderbeck (Eds.), *Families, intergenerationality, and peer group relations* (pp. 1–27). Springer. https://doi.org/10.1007/978-981-4585-92-7_11-2

Coppens, A. D., Corwin, A. I., & Alcalá, L. (2020). Beyond behavior: Linguistic evidence of cultural variation in parental ethnotheories of children's prosocial helping. *Frontiers in Psychology, 11*, 307. https://doi.org/10.3389/fpsyg.2020.00307

Correa-Chávez, M., Mejía-Arauz, R., & Rogoff, B. (Eds.). (2015). *Children learn by observing and contributing to family and community endeavors* (Vol. 49). Academic Press.

Dahl, A. (2017). Ecological commitments: Why developmental science needs naturalistic methods. *Child Development, 11*(2), 79–84. https://doi.org/10.1111/cdep.12217

Danziger, K. (1997). *Naming the mind: How psychology found its language*. Sage.

de Haan, M. (1999). *Learning as cultural practice: How children learn in a Mexican Mazahua community*. Thela Thesis.

Dinishak, J. (2016). The deficit view and its critics. *Disability Studies Quarterly, 36*(4). https://doi.org/10.18061/dsq.v36i4.5236

Dixon-Román, E. J., Jackson, J. L., & de Royston, M. M. (2020). Reconceptualizing the quantitative-qualitative divide. In N. S. Nasir, C. D. Lee, R. D. Pea, & M. M. de Royston (Eds.), *Handbook of the cultural foundations of learning* (pp. 314–329). Routledge.

Du Bois, W. E. B., Battle-Baptiste, W., & Rusert, B. (2018). *W.E.B Du Bois's data portraits: Visualizing Black America*. Princeton Architectural Press.

Eisenberg, N., Spinrad, T. L., & Knafo-Noam, A. (2015). Prosocial development. In R. M. Lerner (Ed.), *Handbook of child psychology and developmental science* (Vol. 3, pp. 610–656). John Wiley & Sons, Inc. https://doi.org/10.1002/9781118963418.childpsy315

Esteban-Guitart, M., & Moll, L. C. (2014). Funds of identity: A new concept based on the funds of knowledge approach. *Culture & Psychology, 20*(1), 31–48. https://doi.org/10.1177/1354067X13515934

Gaskins, S. (1999). Children's daily lives in a Mayan village. In A. Göncü (Ed.), *Children's engagement in the world* (pp. 25–61). Cambridge University Press.

Gharabaghi, K., & Anderson-Nathe, B. (2017). Strength-based research in a deficits-oriented context. *Child & Youth Services, 38*(3), 177–179. https://doi.org/10.1080/0145935X.2017.1361661

González, N., Moll, L. C., & Amanti, C. (2005). *Funds of knowledge: Theorizing practice in households, communities, and classrooms*. L. Erlbaum Associates.

Graham, S. (1992). Most of the subjects were white and middle class: Trends in published research on African Americans in selected APA journals. *American Psychologist, 47*(5), 629–639.

Gutiérrez, K. D., & Jurow, A. S. (2016). Social design experiments: Toward equity by design. *Journal of the Learning Sciences, 25*, 1–34. https://doi.org/10.1080/10508406.2016.1204548

Gutierrez, K., & Larson, J. (2007). Discussing expanded spaces for learning. *Language Arts, 85*(1), 69–77. https://library.ncte.org/journals/la/issues/v85-1

Gutiérrez, K. D., & Rogoff, B. (2003). Cultural ways of learning: Individual traits or repertoires of practice. *Educational Researcher, 32*(5), 19–25.

Hammersley, M. (1992). Deconstructing the qualitative-quantitative divide. In M. Hammersley (Ed.), *What's wrong with ethnography: Methodological explorations* (pp. 159–173). Routledge.
Harding, S. (1992). The afterlife of stories. In G. Rosenwald & R. Ochberg (Eds.), *Storied lives* (pp. 60–75). Yale University Press.
Hart, B., & Risley, T. R. (1995). *Meaningful differences in the everyday experience of young American children*. P.H. Brookes.
Henrich, J., Heine, S. J., & Norenzayan, A. (2010). The weirdest people in the world? *Behavioral and Brain Sciences, 33*(2–3), 61–135. https://doi.org/10.1017/S0140525X0999152X
Jahoda, G. (1986). A cross-cultural perspective on developmental psychology. *International Journal of Behavioral Development, 9*, 417–437.
Johnson, E. J., & Zentella, A. C. (2017). Introducing the language gap. *International Multilingual Research Journal, 11*(1), 1–4. https://doi.org/10.1080/19313152.2016.1258184
Keller, H., & Bard, K. A. (Eds.). (2017). *The cultural nature of attachment: Contextualizing relationships and development*. The MIT Press.
Kessen, W. (1979). The American child and other cultural inventions. *American Psychologist, 34*(7), 815–820.
Kominsky, J. F., Lucca, K., Thomas, A. J., Frank, M. C., & Hamlin, K. (2020). Simplicity and validity in infant research [Preprint]. *PsyArXiv*. https://doi.org/10.31234/osf.io/6j9p3
Koops, W., & Kessel, F. (2015). Common historical roots. Introduction to the special section on recapitulation theory and developmental psychology. *European Journal of Developmental Psychology, 12*(6), 627–629. https://doi.org/10.1080/17405629.2015.1094876
Lancy, D. F. (2020). *Child helpers: A multidisciplinary perspective*. Cambridge University Press. https://doi.org/10.1017/9781108769204
Lave, J. (1997). What's special about experiments as contexts for thinking? In M. Cole, Y. Engeström, & O. Vasquez (Eds.), *Mind, culture, and activity: Seminal papers from the Laboratory of Comparative Human Cognition* (pp. 57–69). Cambridge University Press.
Legare, C. H., & Souza, A. L. (2012). Evaluating ritual efficacy: Evidence from the supernatural. *Cognition, 124*(1), 1–15. https://doi.org/10.1016/j.cognition.2012.03.004
López, A., Rogoff, B., Najafi, B., & Mejía-Arauz, R. (2012). Collaboration and helping as cultural practices. In J. Valsiner (Ed.), *The Oxford handbook of culture and psychology* (Vol. 1–41, pp. 869–884). Oxford University Press.
Marcus, G. E. (1995). Ethnography in/of the world system: The emergence of multi-sited ethnography. *Annual Review of Anthropology, 24*, 95–117. https://doi.org/10.1146/annurev.an.24.100195.000523
Miller, P. J., & Sperry, D. E. (2012). Déjà Vu: The continuing misrecognition of low-income children's verbal abilities. In S. T. Fiske & H. R. Markus (Eds.), *Facing social class: How societal rank influences interaction* (Vol. 1–6, pp. 109–130). Russell Sage Foundation.
Miller, A. L., Stern, C., & Neville, H. (2019). Forging diversity-science-informed guidelines for research on race and racism in psychological science. *Journal of Social Issues, 75*(4), 1240–1261. https://doi.org/10.1111/josi.12356
Modiano, N. (1973). *Indian education in the Chiapas highlands*. Holt, Rinehart and Winston.
Morelli, G., Bard, K., Chaudhary, N., Gottlieb, A., Keller, H., Murray, M., Quinn, N., Rosabal-Coto, M., Scheidecker, G., Takada, A., & Vicedo, M. (2018a). Bringing the real world into developmental science: A commentary on Weber, Fernald, and Diop (2017). *Child Development, 89*(6), e594–e603. https://doi.org/10.1111/cdev.13115
Morelli, G., Quinn, N., Chaudhary, N., Vicedo, M., Rosabal-Coto, M., Keller, H., Murray, M., Gottlieb, A., Scheidecker, G., & Takada, A. (2018b). Ethical challenges of parenting interventions in low- to middle-income countries. *Journal of Cross-Cultural Psychology, 49*(1), 5–24. https://doi.org/10.1177/0022022117746241
Nasir, N. S., Lee, C. D., Pea, R. D., & McKinney de Royston, M. (Eds.). (2020). *Handbook of the cultural foundations of learning*. Routledge.

Nielsen, M., Haun, D., Kärtner, J., & Legare, C. H. (2017). The persistent sampling bias in developmental psychology: A call to action. *Journal of Experimental Child Psychology, 162*, 31–38. https://doi.org/10.1016/j.jecp.2017.04.017

Otto, H., & Keller, H. (Eds.). (2014). *Different faces of attachment*. Cambridge University Press.

Packer, M. J. (2018). *The science of qualitative research* (2nd ed.). Cambridge University Press.

Packer, M. J., & Moreno-Dulcey, F. A. (2022). Theory of puppets?: A critique of the use of puppets as stimulus materials in psychological research with young children. *Cognitive Development, 61*, 101146. https://doi.org/10.1016/j.cogdev.2021.101146

Packer, M., & Moreno-Dulcey, F. A. (2020). *"This puppet will play a game with you": Is it time to take child psychology out of the laboratory?* (Unpublished manuscript).

Packer, M. J., & Tappan, M. B. (2001). *Cultural and critical perspectives on human development*. SUNY Press.

Paradise, R., & Rogoff, B. (2009). Side by side: Learning by observing and pitching in. *Ethos, 37*(1), 102–138.

Paris, D., & Alim, H. S. (Eds.). (2017). *Culturally sustaining pedagogies: Teaching and learning for justice in a changing world*. Teachers College Press.

Plaut, V. C. (2010). Diversity science: Why and how difference makes a difference. *Psychological Inquiry, 21*(2), 77–99. https://doi.org/10.1080/10478401003676501

Quinn, N., & Mageo, J. M. (Eds.). (2013). *Attachment reconsidered: Cultural perspectives on a Western theory*. Palgrave Macmillan.

Revencu, B., & Csibra, G. (2020, December 11). *Opening the black box of early depiction interpretation: From whether to how in the Theory-of-Puppets debate* (submitted). https://doi.org/10.31234/osf.io/u3em8

Rheingold, H. L. (1982). Little children's participation in the work of adults, a nascent prosocial behavior. *Child Development, 53*(1), 114–125. https://doi.org/10.2307/1129643

Rogoff, B. (1995). Observing sociocultural activity on three planes: Participatory appropriation, guided participation, and apprenticeship. In J. V. Wertsch, P. Del Rio, & A. Alvarez (Eds.), *Sociocultural studies of mind* (Vol. 1–6, pp. 139–164). Cambridge University Press.

Rogoff, B. (2003). *The cultural nature of human development*. Oxford University Press.

Rogoff, B., & Coppens, A. D. (in press). The cultural nature of helping without being asked, from the toddler years into middle childhood. In M. Gelfand, C. Chiu, & Y. Hong (Eds.), *Advances in culture and psychology* (Vol. 10). Oxford University Press.

Rogoff, B., Coppens, A. D., Alcalá, L., Aceves-Azuara, I., Ruvalcaba, O., López, A., & Dayton, A. (2017). Noticing learners' strengths through cultural research. *Perspectives on Psychological Science, 12*(5), 876–888. https://doi.org/10.1177/1745691617718355

Rogoff, B., Dahl, A., & Callanan, M. (2018). The importance of understanding children's lived experience. *Developmental Review, 50*, 5–15. https://doi.org/10.1016/j.dr.2018.05.006

Roth, W.-M. (2011). Object/motives and emotion: A cultural-historical activity theoretic approach to motivation in learning and work. In D. M. McInerney, R. A. Walker, & G. A. D. Liem (Eds.), *Sociocultural theories of learning and motivation* (Vol. 1–3, pp. 43–64). Information Age Publishing.

Ryan, R. M., & Deci, E. L. (2000). Self-determination theory and the facilitation of intrinsic motivation, social development, and well-being. *American Psychologist, 55*, 68–78.

Sannino, A., Engeström, Y., & Lemos, M. (2016). Formative interventions for expansive learning and transformative agency. *Journal of the Learning Sciences, 25*(4), 599–633. https://doi.org/10.1080/10508406.2016.1204547

Schmuckler, M. A. (2001). What is ecological validity? *Infancy, 2*(4), 419–436.

Sperry, D. E., Sperry, L. L., & Miller, P. J. (2019). Reexamining the verbal environments of children from different socioeconomic backgrounds. *Child Development, 90*(4), 1303–1318. https://doi.org/10.1111/cdev.13072

Steele, C. M. (2010). *Whistling Vivaldi*. W.W. Norton & Company.

Valencia, R. R. (2010). *Dismantling contemporary deficit thinking*. Routledge.

Valencia, R. R., & Solórzano, D. G. (1997). Contemporary deficit thinking. In R. R. Valencia (Ed.), *The evolution of deficit model thinking* (Vol. 1–6, pp. 160–210). The Falmer Press.

Valsiner, J., & Benigni, L. (1986). Naturalistic research and ecological thinking in the study of child development. *Developmental Review, 6*(3), 203–223. https://doi.org/10.1016/0273-2297(86)90012-2

Vicedo, M. (2014). *The nature and nurture of love*. The University of Chicago Press.

Vossoughi, S., & Gutiérrez, K. D. (2014). Studying movement, hybridity, and change: Toward a multi-sited sensibility for research on learning across contexts and borders. *Teachers College Record, 116*(14), 603–632. https://doi.org/10.1177/016146811411601413

Vossoughi, S., & Zavala, M. (2020). The interview as pedagogical encounter: Nurturing knowledge and relationships with youth. In A. I. Ali & T. L. McCarty (Eds.), *Critical youth research in education* (pp. 136–154). Routledge. https://doi.org/10.4324/9780429277863-11

Whiting, B. B., & Whiting, J. W. M. (1975). *Children of six cultures*. Harvard University Press.

Wortham, S., Mortimer, K., Lee, K., Allard, E., & White, K. D. (2011). Interviews as interactional data. *Language in Society, 40*(1), 39–50. https://doi.org/10.1017/S0047404510000874

Chapter 8
Fertile Ground for Sociocultural Responsivity: Schools and Neighborhoods as Promotive and Inhibiting Environments

Dawn P. Witherspoon, Rebecca M. B. White, Rajni Nair, Tiyobista M. Maereg, and Wei Wei

Schools and neighborhoods are developmental settings outside the home in which children and youth spend a considerable amount of their time. The diversity of these environments, particularly the ethnic-racial composition of these spaces, varies greatly. In this chapter, we focus on race and ethnicity in these settings within the US context because of the historical residential segregation and racial-ethnic stratification of this country. In the United States, residential segregation by race and ethnicity[1] is ubiquitous and functions in unique ways from other Western countries (Meer & Tolsma, 2014). Within this national context, children and youth live in very "separate and unequal" neighborhoods and attend similarly "separate and unequal" schools (https://www.diversitydatakids.org/neighborhoods). Yet, non-white children are the numerical majority of the United States' youth (U.S. Census Bureau, 2018). Further, as the percentage of school age children who are white or Black decreased in the past decade, the proportions of other minoritized ethnic-racial groups (e.g., Latinx, Asian, multiracial) increased in our schools (de Brey et al., 2019). These striking and important population statistics make it critically

[1] We use the term "Black" to refer to individuals of African descent in the United States including the descendants of enslaved Africans, Caribbean Black immigrants and their descendants, and African immigrants and their descendants. We use the term Latinx to refer to individuals of Latin American descent in the United States. We use the term white to refer to individuals of European descent in the United States. We use the term Asian to refer to individuals of Asian descent in the United States.

D. P. Witherspoon (✉) · T. M. Maereg · W. Wei
The Pennsylvania State University, State College, PA, USA
e-mail: dpw14@psu.edu

R. M. B. White · R. Nair
Arizona State University, Tempe, AZ, USA

© The Author(s), under exclusive license to Springer Nature Switzerland AG 2023

D. P. Witherspoon, G. L. Stein (eds.), *Diversity and Developmental Science*,
https://doi.org/10.1007/978-3-031-23163-6_8

necessary for us to understand how the ethnic-racial compositions of where children live and are educated impact their development. In addition to the current demographic shifts that require us to be embrace the varied representation of youth in our neighborhoods and schools, a diverse developmental science is ripe to lead and guide this charge. Extant theories further support such a perspective.

A cultural-contextual developmental perspective is not new to developmental science. There are several pivotal theories that can continue to guide and shape this work. Though we do not go into detail about these theories here, we offer how each theory contributes to our perspective and the focus on ethnic-racial composition within key developmental settings such as neighborhoods and schools. The integrative model of minority child development (García Coll et al., 1996) highlights how social stratification creates systems of oppression that relegate minoritized children and youth to specific environments that have certain resources and risks. One critical element of stratification is race/ethnicity, and two of the important environments are schools and neighborhoods that can promote optimal development or hinder it. Together, the ethnic-racial composition of these environments—teachers, students, residents, youth—create a culture of that environment that shapes development. Related to these tenets, the phenomenological variant of ecological systems theory (PVEST; Spencer, 1995) captures the cultural implications of settings on individuals through its focus on how the realities and experiences within settings shape coping, identities, and behaviors. One's level of vulnerability is a complex mixture of the resources and risks inherent to the individual and the settings they engage. Synergistically, these two cultural-contextual developmental theories highlight the co-action between one's individual characteristics and the elements of the settings to which they are exposed.

As each of these theories provide a very distinctive roadmap for how to craft research questions related to the development of diverse children and youth, our methods need to be equally yoked with our conceptualizations. Phinney et al. (1998) offers a great discussion for how to go beyond universals and think about the use of within and between designs to understand development across different racial and ethnic groups. Within-group designs take an in-group perspective, honing in on the unique experiences of that group, whereas between-group designs focus on the comparison (including both in-group and out-group) between one or more groups to understand developmental universals or highlight differences. Phinney et al. (1998) urges that both types of research designs have utility and import for our science; however, researchers must critically interrogate their research aims and questions to determine the best design. This allows for more nuance—conceptually and methodologically, specifically as it relates to measuring and understanding ethnicity and race as lived experiences across place-based and school contexts.

In our discussion of how neighborhoods and schools are racialized and the impending ethnic-racial compositions of these settings, we propose both within-group (concentration) and between-group (indices, diversity) measures of ethnic-racial composition and how they may be differentially related to youth outcomes. In the sections to follow, we propose a new meta-construct that considers setting-level ethnicity and race, *ethnic-racial compositions*, apply this construct to extant literature focused on neighborhood and school contexts, and reflect on the limits of this approach and offer practical recommendations for scholars engaged and a diverse developmental science that meaningfully considers settings.

Setting-Level Ethnicity and Race in Action

Though there is considerable interest in examining the ethnic and racial makeup of populations within settings, there are also considerable barriers to advancing a comprehensive understanding about how setting race and ethnicity influence development among diverse youth. For example, there is substantial variability in the scholarly terms employed by researchers depending, in part, on the theoretical construct under consideration. Consistent with some theoretical models (Sampson et al., 1997), researchers theorize constructs like homogeneity and segregation, or heterogeneity and diversity. Consistent with other theoretical models (García Coll et al., 1996), researchers theorize access or exposure to a particular group (e.g., access to African Americans). Subsequent operational definitions tend to follow one of two broad approaches: group-specific operationalizations, or multi-group operationalizations. Group-specific operationalizations focus on the proportion of a single ethnic-racial grouping, at any level of specificity, in a setting. We refer to group-specific operationalizations as measures of ethnic-racial concentration, including for example, Black concentration, Indigenous concentration, white concentration, US Mexican concentration, or US Chinese concentration. Multi-group approaches combine information about multiple ethnic-racial groups to produce a single score that takes most or all groups into account. We refer to multi-group operationalizations as ethnic-racial indices, including for example, diversity indices, segregation indices, and others (Budescu & Budescu, 2012; Massey & Denton, 1988).

Though not often discussed, whether a given type of operationalization—concentration vs. index—corresponds well to a given theoretical concept—homogeneity, segregation, heterogeneity, diversity, access, or exposure—often depends upon the underlying distributions of ethnic-racial groups in the setting and/or sample (White et al., 2020). For example, scholars have used single ethnic-racial concentration variables (e.g., % Latinx) to represent: (a) access to specific ethnic-racial groups such as Latinx individuals (e.g., Hong et al., 2014; White et al., 2018a); (b) segregation, isolation, or ethnic-racial homogeneity to represent, for example, the percentage of Latinx individuals relative to another ethnic-racial group (e.g., Massey & Denton, 1988; Rivas-Drake & Witherspoon, 2013; White et al., 2018b); and (c) diversity or heterogeneity that would include the proportion of Latinx individuals compared to other ethnic-racial groups (e.g., Jackson et al., 2016; White et al., 2018a; Witherspoon et al., 2016). Under different underlying conditions (i.e., the number of different groups) and empirical distributions (e.g.,

0–100%), the match between the operational definition and the conceptual definition (e.g., diversity and isolation) may be strong or weak. For example, as White and colleagues have argued (2018a, 2020), in some underlying conditions (2 groups) and sample distributions (e.g., those ranging from 0% to 50%), high scores on an ethnic-racial concentration variable (e.g., % Latinx) may signal heterogeneity and be consistent with diversity (Budescu & Budescu, 2012). In other sample distributions (e.g., those ranging from 0 to 100), regardless of the underlying conditions, high scores on an ethnic-racial concentration variable signal homogeneity. Thus, these high scores are consistent with high access/exposure to a particular group (White et al., 2018b) *and* high isolation or segregation from other groups (e.g., non-Latinxs). For both distributions, low scores on ethnic-racial concentration represent high isolation from the specific group (e.g., Latinx). Under certain conditions, namely two group conditions, low scores on the ethnic-racial concentration variable signal ethnic-racial homogeneity and high access/exposure to one other group. As the number of different ethnic-racial groups in the setting increases, however, low scores on the ethnic-racial concentration variable could signal homogeneity (a higher concentration of one other group) or could signal heterogeneity (a mix of multiple other groups).

Ethnic-racial indices offer a parallel set of complications, also dependent upon the underlying conditions and distributions. These indices can produce identical index values under circumstances that vary considerably relative to any single group's concentration (see White et al., 2018a for a discussion). For example, a setting that has a very high diversity index score may have 30% Latinxs or 5% Latinxs, depending on the underlying conditions and distributions. Thus, a high degree of diversity need not correspond to a high degree of access or exposure to a particular group. Similarly, a setting with a high segregation index score could have 90% Latinxs (and 10% whites) or 10% Latinxs (and 90% whites). A high score on a segregation index, therefore, does not necessarily correspond to a high degree of access or exposure to any group.

Thus, the underlying number of groups and the underlying distributions (a) help determine whether a single ethnic-racial concentration variable (e.g., % Latinx) is signaling ethnic-racial homogeneity/segregation or heterogeneity/diversity and (b) can introduce variability in the ethnic-racial contextual circumstances surrounding a single ethnic-racial index value. Yet, two recent systematic reviews found that very few studies reported the sample distributions of the neighborhood ethnicity and race variables (White et al., 2020; Pasco et al., 2021b) making it difficult to draw conclusions and comparisons across studies. Furthermore, a homogeneous setting could be conceptualized as representing important access to one group or isolation from other group(s), depending on the developmental theory being tested (e.g., integrative model) and/or the framing (e.g., segregation hypotheses). Thus, variability in the way ethnicity and race are conceptualized and operationalized in developmental research impedes scientific progress and compartmentalizes extant literature.

Ethnic-Racial Compositions Meta-Construct

To address these challenges, we define *ethnic-racial compositions* as a meta-construct that captures setting variability in the racial, ethnic, or cultural makeup of (a) populations within settings, (b) group norms and behaviors within settings, and (c) sociocultural signs and symbols within settings. As already inferred above, most prior work focuses on the ethnic-racial makeup of populations within settings using concentration (e.g., % African American) or index (e.g., diversity index) operationalizations. We argue, however, that researchers can also be interested in the compositions of ethnic, racial, and/or cultural group norms and behaviors in settings. For example, researchers may be interested in the proportion of the population in a setting who speak a specific language (e.g., the concentration of Spanish speakers), the number of different languages spoken in a setting (e.g., the diversity of languages spoken), or in norms of racial discrimination in settings (e.g., Russell et al., 2018). They may also be interested in ethnically, racially, or culturally salient signs and symbols within settings. For example, researchers may be interested in the frequency of confederate flags, the frequency of flags from a specific country (concentrations), or in the frequency of flags from a variety of countries of origin (indices) in a neighborhood (Pasco & White, 2020). Similarly, they may be interested in the prevalence of representations (e.g., in media, on billboards, in curricula) of a certain group (ethnic-racial concentrations) or multiple groups (ethnic-racial indices). In this way, we seek to establish a meta-construct that can add both value and meaning to extant work on populations within settings and inform emergent and future work on ethnic, racial, or cultural group norms/behaviors, and sociocultural signs/symbols in settings.

Group Centering of Ethnic-Racial Concentrations

Within the framework provided by the ethnic-racial compositions meta-construct, we argue that it is important to consider whether operationalizations of ethnic-racial concentrations—including those focused on populations, norms/behaviors, or signs/symbols—are (a) centering a within-group (or grouping) or (b) centering an out-group (or grouping). Centering the in-group involves asking how access or exposure to (a) populations that share one's racial, ethnic, or cultural identifications and backgrounds, (b) in-group-salient norms/behaviors, and/or (c) in-group-salient signs/symbols influences development (e.g., Pasco & White, 2020). For example, studies of the effects of African American concentration on youth who identify as African American, or Black concentration on youth who identify as Black, or BIPOC concentration on youth who identify as BIPOC (Benner & Graham, 2013), all examine within-group (at varying levels of specificity) population concentrations on youth who share membership in that group or grouping. Likewise, a study that examined the setting-level concentration of Black Lives Matter signs/symbols on Black youth is centering in-group-salient signs/symbols. Within-group centering should be used when researchers are interested in similarity, sameness, representation, or access to (vs. isolation from) one's own group. Conceptually centering the out-group involves asking how access or exposure to (a) populations that do not share one's racial, ethnic, or cultural background, (b) out-group-salient norms/behaviors, and/or (c) out-group-salient signs/symbols influences development (e.g., Calzada et al., 2020). For example, studies of white concentration on African American youth/families, or Black concentration on Latinx youth/families (Maereg & Witherspoon, 2022; Witherspoon et al., 2016, 2022) examine the effects of an out-group population concentration on youth who do not share membership in that group. Likewise, a study that examined the influence of setting-level prevalence of racial discrimination on BIPOC youth development is centering out-group-salient norms/behaviors because discrimination is defined as dominant racial group members' actions that have differential and negative effects on subordinate, in this case BIPOC, ethnic-racial groups (Williams et al., 2003). Out-group centering should be used when researchers are interested in dissimilarity, difference, or access to (vs. isolation from) another group.

In addition to in-group and out-group conceptual centering, some research employs a colorblind approach to examining ethnic-racial concentrations on development among youth that are heterogeneous on ethnic and racial identifications. For example, it has been common to examine the effects of Black or Latinx concentrations on youth development in samples that are racially and ethnically diverse (e.g., Juvonen et al., 2018). These approaches can be complicit with colorblindness (Neville et al., 2013), in that they can deny racial differences by emphasizing sameness and/or deny racism by emphasizing equal developmental costs/benefits across groups. The approaches also build on the assumption that the concentration of one group (or grouping) means the same thing to all groups. Because we are attempting to encourage researchers to think more critically about culturally and contextually informed approaches to the study of ethnicity and race in settings, we consider these approaches to be outside of the purview of the ethnic-racial compositions meta-construct.

Assessment of Ethnic-Racial Compositions

Ethnic-racial compositions—including both concentration and index operationalizations—can be assessed objectively or subjectively, and the best choice likely depends on the research question. Most objective assessments derive from secondary data or researcher observation (e.g., census, school records) whereas most subjective assessments derive from asking participants directly about ethnic-racial compositions in settings. For example, Stevenson et al. (2005) asked Black adolescents "How many different cultures are there in your neighborhood?" with response choices including *none*, *a few*, or *many* (p. 279). This is a subjective ethnic-racial index because it expects participants to combine information about multiple ethnic-racial groups to produce a single score that takes most or all groups into account. Derlan and Umaña-Taylor (2015) asked Black adolescents "Thinking about your neighborhood, what percentage of the people in your neighborhood do you think are [your specific ethnic group]?" with response choices ranging from *very few [less than 20%],* to *a lot [75–100%]* (p. 3). This is a subjective assessment of the concentration of the within-group population. Others have focused on norms and behaviors, for example, by asking US Mexican parents to indicate the degree to which people in their neighborhoods appreciated Mexican culture, shared Mexican customs, or got involved in activities that support the Mexican American community (Nair et al., 2013; Pasco et al., 2021a). This is a subjective assessment of the concentration of within-group neighborhood norms and behaviors. An out-group concentration sample might consider how the proportion of white teachers in schools affect BIPOC youth's well-being (Goldsmith, 2004). Using principals' reports of the racial composition of teachers and students, categories were created that classified school types and linked those school types to outcomes. Pasco and White et al. (2020) used researcher observations to assess the prevalence of Latinx signs and symbols (e.g., Mexican flags, Mexican murals, Spanish language signage, Latinx businesses) in US Mexican adolescents' neighborhoods. This was an objective assessment of the concentration of within-group signs and symbols that relied on researcher observation. Akani (2016) coded the proportion of images of girls that were Black, White, Latina, Asian, and ambiguous in popular non-fiction books on pubertal development and related those proportions to the proportion of the population of puberty-aged girls sharing the same ethnic-racial identification (e.g., proportion of Asian girl images in the books related to the proportion of US population that is an Asian and female). This was an objective assessment of within-group concentration of ethnic-racial signs and symbols.

In addition, ethnic-racial compositions—including both concentration and index operationalizations—can be assessed categorically rather than continuously. In some cases, for example, researchers take a single ethnic-racial concentration variable (e.g., % Latinx) and simply create ordinal categories like low, moderate, or high Latinx concentration (e.g., Sladek et al., 2020). These types of ordinal categories map easily onto a concentration operationalization and can easily be distinguished relative to in-group or out-group centering. In other cases, however, researchers have operationalized nominal setting categories using information on most or all groups in the setting. For example, Goldsmith (2004) created school categories: white schools, mixed schools, minority schools. Because together, the categories take all or most groups into account, operationalizations like these reflect a multi-group indexed approach.

Finally, within the framework provided by the ethnic-racial compositions meta-construct, it is important to acknowledge that ethnicity, race, and culture are three of several additional salient social position variables (e.g., class, religion, gender, sexuality, ability, documentation status) that, together, situate individuals (Crenshaw, 1989) and settings (Roy, 2018) at diverse locations of power and privilege on a multidimensional space created by numerous systems of inequality. The multiple axes of inequality that create these locations include racism, monoracism

(Johnston-Gurerrero et al., 2020), nativism, xenophobia, classism, religious intolerance, sexism, ableism, and heterosexism. The systems work to marginalize people of color, multiracial people, and immigrants; those who speak languages other than English, are poor, non-Christian (in the United States); and those who identify as lesbian, gay, bisexual, transgender, and/or queer (LGBTQ). Simultaneously, the systems privilege and empower whites, US borns, English speakers, the wealthy, Christians, cis-gender men, and heterosexual people. Incorporating an intersectional lens (Collins, 1989; Crenshaw, 1989), we recognize that individuals and settings exist simultaneously across multiple axes of inequality and, therefore, experience marginalization and/or privilege in multiple, overlapping, and intersecting ways. Thus, although this chapter has a strong focus on conceptualizing and measuring ethnicity and race in settings and recognizes that white people and spaces are privileged over BIPOC people and spaces, we acknowledge and occasionally highlight opportunities to incorporate an intersectional lens. In the sections that follow, we focus on neighborhoods, schools, and peers as key developmental settings and review literature that can be reimagined and reinterpreted using our proposed ethnic-racial compositions meta-construct to capture a more nuanced understanding of the complex interactions between ethnicity and race, people, and places.

Neighborhoods as Fertile Ground: Ethnic-Racial Compositions and Outcomes

Neighborhoods are an important developmental setting for children, youth, and families. Mostly grounded in the Chicago School, the extant neighborhood literature base has followed social disorganization theory (Shaw & McKay, 1942) that suggests that structural, sociodemographic characteristics related to poverty negatively impact youth development. Central to this chapter is social disorganization theory's focus on ethnic-racial composition, namely ethnic-racial heterogeneity, as a debilitating factor for social organization, cohesion, trust, and shared norms and values. More specifically, it is theorized that ethnic-racial heterogeneity erodes the relational fabric of neighborhoods, because residents bring differing cultural experiences, beliefs, and worldviews to the setting. Here, the assumption is that heterogeneity or diversity is troublesome and inhibiting, whereas, homogeneity or ethnic-racial concentration, particularly within-group concentration, is beneficial and promoting. Despite these assumptions, the scholarship supporting this perspective is limited and mixed. Using the ethnic-racial compositions meta-construct, we aim to summarize the literature using this perspective to further explicate how the framing of ethnic-racial composition—population concentration compared to population diversity—has important implications for interpretation of research findings.

As social disorganization theory squarely focuses on ethnic-racial heterogeneity, conversely, collective socialization perspectives (Jencks & Mayer, 1990) such as collective efficacy (Sampson et al., 1997) inherently focuses on similarity or concentration as a promotive factor for youth development. Implicitly, the assumption is that neighborhood settings will be more cohesive with greater social capital and positive relational experiences when residents share complementary views, particularly as it relates to acceptable behavior. This social fabric is more likely when there is population, in-group concentration. However, some scholars have shown a non-linear association between group concentration and collective efficacy (i.e., cohesion and trust; informal social control) (Browning et al., 2016). Therefore, more clarity is needed in how ethnic-racial compositions impact important neighborhood-level social processes.

As noted in the previous section of the paper, there are two major ways to approach the meta-construct of ethnic-racial compositions, including the group-specific approach (i.e., ethnic-racial concentration) and the multi-group approach (i.e., ethnic-racial indexes). These two approaches can be measured either objectively and subjectively and can be within-group centered or out-group centered. In the following sections, we provide an overview of the place-based literature's assessments of ethnic-racial composition and then present some empirical studies to illustrate how different methodological approaches were utilized to measure ethnic-racial compositions in neighborhoods. We discuss the studies to highlight how different assessments fit with our meta-construct, what they were measuring conceptually, and their empirical findings for child and youth development. Doing so, we hope to encourage researchers to be more conscientious on the choice of measurement and what they are measuring in different contexts. This review is not intended to be exhaustive but to provide examples of methods of the meta-construct and their associations with child and youth outcomes. Please see White et al. (2020) for a more exhaustive review on the associations between ethnic-racial compositions and various aspects of youth outcomes across different ethnic-racial groups.

To date, by far most place-based research has focused on residential neighborhood settings, often operationalized as an administratively defined static geographic space (i.e., census tract) and by extension the ethnic-racial composition of this important developmental setting. However, the residential neighborhood does not capture all the areas experienced that impact behavior. Therefore, in developmental research, more recent investigations have begun to explore how the daily locations that individuals, youth, and families experience (i.e., activity space) contribute to youth development. Often, youth spend a considerable amount of time outside of and navigate beyond their residential neighborhood to meet their needs (Mason & Korpela, 2009; Witherspoon & Hughes, 2014). In these different spaces, youth may encounter vastly different or strikingly similar ethnic-racial compositions. Similar to neighborhood research, most activity space studies depend upon social disorganization theory but focus almost equally on ethnic-racial heterogeneity (i.e., diversity) and ethnic-racial homogeneity (i.e., in-group and out-group concentration). The extension from activity space methodology is the ability to investigate fluctuations in ethnic-racial compositions across key settings within and across time (e.g., day, weekday vs. weekend, week). Below, we selectively synthesize the extant neighborhood and activity space literature to highlight how ethnic-racial compositions have been investigated in studies of youth development. When possible, we highlight important and robust trends related to the various definitions of ethnic-racial compositions.

Group-Specific Ethnic-Racial Concentration

In the neighborhood literature, the bulk of the research uses ethnic-racial concentration (i.e., the percentage of a group in the neighborhood) to assess ethnic-racial compositions in neighborhoods and their association with various child and youth outcomes (e.g., mental well-being, academic outcomes, and culture-related processes, stressors, and parenting strategies (i.e., ethnic-racial socialization). According to social disorganization theory (Shaw & McKay, 1942), a greater within-group concentration can serve as a promoting factor for youth development. However, the findings have not been consistent and vary by study outcomes and the sample characteristics.

Most of the existing studies that measure ethnic-racial concentration used an objective in-group centered concentration measure. In these studies, the percentage of the neighborhood residents of a particular group was drawn from an official record (e.g., census data) and its association with youth development was studied in the same group. Overall, using the objective in-group centered measure of ethnic-racial concentration, studies found that greater in-group concentration was promotive for youth development, especially for socioemotional well-being and positive feelings about their ethnic-racial group with some exceptions (e.g., Lara-Cinisomo et al., 2013; Tam & Freisthler, 2015). For instance, Hurd and colleagues (2013) found that Black concentration predicted Black youth's better mental well-being via increased social support and social cohesion. Studies also showed that, among Latinx youth, an increase in Latinx concentration was associated with youth's higher levels of ethnic-racial identity (e.g., Feinauer & Whiting, 2012; White et al., 2018b), lower levels of externalizing (Basáñez et al., 2013; Martinez & Polo, 2018) and internalizing symptoms (White et al., 2018b), less peer discrimination (White et al., 2018a), and more prosocial behaviors (White et al., 2018b), either directly or indirectly. Studies also show that within-group concentration can serve as a protective factor in the associations between other risk factors and youth mental well-being (Basáñez et al., 2013; White et al., 2012, 2013). For example, Latinx concentration attenuated the negative impact of discrimination on youth mental health among Latinx youth (Basáñez et al., 2013). However, researchers must be careful when choosing a specific in-group concentration measurement and consider who the in-group is as there are different levels of specificity regarding who is defined as the in-group (i.e., specific ethnic groups, pan-ethnic groups, or even more broadly BIPOC). For instance, in a pan-Latinx sample including both immigrant and non-immigrant youth, researchers (Lee & Liechty, 2015) operationalized the in-group concentration as Latinx immigrant density which was the percent of Hispanic-origin foreign-born immigrants who did not speak English well or not at all. They found that in-group concentration (i.e., Latinx immigrant density) was only associated with lower levels of depressive symptoms for *immigrant* youth but not for *non-immigrant* youth, which suggests that the level of in-group specificity matters. These findings, in whole, support the theoretical assumptions that a greater proportion of co-ethnic neighbors can positively affect child and youth development given the strong social ties and shared norms that are more easily formed among neighbors from similar backgrounds.

Although studies have suggested that ethnic-racial concentration is beneficial for youth mental health and ethnic-racial identity, when it comes to academic outcomes, ethnic-racial concentration seems to serve as a risk factor. Bennett (2011) found that, compared to Black youth living in lower Black concentrated neighborhood (with lower than 25% Black residents), living in neighborhoods with greater Black concentration predicted lower reading scores, especially for Black male youth. Similar findings were observed with Chinese American youth in which in-group centered Chinese American concentration predicted lower academic achievement (Wei et al., 2021). These findings may suggest that ethnic-racial concentration can be a proxy for neighborhood risk factors (e.g., low SES, physical disorders, and limited institutional resources). Due to reasons such as limited economic opportunities, ethnic minority groups may choose to live in a certain area, which creates economically and socially segregated ethnic enclaves that may not have access to high-quality schools and teachers. Also, with stressors and risk factors prevalent in these segregated neighborhoods, parents may not have the energy, time, and resources to support youth achievement. These realities and consequences of social stratification contribute to the segregation paradox.

Other studies have also used out-group centered ethnic-racial concentration to assess how youth's access or exposure to neighbors who are not of their group matter for their development. For example, in a sample of Black emerging adults, researchers found that out-group centered ethnic-racial concentration (i.e., the percent of white residents) was positively related to perceived discrimination and white concentration also worsened the relationship between discrimination and physiological stress responses measured by cortisol functioning (Lee et al., 2018).

While most of the studies reviewed used objective measures of ethnic-racial concentration, some studies have incorporated subjective measures of ethnic-racial concentration and found that the subjective measure has additional predictive value. For example, Juang and Alvarez (2011) measured both objective and subjective in-group centered Chinese concentration and found that only subjective Chinese concentration was related to discrimination. In another study among Chinese Americans, researchers found that subjective Chinese concentration was positively related to ethnic-racial identity (Juang & Nguyen, 2010). Conversely, Black concentration (subjective and objective) was not directly related to Black youth's ethnic-racial identity (Derlan & Umaña-Taylor, 2015; Rucker et al., 2014; Seaton & Carter, 2018).

Multi-Group Ethnic-Racial Indexes

Another frequent way to study neighborhood ethnic-racial compositions is to measure it with an objective ethnic-racial index that considers multiple groups. A commonly used index is Simpson's Diversity Index, which takes into account the number and proportion of all the ethnic-racial groups in the neighborhood. Different from ethnic-racial concentration, based on social disorganization theory, ethnic-racial diversity decreases the likelihood that neighbors from different ethnic backgrounds form strong social ties and develop shared norms and thus may serve as a risk factor for child and youth development.

Empirical studies have yielded mixed findings. Among Black youth, neighborhood diversity was unrelated to expectations of racially discriminatory treatment (Witherspoon et al., 2016) and unrelated to depression (Isom Scott & Seal, 2019). For Asian American youth, neighborhood diversity was differentially associated with youth achievement depending on whether youth have other risk factors. More specifically, among US Hmong adolescents, higher levels of neighborhood diversity were related to lower grades among those who had other financial, cultural, or educational risks, but were associated with higher grades for their lower-risk counterparts (Lee et al., 2017). For Latinx youth, especially those who lived in non-traditional immigrant destinations, neighborhood diversity was linked to Latinx parents' greater endorsement of the importance of ethnic-racial socialization and was indirectly associated with Latinx youth's ERI (Witherspoon et al., 2022).

Whereas most studies either used group-specific concentration or a multi-group index, studies have started to employ both assessments to measure different aspects of neighborhood ethnic-racial compositions. For example, Wang et al. (2020) measured both the out-group centered ethnic-racial concentration (i.e., numeric marginalization--percent of neighbors or students at school that are not child's ethnic-racial group) and neighborhood diversity to explore how diverse neighborhoods and schools matter for parent's cultural socialization and young child's social competence. They found that, compared to numeric marginalization, diversity had more predictive value for minority parents' cultural socialization. Their findings also suggested that the associations between cultural socialization and child social competence was only significant for young children who lived in neighborhoods that were high in diversity and they were the numerically marginalized group. Another study conducted by Witherspoon et al. (2022) employed both a subjective measure of the out-group centered norms and behaviors (mean perceived discrimination in neighborhood) and objective neighborhood diversity. They found that only the subjective mean discrimination experience was positively associated with Latinx and Black parents' cultural socialization behavior but not the objective neighborhood diversity measure.

In this section we showed how neighborhood research and place-based literature has explored ethnic and racial composition using both objective and subjective accounts, diversity, and concentration to demonstrate the impact of racialized spaces on youth development. In the next section, we explore how schools' and peers' ethnic-racial compositions may impact a variety of youth outcomes in multifaceted ways.

Schools and Peers as Fruitful and Productive Ground: Ethnic-Racial Concentration and Outcomes

Similar to neighborhoods, racial and ethnic minoritized youth disproportionally attend high poverty and more segregated schools (Francies & Kelley, 2021). Since the hallmark ruling of Brown vs. the Board of Education in 1954, scholars, policy-makers, and school officials have participated in a decades long ongoing scholarly and legal debate about the role of ethnic-racial composition in the educational and psychosocial lives of US students (Rothstein, 2004). Despite these longstanding debates, many questions remain about the impact of ethnic-racial composition on student outcomes and have resulted in continued policy shifts focused on integration, re-segregation, and resource allocation (Francies & Kelley, 2021; Diem et al., 2022; Williams & Hamm, 2018). Theoretical literature on schools, paralleling some neighborhood literature, provides several competing perspectives from which scholars theorize, operationalize, and examine the impact of the ethnic-racial compositions meta-construct. First, given social stratification in the United States, the United States promotes investments in some schools due to whiteness that result in human, material, and social resources. Also, there are contentions that predominantly white, lower ethnic-racial minority, and lower-poverty schools bestow advantages onto student populations by way of access to resources such as high-quality teachers, higher teacher retention, and access to newer curricula, technology, and materials. We will refer to this as the *access to structures of privilege hypothesis*. In contrast, work situated within cultural developmental frameworks (García Coll et al., 1996) and belonging approaches (e.g., person-/stage-environment fit, Caplan & Van Harrison, 1993; Eccles et al., 1993) theorize that minoritized youth experience increased sense of belonging when situated in environments in which they encounter youth of similar backgrounds. In this body of work, scholars suggest that settings characterized by high levels of co-ethnics or, ethnic/racially consonant settings, can more easily promote cohesion, connection, and well-being, whereas a portion of extant literature refers to this as the segregation paradox (Ackert, 2018), we take a more strengths-based approach and refer to this as the *cultural resilience/racial consonance hypothesis*. Finally, there remains a third perspective. Situated within cognitive (Piaget, 2000) and intercultural contact (Allport, 1954) theoretical models, other scholars contend that higher levels of heterogeneity (diversity) in setting-level ethnicity and race promotes well-being and fosters intergroup relations (Allport, 1954; Pettigrew, 2008) as this diversity provides youth with opportunities to engage in sustained, meaningful contact with ethnic-racial "others" which can build empathy and reduce bias and racist attitudes (Tropp et al., 2022). We refer to this as the *multiculturalism coalition building/integration* hypothesis. Importantly much of this work has been situated within the context of white youth (e.g., how does diversity affect the views of white youth; White et al. (2009)). Researchers interested in the impact of diversity on minoritized youth, however, suggest that a more equitable representation of multiple ethnic groups within the school context can help to balance power and privilege within a school thereby shifting group dynamics and decreasing vulnerability (Graham et al., 2014). Below we highlight how varied operationalizations of the meta-construct (e.g., group specific or concentration vs. multi-group or index) combined with particular types of centering (in-group/out-group), and objective vs. subjective measurement have led to the creation and sustained use of these competing narratives.

It is also important to note that within this section, we will also highlight the work with peers, as much of the peer literature is based within school settings. Peers play a vital role in youth's development, especially as they enter their adolescence (Benner & Wang, 2017). *Peers* is used as a broad descriptor to describe other youth with whom the target child or adolescent shares space (e.g., classroom or school generally) (Carter et al., 2017; Douglass et al., 2017). Peers with whom youth share an existing relationship or closeness with are further defined as *friends* (Douglass et al., 2017). When examining ethnic-racial composition of peers and friends, the literature is situated within the ethnic homophily (Bellmore et al., 2007), propinquity, and/or intergroup contact theory (Allport, 1954). Homophily is often used in friendship literature to discuss why youth form or have same-ethnic and -race friends. This theory suggests that youth will form friendships with those who are like them so youth will more likely have friends who share their ethnic-racial background (Aboud & Mendelson, 1996). This theory lends itself to use of group-specific, within-group centered concentration. Propinquity posits that friendships are formed with those who are within proximity—physical or functional (Echols & Graham, 2013; Mouw & Entwisle, 2006). This theory is associated with our meta construct of the ethnic-racial compositions as it relates to exposure within particular settings; settings, proximal, or distal to the focal child, may differ dramatically in ethnic-racial compositions. These differences in key peer settings may shape the peer groups and friendships youth form. For example, in racially and ethnically diverse schools, adolescents are presumed to have more cross-ethnic or -race friendships due to the availability of youth outside of one's ethnic-racial in-group (Graham et al., 2014; Knifsend & Juvonen, 2014). Intergroup contact theory posits that contact with individuals from an out-group will lessen bias and prejudice and improve intergroup attitudes (Allport, 1954), which provides evidence for how exposure to out-group concentration impacts youth. From an intergroup contact perspective, the ethnic-racial compositions could be ethnic-racial concentration or a multi-group index with out-group centering. Each of these vantage points of ethnic-racial concentration may offer different hypotheses about impact on child and youth outcomes.

Since most literature is situated within the school context, this section will focus on empirical examples of methodologies examining ethnic-racial composition of peers and friends and their fit with this chapter's meta-construct of ethnic-racial composition. Moreover, one study can utilize multiple approaches listed in this chapter to examine ethnic-racial compositions. For example, Benner and Wang (2017) utilized both a subjective ethnic-racial index and within-group centered concentration to understand the role of cross-ethnic friendships and friend's experiences of discrimination in the relationship between racial-ethnic discrimination and well-being for adolescents. The authors focus on the impact of cross-ethnic racial friendship automatically categorizes this as a multi-group approach (Benner &

Wang, 2017). Additionally, this study how same-ethnic friendships impacted the link between adolescents' reports of discrimination and well-being. This study subjectively assessed both an index and concentration measure of ethnic-racial compositions.

School and Peer Ethnic-Racial Structuring: Populations

Group Specific Ethnic-Racial Concentration Constructs Like neighborhoods researchers, scholars interested in school settings and peer groups have often employed the use of group specific in-group and out-group centered ethnic-racial concentration variables and evaluated their role in BIPOC youth's academic and socioemotional well-being. In the peer literature, exploring *subjective within-group centered concentration* allows for the possibility that youth may not be exposed to or engage with all peers in their schools/immediate environments. For example, Graham and Morales-Chicas (2015) examined how perception of same-ethnic representation in participants math class and perception of ethnic climate influenced participants' attitude towards their 9th grade math class and found that the perception of same-ethnic peers played a significant role toward feelings of belonging and competence in class when ethnic climate was perceived to be low or average (Graham & Morales-Chicas, 2015). In the friendship literature, within-group concentration for Black or African American adolescents is positively associated with school belonging and ethnic-racial identity and negatively associated with externalizing behaviors (Carter et al., 2017; Derlan & Umaña-Taylor, 2015; Douglass et al., 2017; Graham et al., 2014). Investigating the relationship between pubertal timing, same-ethnicity friends, and psychosocial outcomes with Black girls, Carter and colleagues (2017) found that the number of same-race peers was positively associated with high school connectedness and negatively associated with externalizing behaviors for early-developing girls. For late-developing girls, having fewer same-race friends was associated with a stronger sense of school belonging (Carter et al., 2017). Additionally, subjective within-group centered concentration is also positively associated with ethnic-racial identity constructs such as affirmation-belonging and centrality (Derlan & Umaña-Taylor, 2015; Douglass et al., 2017). Further, within-group concentration can serve as protective factor against discrimination for adolescents. In a subsample of youth that indicated having same-ethnic friendships, Benner and Wang (2017) found that same-ethnic friends' discrimination (as reported by the friend) moderated the relation between the target adolescent's reports of teacher-perpetrated discrimination and school engagement such that when friends report less teacher discrimination, students' experiences with discrimination were related to lower school engagement (Benner & Wang, 2017). This relation did not exist for student's whose friends reported higher levels of discrimination (Benner & Wang, 2017). These findings suggest that among adolescents, homophily (in experiences and ethnicity-race) of peers have important implications for the negative impact of bias on academics; having same-ethnic friends with similar discrimination experiences may "normalize" the experience or situate the problematic behavior in the actor (e.g., teacher in this study) resulting in less negative impact on academic behaviors.

In contrast, in a study of a heterogenous sample of Latinx elementary school students, Lee and Klugman (2013) found that Latinx concentration, *based on archival records*, had a positive effect on math test scores only for children of immigrant Latinx parents. However, using the same *objective* measure, researchers showed that Latinx concentration was negatively related to general knowledge test scores only for Latinx children of US born parents. These findings follow tenets of the integrative model (García Coll et al., 1996) and suggests that setting-level ethnic-racial compositions constructs may have differential impacts on various academic outcomes and that other individual cultural variables may modify these associations. These same researchers (Lee & Klugman, 2013) also attempted an out-group centered (% Black) approach and found that Black concentration did not impact Latinx students' math or general knowledge scores but did affect reading scores. Further, the negative effect on reading scores was only found for Latinx children of US born parents. Additionally, these researchers used an out-group centered approach (% Latinx) for Black students and found that Latinx student concentration did not have a significant effect on math, reading, or general knowledge test results for Black students.[2] As a whole these results support the *cultural resilience/racial consonance* hypothesis demonstrating that school settings with high levels of in-group centered concentration become transformed into culturally supportive and safe spaces allowing youth opportunities for connections and community that in turn enhances their socio-emotional and academic well-being. Additionally, these results suggest that the in-group versus out-group centering of the meta-construct is critically important to understand impact on developmental outcomes. Further, these studies show that the findings may be further qualified by in-group variability components (e.g., nativity) or other measures of ethnic-racial compositions.

Another study used a group specific objective (% Latinx concentration) operationalization of the meta-construct to create a typology of schools (e.g., high, moderate or low Latinx enrollment) and investigated whether school type moderated associations between bicultural stress and subjective sleep problems among Latinx youth (Sladek et al., 2020). This study found that the relation between stress and sleep problems and sleep duration was moderated by school type such that, for students in high Latinx enrollment schools increases in average stress were related to increases in sleep problems and increases in daily bicultural stress which led to decreases in sleep duration. However, it is important to note that average and daily stress reports were unrelated to sleep problems or sleep duration for Latinx students in average or low Latinx enrollment schools. At face value, this study's findings are complicated and seem to offer support for the *multiculturalism coalition building/integration* hypothesis.

[2] It is important to note that within this study, out-group centering with Latinx concentration did not matter for majority, white and minoritized, Black students.

Looking at the peer context within school, the *objective within-group centered concentration approach* is concerned with how exposure to same-ethnicity and -race peers and friends affect the outcomes of youth's development. Additionally, this approach utilizes an objective measure such as an official record (e.g., Department of Education data on ethnic-racial breakdown of schools) or self-report of ethnic-racial background. Benner and Graham (2007) use this approach to understand how ethnic congruence from middle school to high school affects adolescents' school outcomes (i.e., school climate, school worries, and Grade Point Average (GPA)). The study utilized school-level ethnicity/race data from the California Department of Education to construct the student's ethnic congruence from middle school to high school to understand its effect on youth's school outcome. Ethnic incongruence, or decline in same-ethnic peers, across the transition from middle school to high school was related to a decline in feelings of belonging for African American youth, whereas ethnic congruence across the transition was related to fewer reports of academic worries from middle school to high school for all the adolescents in the sample (Benner & Graham, 2007). This study shows the importance and protective role of consistency of exposure to in-group concentration across a developmental and ecological transition. Douglass and colleagues (2017) offer another example that utilizes an objective within-group centered concentration measure to understand its effects on ethnic identity (i.e., centrality) over time. Using Department of Education data, a value was created for each participant that represented the proportion of same-ethnic/race peers, and the authors reported that although having a high proportion same-ethnic/race peers in schools was related to higher reports of centrality initially, but over time, a higher proportion of ethnic/race peers was associated with a decline in centrality. Taken together with Benner and Graham (2007), these studies show that within-group centered concentration plays an important role during periods of transition for adolescents.

Within the peer literature, when using an *objective out-group centered concentration,* the main goal is to understand how exposure or involvement with peers or friends outside of one's ethnic or racial group affects youth's outcomes. Rastogi and Juvonen (2019) is a great exemplar of using this approach in understanding the relation between cross-ethnic friendships (i.e., number and stability) and intergroup attitudes with self-identifying Black and Latinx youth. Cross-race friendships were coded when nominated friends were part of the racial-ethnic out-group (as self-identified by the nominated friend). The authors found that differences in the number of Black-Latinx friendships reported over the course of their middle school years and that the presence of at least one stable cross-ethnic friendship was related to positive attitudes toward the target out-group (Rastogi & Juvonen, 2019). This supports intergroup contact theory such that exposure to out-group through friendship may mitigate prejudice and bias.

However, other school studies have found that the distinction between in-group and out-group centering did not matter. For example, in a multi-ethnic sample of high school students, researchers found that objective measures of ethnic-racial concentration that were out-group centered for some members of the sample and in-group centered for others (i.e., % white), exerted a similar effect on Latinx,

Black, and white youth's engagement such that greater white concentration was associated with lowered reports of liking school and higher levels of course engagement for all students (Ackert, 2018). Essentially in this example for Latinx and Black youth the operationalization of ethnic-racial concentration (% white) represents an out-group centered approach; whereas, for white youth the operationalization of ethnic-racial concentration (% white) is in-group centered. Such findings provide support for two of the theoretical hypotheses; results linking white concentration to course engagement provide support for the *access to privileged structures* hypothesis whereas findings for liking school support the *cultural resilience* hypothesis in which lack of access to co-ethnic students seems to hinder social ties and inhibit overall connectedness.

Yet this research as described above has been further complicated as researchers have begun to note that relations between group specific measures and outcomes may be qualified by race. Using a group specific concentration measure, some studies indicate that relations between group specific concentration measures and academic outcomes remain consistent across racial/ethnic groups (Bankston III & Caldas, 1998; Lee & Klugman, 2013), whereas other studies have indicated that Black students, compared with white students, experience more negative impacts on their academics when they attended segregated (e.g., 75% or more minority) schools (Bankston III & Caldas, 1998; Gurin et al., 2002; Kainz & Pan, 2014), Benner and Crosnoe (2011) found that white students, but not minority students, performed better academically with more equal representation of racial groups within the school. In contrast, Clayton (2011) found that compared with Black and Latinx students, white students were more negatively affected academically with more equal representation. These two latter examples could also be considered multi-group ethnic-racial indices with out-group centering.

Multi-Group Ethnic-Racial Indices Similar to the neighborhood literature, examinations of the meta-construct in schools have also utilized multi-group ethnic-racial indices. The vast majority of this work is situated in examining the *multiculturalism coalition building/integration* hypothesis; yet the research here is far from equivocal. For example across a series of studies among multi-ethnic middle school students, researchers using multi-group objective ethnic-racial indices (e.g., school diversity/classroom diversity) found that greater school ethnic-racial diversity was related to a range of psychosocial outcomes, including increased perceptions of school safety, decreased perceptions of victimization and loneliness (Graham, 2018; Juvonen et al., 2006, 2018), and linked to more positive perceptions of teachers' treatment and lower desire to associate with only in-group members (Juvonen et al., 2018). These findings provide support for the *multiculturalism coalition building/ integration*. Yet, other work has shown opposing effects linking greater school and teacher ethnic-racial diversity to greater school suspensions and conflict (Eitle & Eitle, 2004; Pitts, 2005). In a study of Black adolescents, Seaton and Yip (2009) linked an objectively measured school ethnic-racial diversity (e.g., diversity index score) to increased perceptions of discrimination and found that collective/institutional discrimination negatively impacted self-esteem and was strongest for students in high diversity (combined neighborhood and school diversity) settings. These findings challenge the *multiculturalism coalition building/integration* hypothesis. However, partial support for this hypothesis is found such that the negative impact of discrimination on life satisfaction was only present in low diversity settings. These findings provide competing evidence challenging oversimplified explanations of how diversity and multi-group representation impact youth development. We discuss these findings and ideas in a more detailed fashion in our peer section.

Similar to the broader school literature, work that focuses on the peer environment shows similar patterns. The peer literature utilizing *objective ethnic-racial indices* often incorporate data from official records (e.g., school-level ethnicity-race data from Department of Education) to derive a percentage or proportion of cross-ethnic and -race peers (Benner & Graham, 2007; Knifsend & Juvonen, 2014). Additionally, some utilize self-report of ethnicity/race to determine the existence of cross-ethnic and -race friendships, particularly in studies that focus on reciprocal nominations (Kelleghan et al., 2019). Students in ethnically/racially diverse classrooms reported greater sense of safety, fewer reports of victimization and lower sense of loneliness (Graham et al., 2014). Additionally, cross-ethnic friendships were related to greater displays of empathy, greater sense of safety, less perceived victimization, decreased sense of loneliness, and greater levels of inclusion and leadership (Graham et al., 2014; Jugert et al., 2013; Kawabata & Crick, 2008; Munniksma & Juvonen, 2012). As it relates to stability and quality of friendships, cross-ethnic friendships yield mixed results such that some studies indicate that cross-ethnic friendships are similar or better than those of same-ethnic and -race friendships in some quality indicators (Aboud et al., 2003; Graham et al., 2014). Just as ethnic-racial indices can be measured objectively, they can also be measured *subjectively* through individuals' perceptions. However, it is important to note that we found very few examples of this in the extant peer literature. A few notable exceptions that considered multi-group samples and participants' reports of their friends' ethnicities/races showed that cross ethnic/race friendships played a protective role against discrimination (Benner & Wang, 2017) and impacted the stability and quality of friendships (Lessard et al., 2019).

In addition to examining school ethnic-racial diversity with respect to students, some scholars examine school ethnic-racial diversity with respect to teachers. A longitudinal study (across a two-year period in high school) of Latinx adolescents used objective measures of student and teacher diversity (e.g., diversity score) to evaluate the unique impact of diversity in the student body and diversity in teaching staff on Latinx youth experiences of discrimination (Benner & Graham, 2011). This study found that as student ethnic-racial diversity increased student reports of discrimination increased, following the *cultural resilience* hypothesis (Benner & Graham, 2011). In contrast, as teacher ethnic-racial diversity increased, students were less likely to report discrimination, supporting the *multiculturalism coalition building/integration*. Here again, findings provide competing support for school ethnic-racial compositions hypotheses. Another important and alternative interpretation of these findings is that increasing diversity among those with power (i.e., teachers) can be promotive for youth, but the diversity in the student body, among peers, may promulgate racism, stereotypes, and white supremacy.

Other work has taken a slightly different operationalization to the multi-group ethnic racial index, using a categorical, school level variable/to describe the setting (e.g., majority white/majority Black/no majority). Oftentimes, scholars utilizing this approach are interested in understanding the ways in which the meta-construct may moderate or qualify relations between risk factors and developmental outcomes. For example, using a multi-ethnic sample of eighth graders, Goldsmith (2004) examined the effect of attending separate white (i.e., predominantly majority white students), mixed schools (i.e., predominantly white teachers), and separate-minority schools[3] (i.e., predominantly BIPOC students and teachers) on a students' educational beliefs (including occupational expectations, educational aspirations, and attitudes about teaching, math and science, and English and history). It is important to note here that this is one of the few studies in which the underlying concentration of students *and* teachers were used to create the discrete categories of types of schools. Results indicated that for Black and Latinx students being in separate minority schools (e.g., less than 50% white students *and* teachers) and to a lesser degree mixed schools (e.g., less than % white students but predominantly white teachers) was related to higher overall educational beliefs as compared to Black and Latinx students in majority white schools (Goldsmith, 2004). In contrast school type only influenced white students' educational aspirations suggesting that white students who attended separate minority schools had increased odds of aspiring to go to college than white students who attended separate white schools (Goldsmith, 2004). Together these findings offer support for two hypotheses. For Black and Latinx students, results coalesce in support of the *cultural resilience/racial consonance* hypothesis, whereas for white students, results provide support for *multiculturalism coalition building/integration* hypothesis. Finally, some studies have evaluated diversity in classrooms rather than school-wide settings. Using a multi-group racial index (e.g., diversity score), Benner and Yan (2015) found that the impact of classroom diversity on parental involvement was qualified by a within-group centered ethnic racial concentration variable (e.g., proportion of same race/same ethnicity peers). Researchers found that classroom diversity had a positive impact on parental involvement but only when there was a high-proportion within-group concentration. These findings offer support for both the *multiculturalism coalition building/integration* and the *cultural resilience/racial consonance hypothesis*. This study is also particularly interesting in that it used both a concentration and diversity index together and evaluated the impact of the racial and ethnic

[3] In this study, the authors were interested in the categorizing schools based on racial-ethnic concentration of white students and teachers. Principals' reports of the racial-ethnic identification of students and teachers were used to determine proportions of each group. Separate-white schools are those with a majority (more than half) of the students in the school are white. Mixed schools include were those with half or less of the students were white whereas the majority (more than half) of the teachers were white. In separate-minority schools, half or less of the students were white and half or less of the teachers were white.

structuring of a school on parental behavior. A note of the subjective perceptions of ethnic-racial concentrations and multi-group indices at the school level. Whereas the neighborhood and peer literature (see next section) also utilize subjective (e.g., perception based) measures, the use of subjective perceptions for school level racial ethnic constructs is rare.

School Ethnic-Racial Structuring: Norms, Signs, and Symbols

Whereas much of the literature for schools has devoted attention to evaluating the racial and ethnic structuring of the populations found within these settings, less evaluated the ethnic -racial structuring of other salient facets of schools such as signs and symbols (e.g., curricula). A limited body of work has focused on textual analysis of social studies curricula and demonstrates the utility of understanding the impact of racial structuring of the instruments of students. For example, in a qualitative study of Mexican-American students, Almarza (2001) found students attached little significance and reported decreased motivation for the study of American history due to its devaluing their culture and people and associated their teacher with this same act of devaluation because they perpetuated a hegemonic curriculum as opposed to incorporating culturally diverse counter-narratives. Similar to these findings several researchers found that dominant perspectives in addition to teachers' lack of counter-discourse contributed to negative perceptions of US history held by Korean immigrant middle and high school students (An et al., 2014; Choi et al., 2011), African American (Epstein, 2000), and Latinx students (Almarza, 2001; Salas et al., 2013). Studies of multi-ethnic minority youth have also indicated that youths' ethnic, linguistic, and cultural identities influence the way they learn and subsequently perceive social studies, mainly finding history class to be an oppressive experience (Harris & Reynolds, 2014). It is important to note that much of this literature has focused specifically on connection between student's exposure to representation in social studies curricula and their interest and perception of the subject of social studies, and we found no work that has linked objective or subjective ratings of curriculum representation (e.g., group specific concentrations or multi-racial indices) to broader youth academic or socio-emotional outcomes. Yet, a fast-growing (and largely qualitative) research literature in education focused on classroom pedagogy and the importance of "culturally relevant pedagogy" has identified representative curricula as a potential mechanism to increase the academic well-being and enhance connection and motivation among marginalized youth (e.g., Ladson-Billings, 1995). For example, recent work evaluating the impact of a Mexican-American studies program in high schools linked participation in this curriculum to increased probability of graduation and passing rates of standardized exam (Cabrera et al., 2014). Taken together, this work lays the foundation for our contention that though underutilized, objective and subjective assessments of the racial/ethnic structuring of salient school signs and symbols in school settings (e.g., curricula) represent an important avenue for school-based research.

Limitations and Recommendations

Through this chapter, we have proposed a new way to think about the interaction between race/ethnicity and settings suggesting a new meta-construct, *ethnic-racial compositions*. This meta-construct is meant to be flexible, increasing the potential for use and inclusion in research. We defined *ethnic-racial compositions* as a meta-construct that captures setting variability in the racial, ethnic, or cultural makeup of (a) populations within settings, (b) group norms and behaviors within settings, and (c) sociocultural signs and symbols within settings. We provided examples of how this new construct can be operationalized in extant research. From our review of literature focused on neighborhood and school effects on youth development, we found examples of ethnic-racial compositions of populations within samples using a concentration or index approach with objective or subjective measurement. However, we found relatively fewer examples of group norms and behaviors within settings or sociocultural signs and symbols within settings. We urge scholars to embrace these types of representations of ethnic-racial composition in neighborhoods and schools to advance our understanding of how race and ethnicity create meaning within settings. For example, given the pervasive and growing nature of discrimination in neighborhoods and schools following the height of the global pandemic and the senseless murders of BIPOC individuals from authority figures or disproportionate suspensions/expulsions of Black and Brown children in schools, researchers can explore collective experiences of discrimination across neighborhoods or within classrooms, tracks, or programs at schools to determine how these shared experiences impact well-being. Also, there were very few studies that examined norms, signs, and symbols as cultural manifestations of ethnic-racial compositions. Following PVEST (Spencer, 1995), the meaning making associated with these identities within settings is facilitated and experienced through these cultural artifacts. With the rise of participatory methods, inclusion of youth voice, and use of methods like Photovoice (see Chap. 9 of this book), diversity developmental science scholars can capitalize on these approaches and methods to advance our understanding of how these elements communicate meaningful messages to children and youth.

As the world and United States is becoming more diverse, it is imperative that we bring an intersectionality lens to this work that includes a critical examination of power and privilege in these settings. Scholars can go beyond concentration and diversity indexes to grapple with minoritization and marginalization as extensions and complements to our ethnic-racial compositions construct. Combining subjective accounts of concentration with objective accounts of numerical marginalization may offer deep insight into how neighborhoods' and schools' ethnic-racial representation shape the lives of diverse children, youth and families. Further, we must consider the multiple aspects of youth's identification—ethnic-racial, gender, socioeconomic status, immigrant status, sexual orientation—as these identities contribute to the meaning making that goes into their understanding, beliefs, and behaviors. Also, stratification along these social positions is rampant within the United States and across the world, creating multiplicative systems of oppression—xenophobia, classism, heterosexism, and sexism. Diversity developmental science can bear fruit in this domain and provide practical, translational, and transformative work to improve the lives of diverse, marginalized, children and youth.

Last, it is important to acknowledge and address the representation of specific ethnic-racial groups in this work and how we create ethnic-racial categories. We must address the underrepresentation of US Asian concentrations and Indigenous concentrations in work on setting-level ethnicity and race as well as multiracial or multiethnic populations (see Chap. 6 of this volume). As Asian and multiracial populations are growing rapidly, and among those with the greatest increase in the United States, it is imperative that our work is representative of these children and youth's experiences. Scholars should now begin to understand the unique and overlapping experiences of these groups within settings that may have unbalanced social (Seidman, 1990) and racial (Hughes & Watford, 2021) regularities. Also, how we create ethnic-racial categories will impact how *ethnic-racial compositions* measurement works to predict children and youth's outcomes. For example, there is not always the same level of specificity at individual and setting levels. If we combine all children and youth of color, we devise a pan-ethnic-racial category; often we do not consider how the pan-ethnic or pan-racial grouping impacts the outcomes of that pan-ethnic or pan-racial group (e.g., effect of BIPOC concentration on Black youth). Further, do we consider such a measure group-centered concentration on the within-group or another operationalization. What is lost if we do that? On the other hand, if we consider the percent BIPOC youth on Black youth, then this becomes an index that takes multiple groups into account. As we move to more specificity and nuance, the lines may become less clear; however, the important message is to do the exercise of thinking about what type of setting-level ethnic-racial composition variable is most important for the developmental phenomenon under consideration. Utilizing such an appropriate can only advance our diversity developmental science.

References

Aboud, F. E., & Mendelson, M. J. (1996). Determinants of friendship selection and quality: Developmental perspectives. In W. M. Bukowski, A. F. Newcomb, & W. W. Hartup (Eds.), *The company they keep: Friendship in childhood and adolescence* (pp. 87–112). Cambridge University Press.

Aboud, F. E., Mendelson, M. J., & Purdy, K. T. (2003). Cross-race peer relations and friendship quality. *International Journal of Behavioral Development, 27*(2), 165–173. https://doi.org/10.1080/01650250244000164

Ackert, E. (2018). Segregation paradox? School racial/ethnic and socioeconomic composition and racial/ethnic differences in engagement. *Social Science Research, 70*, 144–162. https://doi.org/10.1016/j.ssresearch.2017.10.010

Akani, B. C. (2016). *Portrayals of Black girls in books on puberty for girls* (Unpublished honors thesis). University of Michigan.

Allport, G. W. (1954). *The nature of prejudice*. Addison-Wesley.

Almarza, D. J. (2001). Contexts shaping minority language students' perceptions of American history. *Journal of Social Studies Research, 25*(2), 4–22.

An, S., Kim, J. M., & Delacruz, S. (2014). U.S. history through the eyes of South Korean youth. *The Social Studies, 105*(4), 165–174. https://doi.org/jfsv

Bankston, C. L., III, & Caldas, S. J. (1998). Family structure, schoolmates, and racial inequalities in school achievement. *Journal of Marriage and the Family*, 715–723.

Basáñez, T., Unger, J. B., Soto, D., Crano, W., & Baezconde-Garbanati, L. (2013). Perceived discrimination as a risk factor for depressive symptoms and substance use among Hispanic adolescents in Los Angeles. *Ethnicity & Health, 18*(3), 244–261.

Bellmore, A. D., Nishina, A., Witkow, M. R., Graham, S., & Juvonen, J. (2007). The influence of classroom ethnic composition on same- and other-ethnicity peer nominations in middle school. *Social Development, 16*(4), 720–740. https://doi.org/10.1111/j.1467-9507.2007.00404.x

Benner, A. D., & Crosnoe, R. (2011). The racial/ethnic composition of elementary schools and young children's academic and socioemotional functioning. *American Educational Research Journal, 48*(3), 621–646. https://doi.org/10.3102/0002831210384838

Benner, A. D., & Graham, S. (2007). Navigating the transition to multi-ethnic urban high schools: Changing ethnic congruence and adolescents' school-related affect. *Journal of Research on Adolescence, 17*(1), 207–220. https://doi.org/cg4jpb

Benner, A. D., & Graham, S. (2011). Latino adolescents' experiences of discrimination across the first 2 years of high school: Correlates and influences on educational outcomes. *Child Development, 82*(2), 508–519. https://doi.org/10.1111/j.1467-8624.2010.01524.x

Benner, A. D., & Graham, S. (2013). The antecedents and consequences of racial/ethnic discrimination during adolescence: Does the source of discrimination matter? *Developmental Psychology, 49*(8), 1602. https://doi.org/10.1037/a0030557

Benner, A. D., & Wang, Y. (2017). Racial/ethnic discrimination and adolescents' well-being: The role of cross-ethnic friendships and friends' experiences of discrimination. *Child Development, 88*(2), 493–504. https://doi.org/10.1111/cdev.12606

Benner, A. D., & Yan, N. (2015). Classroom race/ethnic composition, family-school connections, and the transition to school. *Applied Developmental Science, 19*(3), 127–138. https://doi.org/10.1080/10888691.2014.983028

Bennett, P. R. (2011). The relationship between neighborhood racial concentration and verbal ability: An investigation using the institutional resources model. *Social Science Research, 40*(4), 1124–1141.

Browning, C. R., Dirlam, J., & Boettner, B. (2016). From heterogeneity to concentration: Latino immigrant neighborhoods and collective efficacy perceptions in Los Angeles and Chicago. *Social Forces, 95*(2), 779–807. https://doi.org/f9pwb7

Budescu, D. V., & Budescu, M. (2012). How to measure diversity when you must. *Psychological Methods, 17*(2), 215–227. https://doi.org/f32r4w

Cabrera, N. L., Milem, J. F., Jaquette, O., & Marx, R. W. (2014). Missing the (student achievement) forest for all the (political) trees: Empiricism and the Mexican American studies controversy in Tucson. *American Educational Research Journal, 51*(6), 1084–1118. https://doi.org/chxc

Calzada, E. J., Roche, K. M., White, R. M. B., Partovi, R., & Little, T. D. (2020). Family strengths and Latinx youth externalizing behavior: Modifying impacts of an adverse immigration environment. *Journal of Latinx Psychology, 8*, 332–248. https://doi.org/frm2

Caplan, R. D., & Van Harrison, R. (1993). Person-environment fit theory: Some history, recent developments, and future directions. *Journal of Social Issues, 49*(4), 253–275.

Carter, R., Leath, S., Butler-Barnes, S. T., Bryd, C. M., Chavous, T. M., Caldwell, C. H., & Jackson, J. S. (2017). Comparing associations between perceived puberty, same-race friends and same-race peers, and psychosocial outcomes among African American and Caribbean Black girls. *Journal of Black Psychology, 43*(8), 836–862. https://doi.org/10.1177/0095798417711024

Choi, Y., Lim, J. H., & An, S. (2011). Marginalized students' uneasy learning: Korean immigrant students' experiences of learning social studies. *Social Studies Research and Practice, 6*(3), 1–17. https://doi.org/10.1108/SSRP-03-2011-B0001

Clayton, J. K. (2011). Changing diversity in U.S. schools: The impact on elementary student performance and achievement. *Education and Urban Society, 43*(6), 671–695. https://doi.org/10.1177/0013124510380909

Collins, P. H. (1989). The social construction of black feminist thought. *Signs: Journal of Women in Culture and Society, 14*(4), 745–773.

Crenshaw, K. (1989). Demarginalizing the intersection of race and sex: A black feminist critique of antidiscrimination doctrine, feminist theory and antiracist politics. *University of Chicago Legal Forum*, 139–167.

de Brey, C., Musu, L., McFarland, J., Wilkinson-Flicker, S., Diliberti, M., Zhang, A., Branstetter, C., & Wang, X. (2019). *Status and trends in the education of racial and ethnic groups 2018* (NCES 2019-038). U.S. Department of Education. Retrieved from https://nces.ed.gov/pubsearch/

Derlan, C. L., & Umaña-Taylor, A. J. (2015). Contextual predictors of African American adolescents' ethnic-racial identity affirmation-belonging and resistance to peer pressure. *Journal of Adolescence, 41*, 1–6. https://doi.org/f7fk9b

Diem, S., Good, M., Smotherson, B., Walters, S. W., & Bonney, V. N. A. (2022). Language and power dynamics: A critical policy analysis of racial and choice discourses in school integration policies. *Education Policy Analysis Archives, 30*(12), 10.14507/epaa.30.6995.

Douglass, S., Mirpuri, S., & Yip, T. (2017). Considering friends within the context of peers in school for the development of ethnic/racial identity. *Journal of Youth and Adolescence, 46*(2), 300–316. https://doi.org/10.1007/s10964-016-0532-0

Eccles, J. S., Midgley, C., Wigfield, A., Buchanan, C. M., Reuman, D., Flanagan, C., & Mac Iver, D. (1993). Development during adolescence: The impact of stage-environment fit on young adolescents' experiences in schools and in families. *American Psychologist, 48*(2), 90–101. https://doi.org/dk6x49

Echols, L., & Graham, S. (2013). Birds of a different feather: How do cross-ethnic friends flock together? *Merrill-Palmer Quarterly, 59*(4), 461–488. https://doi.org/gmx5w6

Eitle, T. M., & Eitle, D. J. (2004). Inequality, segregation, and the overrepresentation of African Americans in school suspensions. *Sociological Perspectives, 47*(3), 269–287. https://doi.org/10.1525/sop.2004.47.3.269

Epstein, T. (2000). Adolescents' perspectives on racial diversity in U.S. history: Case studies from an urban classroom. *American Educational Research Journal, 37*(1), 185–214. https://doi.org/10.3102/00028312037001185

Feinauer, E., & Whiting, E. F. (2012). Examining the sociolinguistic context in schools and neighborhoods of pre-adolescent Latino students: Implications for ethnic identity. *Journal of Language, Identity & Education, 11*(1), 52–74.

Francies, C., & Kelley, B. (2021). *State policy options to promote k-12 integration*. Education Commission of the States. https://files.eric.ed.gov/fulltext/ED613900.pdf

García Coll, C., Lamberty, G., Jenkins, R., McAdoo, H. P., Crnic, K., Wasik, B. H., & Vázquez García, H. (1996). An integrative model for the study of developmental competencies in minority children. *Child Development, 67*, 1891–1914. https://doi.org/10.2307/1131600

Goldsmith, P. A. (2004). Schools' racial mix, students' optimism, and the Black-White and Latino-White achievement gaps. *Sociology of Education, 77*(2), 121–147. https://doi.org/10.1177/003804070407700202

Graham, S. (2018). Race/ethnicity and social adjustment of adolescents: How (not if) school diversity matters. *Educational Psychologist, 53*(2), 64–77. https://doi.org/10.1080/00461520.2018.1428805

Graham, S., & Morales-Chicas, J. (2015). The ethnic context and attitudes toward 9th grade math. *International Journal of Educational Psychology, 4*(1), 1–32.

Graham, S., Munniksma, A., & Juvonen, J. (2014). Psychosocial benefits of cross-ethnic friendships in urban middle schools. *Child Development, 85*(2), 469–483. https://doi.org/10.1111/cdev.12159

Gurin, P., Dey, E., Hurtado, S., & Gurin, G. (2002). Diversity and higher education: Theory and impact on educational outcomes. *Harvard Educational Review, 72*(3), 330–367.

Harris, R., & Reynolds, R. (2014). The history curriculum and its personal connection to students from minority ethnic backgrounds. *Journal of Curriculum Studies, 46*(4), 464–486. https://doi.org/jfsw

Hong, S., Zhang, W., & Walton, E. (2014). Neighborhoods and mental health: Exploring ethnic density, poverty, and social cohesion among Asian Americans and Latinos. *Social Science & Medicine, 111*, 117–124. https://doi.org/10.1016/j.socscimed.2014.04.014

Hughes, D. L., & Watford, J. A. (2021). Racial regularities: Setting-level dynamics as a source of ethnic-racial socialization. *American Journal of Community Psychology, 70*, 3–17. https://doi.org/10.1002/ajcp.12565

Hurd, N. M., Sellers, R. M., Cogburn, C. D., Butler-Barnes, S. T., & Zimmerman, M. A. (2013). Racial identity and depressive symptoms among Black emerging adults: The moderating effects of neighborhood racial composition. *Developmental Psychology, 49*(5), 938–950. https://doi.org/10.1037/a0028826

Isom Scott, D. A., & Seal, Z. T. (2019). Disentangling the roles of negative emotions and racial identity in the theory of African American offending. *American Journal of Criminal Justice, 44*(2), 277–308. https://doi.org/10.1007/s12103-018-9453-7

Jackson, A., Browning, C., Krivo, L., Kwan, M. P., & Washington, H. M. (2016). The role of immigrant concentration within and beyond residential neighborhoods in adolescent alcohol use. *Journal of Youth and Adolescence, 45*, 17–34. https://doi.org/f76nhm

Jencks, C., & Mayer, S. (1990). The social consequences of growing up in a poor neighborhood. In L. E. Lynn & M. F. H. McGeary (Eds.), *Inner-city poverty in the United States* (pp. 111–186). National Academy Press.

Johnston-Gurerrero, M. P., Tran, V., & Combs, L. (2020). Multiracial identities and monoracism: Examining the influence of oppression. *Journal of College Student Development, 61*(1), 18–33. https://doi.org/ftk6

Juang, L. P., & Alvarez, A. N. (2011). Family, school, and neighborhood: Links to Chinese American adolescent perceptions of racial/ethnic discrimination. *Asian American Journal of Psychology, 2*(1), 1–12. https://doi.org/10.1037/a0023107

Juang, L. P., & Nguyen, H. H. (2010). Ethnic identity among Chinese-American youth: The role of family obligation and community factors on ethnic engagement, clarity, and pride. *Identity: An International Journal of Theory and Research, 10*(1), 20–38.

Jugert, P., Noack, P., & Rutland, A. (2013). Children's cross-ethnic friendships: Why are they less stable than same-ethnic friendships? *European Journal of Developmental Psychology, 10*(6), 649–662. https://doi.org/10.1080/17405629.2012.734136

Juvonen, J., Nishina, A., & Graham, S. (2006). Ethnic diversity and perceptions of safety in urban middle schools. *Psychological Science, 17*(5), 393–400. https://doi.org/10.1111/j.1467-9280.2006.01718.x

Juvonen, J., Kogachi, K., & Graham, S. (2018). When and how do students benefit from ethnic diversity in middle school? *Child Development, 89*(4), 1268–1282. https://doi.org/10.1111/cdev.12834

Kainz, K., & Pan, Y. (2014). Segregated school effects on first grade reading gains: Using propensity score matching to disentangle effects for African-American, Latino, and European-American students. *Early Childhood Research Quarterly, 29*(4), 531–537.

Kawabata, Y., & Crick, N. R. (2008). The role of cross-racial/ethnic friendships in social adjustment. *Developmental Psychology, 44*(4), 1177–1183. https://doi.org/10.1037/0012-1649.44.4.1177

Kelleghan, A., Mali, L., Malamut, S., Badaly, D., Duong, M., & Schwartz, D. (2019). Cross-ethnic friendships, intergroup attitudes, intragroup social costs, and depressive symptoms among Asian American and Latino American youth. *Journal of Youth and Adolescence, 48*(11), 2165–2178. https://doi.org/10.1007/s10964-019-01143-7

Knifsend, C. A., & Juvonen, J. (2014). Social identity complexity, cross-ethnic friendships, and intergroup attitudes in urban middle schools. *Child Development, 85*(2), 709–721. https://doi.org/10.1111/cdev.12157

Ladson-Billings, G. (1995). Toward a theory of culturally relevant pedagogy. *American Educational Research Journal, 32*(3), 465–491. https://doi.org/dx3gcj

Lara-Cinisomo, S., Xue, Y., & Brooks-Gunn, J. (2013). Latino youth's internalising behaviours: Links to immigrant status and neighbourhood characteristics. *Ethnicity & Health, 18*(3), 315–335. https://doi.org/10.1080/13557858.2012.734278

Lee, J. C., & Klugman, J. (2013). Latino school concentration and academic performance among Latino children. *Social Science Quarterly, 94*, 977–1015. https://doi.org/10.1111/j.1540-6237.2012.00935.x

Lee, M. J., & Liechty, J. M. (2015). Longitudinal associations between immigrant ethnic density, neighborhood processes, and Latino immigrant youth depression. *Journal of Immigrant and Minority Health, 17*(4), 983–991. https://doi.org/10.1007/s10903-014-0029-4

Lee, M., Lam, B. O. Y., & Madyun, N. I. (2017). Effects of different-race exposure in school and neighborhood on the reading achievement of Hmong students in the United States. *Urban education, 52*(10), 1255–1283. https://doi.org/10.1016/j.ecresq.2014.06.005

Lee, D. B., Eisman, A. B., Stoddard, S. A., Peckins, M. K., Goldstick, J. E., Hsieh, H. F., et al. (2018). Racial discrimination and cortisol in African American emerging adults: The role of neighborhood racial composition. *Cultural Diversity and Ethnic Minority Psychology, 24*(4), 521.

Lessard, L. M., Kogachi, K., & Juvonen, J. (2019). Quality and stability of cross-ethnic friendships: Effects of classroom diversity and out-of-school contact. *Journal of Youth and Adolescence, 48*(3), 554–566. https://doi.org/10.1007/s10964-018-0964-9

Maereg, T. M., & Witherspoon, D. P. (2022). *The interactive relationship between ethnic-racial identity, academic expectations, and neighborhood ethnic-racial concentration: Variation by gender and ethnicity-race* (Unpublished masters thesis). Pennsylvania State University.

Martinez, W., & Polo, A. (2018). Neighborhood context, family cultural values, and Latinx youth externalizing problems. *Journal of Youth and Adolescence, 47*, 2440–2452. https://doi.org/gd8d3v

Mason, M. J., & Korpela, K. (2009). Activity spaces and urban adolescent substance use and emotional health. *Journal of Adolescence, 32*(4), 925–939. https://doi.org/dzd4c6

Massey, D. S., & Denton, N. A. (1988). The dimensions of residential segregation. *Social Forces, 67*, 281–315.

Meer, T. V. D., & Tolsma, J. (2014). Ethnic diversity and its effects on social cohesion. *Annual Review of Sociology, 40*, 459–478. https://doi.org/10.1146/annurev-soc-071913-043309

Mouw, T., & Entwisle, B. (2006). Residential segregation and interracial friendship in schools. *American Journal of Sociology, 112*(2), 394–441. https://doi.org/10.1086/506415

Munniksma, A., & Juvonen, J. (2012). Cross-ethnic friendships and sense of social-emotional safety in a multiethnic middle school: An exploratory study. *Merrill-Palmer Quarterly, 58*(4), 489–506. https://doi.org/10.1353/mpq.2012.0023

Nair, R. L., White, R. M. B., Roosa, M. W., & Zeiders, K. H. (2013). Cultural stressors and mental health symptoms among Mexican Americans: A prospective study examining the impact of family and neighborhood context. *Journal of Youth and Adolescence, 42*, 1611–1623. https://doi.org/ftk7

Neville, H. A., Awad, G. H., Brooks, J. E., Flores, M. P., & Bluemel, J. (2013). Color-blind racial ideology: Theory, training, and measurement implications in psychology. *American Psychologist, 68*(6), 455. https://doi.org/10.1037/a0033282

Pasco, M. C., & White, R. M. B. (2020). A mixed methods approach to examining Mexican-origin adolescents' use of ethnic-racial labels in neighborhood contexts. *Journal of Adolescent Research, 35*, 489–520. https://doi.org/ftk8

Pasco, M. C., White, R. M. B., Iida, M., & Seaton, E. K. (2021a). A prospective examination of neighborhood social and cultural cohesion and parenting processes on ethnic-racial identity among U.S. Mexican adolescents. *Developmental Psychology, 57*(5), 783–795. https://doi.org/10.1037/dev0001170

Pasco, M. C., White, R. M. B., & Seaton, E. K. (2021b). A systematic review of neighborhood ethnic–racial compositions on cultural developmental processes and experiences in adolescence. *Adolescent Research Review, 6*, 229–246. https://doi.org/jfxz

Pettigrew, T. F. (2008). Future directions for intergroup contact theory and research. *International Journal of Intercultural Relations, 32*(3), 187–199. https://doi.org/bnv9w8

Phinney, J. S., Madden, T., & Santos, L. J. (1998). Psychological variables as predictors of perceived ethnic discrimination among minority and immigrant adolescents. *Journal of Applied Social Psychology, 28*(11), 937–953. https://doi.org/fd9cs5

Piaget, J. (2000). Piaget's theory of cognitive development. *Childhood Cognitive Development: The Essential Readings, 2*, 33–47.

Pitts, D. W. (2005). Diversity, representation, and performance: Evidence about race and ethnicity in public organizations. *Journal of Public Administration Research and Theory, 15*(4), 615–631.

Rastogi, R., & Juvonen, J. (2019). Interminority friendships and intergroup attitudes across middle school: Quantity and stability of Black-Latino ties. *Journal of Youth & Adolescence, 48*, 1619–1630. https://doi.org/10.1007/s10964-019-01044-9

Rivas-Drake, D., & Witherspoon, D. (2013). Racial identity from adolescence to young adulthood: Does prior neighborhood experience matter? *Child Development, 84*, 1918–1932. https://doi.org/f5gv9f

Rothstein, R. (2004). *Class and schools: Using social, economic, and educational reform to close the Black-White achievement gap*. Teachers College Press.

Roy, A. L. (2018). Intersectional ecologies: Positioning intersectionality in settings-level research. *New Directions for Child and Adolescent Development, 2018*(161), 57–74. https://doi.org/ftk9

Rucker, J. M., Neblett, E. W., Jr., & Anyiwo, N. (2014). Racial identity, perpetrator race, racial composition of primary community, and mood responses to discrimination. *Journal of Black Psychology, 40*(6), 539–562. https://doi.org/10.1177/0095798413499371

Russell, D. W., Clavél, R. D., Cutrona, C. E., Abraham, W. T., & Burzette, R. G. (2018). Neighborhood racial discrimination and the development of major depression. *Journal of Abnormal Psychology, 127*, 150–159. https://doi.org/gc63nv

Salas, S., Jones, J. P., Perez, T., Fitchett, P. G., & Kissau, S. (2013). Habla con ellos—Talk to them: Latinas/os, achievement, and the middle grades. *Middle School Journal, 45*, 18–23. https://doi.org/10.1080/00940771.2013.11461878

Sampson, R. J., Raudenbush, S. W., & Earls, F. (1997). Neighborhoods and violent crime: A multilevel study of collective efficacy. *Science, 277*(5328), 918–924.

Seaton, E. K., & Carter, R. (2018). Pubertal timing, racial identity, neighborhood, and school context among Black adolescent females. *Cultural Diversity and Ethnic Minority Psychology, 24*(1), 40–50. https://doi.org/10.1037/cdp0000162

Seaton, E. K., & Yip, T. (2009). School and neighborhood contexts, perceptions of racial discrimination, and psychological well-being among African American adolescents. *Journal of Youth and Adolescence, 38*, 153–163. https://doi.org/10.1007/s10964-008-9356-x

Seidman, E. (1990). Pursuing the meaning and utility of social regularities for community psychology. In P. Tolan, C. Keys, F. Chertok, & L. A. Jason (Eds.), *Researching community psychology: Issues of theory and methods* (pp. 91–100). American Psychological Association. https://doi.org/10.1037/10073-008

Shaw, C. R., & McKay, H. D. (1942). *Juvenile delinquency and urban areas*. University of Chicago Press.

Sladek, M. R., Doane, L. D., & Park, H. (2020). Latino adolescents' daily bicultural stress and sleep: Gender and school context moderation. *Health Psychology, 39*(3), 179–189. http://dx.doi.org.ezproxy1.lib.asu.edu/10.1037/hea0000824

Spencer, M. B. (1995). Old issues and new theorizing about African American youth: A phenomenological variant of ecological systems theory. In R. L. Taylor (Ed.), *African-American youth: Their social and economic status in the United States* (pp. 37–69). Praeger.

Stevenson, H. C., McNeil, J. D., Herrero-Taylor, T., & Davis, G. Y. (2005). Influence of perceived neighborhood diversity and racism experience on the racial socialization of black youth. *Journal of Black Psychology, 31*(3), 273–290. https://doi.org/c9ztvz

Tam, C., & Freisthler, B. (2015). An exploratory analysis of linguistic acculturation, neighborhood, and risk behaviors among children of Southeast Asian immigrants. *Child and Adolescent Social Work Journal, 32*(4), 383–393. https://doi.org/ggk5jf

Tropp, L. R., White, F., Rucinski, C. L., & Tredoux, C. (2022). Intergroup contact and prejudice reduction: Prospects and challenges in changing youth attitudes. *Review of General Psychology, 26*(3), 342–360. https://doi.org/10.1177/10892680211046517

U.S. Census Bureau. (2018, March 15). *New Census Bureau report analyzes U.S. population projections*. Retrieved from https://www.census.gov/newsroom/releases/archives/population/cb12-243.html

Wang, Y., Benner, A. D., & Boyle, A. E. (2020). Family cultural socialization in childhood: Navigating ethnic/racial diversity and numeric marginalization in school and neighborhood settings. *Cultural Diversity and Ethnic Minority Psychology, 28*, 449. https://doi.org/10.1037/cdp0000435

Wei, W., Witherspoon, D. P., Hughes, D., & Way, N. (2021). Chinese American adolescents' academic performance and psychological adjustment: The role of neighborhood and family. *Asian American Journal of Psychology, 12*(1), 26–40. https://doi.org/10.1037/aap0000215

White, F. A., Wooten, B., Man, J., Diaz, H., Rasiah, J., Swift, E., & Wilkinson, A. (2009). Adolescent racial prejudice development: The role of friendship quality and inter-racial contact. *International Journal of Intercultural Relations, 33*(6), 524–534. https://doi.org/10.1016/j.ijintrel.2009.06.008

White, R. M., Deardorff, J., & Gonzales, N. A. (2012). Contextual amplification or attenuation of pubertal timing effects on depressive symptoms among Mexican American girls. *Journal of Adolescent Health, 50*(6), 565–571. https://doi.org/g7tm

White, R. M., Deardorff, J., Liu, Y., & Gonzales, N. A. (2013). Contextual amplification or attenuation of the impact of pubertal timing on Mexican-origin boys' mental health symptoms. *Journal of Adolescent Health, 53*(6), 692–698. https://doi.org/f5hm8z

White, R. M. B., Knight, G. P., Jensen, M., & Gonzales, N. A. (2018a). Ethnic socialization in neighborhood contexts: Implications for ethnic attitude and identity development among Mexican-origin adolescents. *Child Development, 89*(3), 1004–1021. https://doi.org/gdkgpb

White, R. M. B., Zeiders, K. H., & Safa, M. D. (2018b). Neighborhood structural characteristics and Mexican-origin adolescents' development. *Development and Psychopathology, 30*, 1679–1698. https://doi.org/d7ph

White, R. M. B., Pasco, M. C., Korous, K. M., & Causadias, J. M. (2020). A systematic review and meta-analysis of the association of neighborhood ethnic-racial concentrations and adolescent behaviour problems in the U.S. *Journal of Adolescence, 78*, 73–84. https://doi.org/ftmc

Williams, J. L., & Hamm, J. V. (2018). Peer group ethnic diversity and social competencies in youth attending rural middle schools. *The Journal of Early Adolescence, 38*(6), 795–823. https://doi.org/10.1177/0272431617699945

Williams, D. R., Neighbors, H. W., & Jackson, J. S. (2003). Racial/ethnic discrimination and health: Findings from community studies. *American Journal of Public Health, 93*, 200–208. https://doi.org/dtvcm6

Witherspoon, D. P., & Hughes, D. L. (2014). Early adolescent perceptions of neighborhood: Strengths, structural disadvantage, and relations to outcomes. *Journal of Early Adolescence, 34*(7), 866–895. https://doi.org/10.1177/0272431613510404

Witherspoon, D. P., Seaton, E. K., & Rivas-Drake, D. (2016). Neighborhood characteristics and expectations of racially discriminatory experiences among Black adolescents. *Child Development, 87*(5), 1367–1378. https://doi.org/f88h6g

Witherspoon, D. P., Smalls, C., Wei, W., & Hughes, D. (2022). Neighborhood-level predictors of African American and Latinx parents' ethnic-racial socialization. *American Journal of Community Psychology, 69*, 183–200. https://doi.org/10.1002/ajcp.12555

Chapter 9
Centering Youth Voice in Developmental Science: A Research Roadmap for Partnerships with Latinx Youth

Josefina Bañales, Bernardette J. Pinetta, Sukhmani Singh, Alfred J. Rodriguez, Adriana Aldana, and Felicia J. Gutierrez

Developmental science with youth has traditionally taken a "top-down" approach to research, wherein developmentalists—who are often White adults with PhDs—create and implement research agendas based on their own scientific and personal interests without meaningful collaboration with youth on issues and solutions they care about (Aldana et al., 2019; Ozer, 2016). By engaging in this traditional approach with youth and communities marginalized by oppression, researchers exploit, albeit unintentionally, communities in order to advance their careers, needs, and interests of academic institutions. The research approach of conducting research *on* youth marginalized by oppression contributes to distrust between researchers and youth, and resultant programs of research that do not result in systemic, anti-oppressive change, and produces research that is decontextualized of the complexity of youths' experiences and voices (see Fine, 2018; Rivas-Drake et al., 2016; Teixeira et al., 2021). Instead, developmental science may function as transformative research, or work that is guided by a desire to preserve the dignity of youth who have been historically exploited by researchers who are systems actors in institutions and actively seeks to rectify this marginalization (Ginwright & Cammarota, 2002).

In order to vision and enact a more transformative developmental science, the current chapter proposes a "research roadmap" that offers considerations and recommendations to developmentalists—defined as adult researchers in institutions

J. Bañales (✉) · A. J. Rodriguez · F. J. Gutierrez
University of Illinois, Chicago, Chicago, IL, USA
e-mail: jbanal2@uic.edu

B. J. Pinetta
University of Michigan, Ann Arbor, MI, USA

S. Singh
University of Connecticut, Hartford, CT, USA

A. Aldana
California State University, Dominguez-Hills, Carson, CA, USA

© The Author(s), under exclusive license to Springer Nature Switzerland AG 2023
D. P. Witherspoon, G. L. Stein (eds.), *Diversity and Developmental Science*,
https://doi.org/10.1007/978-3-031-23163-6_9

(e.g., higher education institutions, national and local research organizations, youth serving organizations) who study the development of social, cognitive, and biological phenomena among youth—on how Latinx youths' voices may be elevated throughout the entire research process. This roadmap is grounded in the principles of youth participatory action research (YPAR) and other critical research and methodological frameworks (Domínguez, 2021; Ginwright & Cammarota, 2002; Rivas-Drake et al., 2016; Ozer, 2016). First, we discuss assumptions and approaches in mainstream developmental science and the larger scientific enterprise, many of which are grounded in white supremacy. Second, we describe the importance of youth voice and the need to center Latinx youths' voices in developmental science to counter white supremacy cultural assumptions and practices in the field. Finally, we offer concrete recommendations on how to elevate Latinx youths' voices throughout the entire research process.

This research is grounded in the experiences of Latinx youth and communities because we all share a commitment to the advancement of Latinx youth, which is informed by our social identities as Latina/o/x or as people of color, in general. This chapter aims to inspire research that centers Latinx youths' voices in developmental science to support the positive development of youth and their communities. Additionally, this chapter aims to push developmental science forward as a field that works toward the liberation of Latinx youth and other communities marginalized by systems of oppression.

Latinx Youth and Developmental Science

Despite recent growth in research on Latinx youths' psychosocial and behavioral development, there is a continued need to understand the complexities of youths' diverse racial, ethnic, cognitive, biological, cultural, and sociopolitical experiences (Azmitia, 2021; Bañales & Rivas-Drake, 2022). The underrepresentation of Latinx youth in developmental science is concerning because the field is a socializing agent that shapes how adults in power (e.g., developmentalists, teachers, youth development practitioners, government officials) view, treat, and allocate resources to youth and their communities. Although inroads have been made to better include Latinx youth in developmental science (see Azmitia, 2021) and YPAR work (see Anyon et al., 2018), they are often considered as a monolithic group without critical attention to how their diverse social identities and experiences inform their psychological and behavioral development. Indeed, Latinx youth and their communities are diverse along multiple axes of social identity and are differentially positioned in the U.S.'s power hierarchies. Latinx youth speak English, Spanish, Spanglish, or Indigenous languages, are first generation immigrants or do not identity as immigrants at all, may be undocumented or documented in the U.S., identify as Afro-Latinx, Latinx/o/a, Asian, or White, have different skin tones, come from different ethnic backgrounds, and live in different regions and areas of the U.S. (Kiang et al., 2020; Salas Pujols, 2020; Suárez-Orozco et al., 2011; Umaña-Taylor et al., 2002).

Given this diversity, theoretical and empirical work that aims to advance the lives of Latinx youth, including YPAR, must be intentional with who is included in the work in order to engage in research that is contextually grounded and relevant to youths' lives. In doing so, such work can be better leveraged by youth and institutions to transform unequal systems that negatively affect their lived experiences.

There are deficit assumptions about Latinx youth and their communities (Rivas-Drake et al., 2016; Rodriguez & Morrobel, 2004), which are informed by the larger scientific enterprise and societal assumptions on work ethic, merit, worthiness, motivation, and intelligence (Lee et al., 2021; Ponterotto, 2005). Deficit perspectives enacted in research focus overly on what is going wrong in youths' lives (e.g., an overemphasis on negative outcomes such as antisocial behavior) as opposed to what is going right in youths' lives (e.g., outcomes related to psychological well-being and sense of purpose) (García Coll et al., 1996; Spencer, 2006). Deficit perspectives often coincide with acritical perspectives that do not consider the role structural racism and other forms of interconnected oppressions play in shaping the psychological and behavioral development of youth marginalized by systems of oppression (García Coll et al., 1996; Spencer, 2021; Singh et al., 2018). Consistent with mainstream U.S. American ideals that value individualism, constant individual work ethic, and other principles of classical liberalism that perpetuate color-evasive ideology (Bonilla-Silva, 2015; Jost & Hunyady, 2005), deficit, acritical perspectives place blame and responsibility of the causes of negative outcomes in youths' lives on youth and their communities, and not the systems of oppression that fuel negative youth outcomes. For example, deficit, acritical perspectives on Latinx youths' academic achievement may attribute youths' academic motivation (or assumed lack thereof) solely to individual factors (e.g., determination to excel in school), as opposed to policy that does not provide predominantly Latinx schools with sufficient funds to purchase accurate and updated textbooks or culturally responsive teachers who engage youth and their families in the learning process (Rios & Vigil, 2017; Valencia et al., 2020).

The American, color-evasive ideals that emphasize that individuals should have sole responsibility over their life experiences are also expressed in remarks about Latinx communities by researchers, such as, "*they* don't understand or speak English so they do not want to engage with English-speaking researchers," "they don't trust us, so they don't come into the research lab for studies," or "the Latinx community is too small to focus on for research." Indeed, low-income Latinx youth, Spanish speaking youth, recent immigrants, and undocumented youth in the U.S. can be more difficult to recruit and retain in research due to systemic barriers that gatekeep access to research (Hernández et al., 2013; Knight et al., 2009; Rivas-Drake et al., 2016). Yet, acritical perspectives contribute to false beliefs that Latinx youth and families are to blame for their lack of representation in research and fail to critique how the structure, process, and assumptions of developmental science might contribute to how communities marginalized by systems of oppression are engaged throughout the research process (Aldana & Richards-Schuster, 2021; Hughes & Seidman, 2002; Rivas-Drake et al., 2016). A developmental science that facilitates and centers Latinx youths' voices in research results in co-created research knowledge and tools that promote the collective well-being and liberation of Latinx communities.

White Supremacy in Developmental Science

Aldana and Richards-Schuster (2021) identify and critique characteristics of white supremacy culture ingrained in developmental science with youth. They argue that white supremacy manifests in the field through paternalism, quantity over quality, assumed objectivity, individualism, power hoarding, worship of the written word, and dichotomous thinking. Paternalism is the notion that people with power (e.g., developmentalists) make decisions for individuals with less power (e.g., youth, racialized youth, girls and femmes), which is a perspective and practice grounded in white colonial patriarchy (Aldana & Richards-Schuster, 2021; Okun, 2021). Paternalism in research appears when researchers conduct research without youths' authentic collaboration or develop programs without considering if youth wanted the program in the first place. Adultism—or a system of oppression grounded in negative attitudes toward young people that contributes to laws and practices that minimize or ignore the perspectives of youth, while the perspectives of adults are centered (Bettencourt, 2020) is closely related to paternalism (Aldana & Richards-Schuster, 2021). Adultism in developmental science often appears when developmentalists—who are adults—conduct research on youth development but do not meaningfully collaborate with youth in the development and enactment of research. This adult centered approach to research limits an understanding of the nature and scope of research questions that are most pressing to youths' lives and limits the development of practical and creative solutions to address complex social issues that affect youth and their communities (Teixeira et al., 2021).

Quantity over quality refers to the need to produce a large amount of published research articles rather than engaging in work that directly impacts communities' daily lives. This assumption is evinced in the expectation that developmentalists who seek promotion must "hit the ground running" in their research agendas, creating a sense of urgency to engage in fast-paced research that may result in work that is not responsive to the needs of youth and their communities (Aldana & Richards-Schuster, 2021). Smith (2012) unpacks the positivist paradigm and its epistemological insistence that science is "objective" and researchers are, by virtue of their positions and privilege, neutral and outside of the research context. The idea of assumed objectivity is closely linked to Enlightenment-based, Western notions of individualism that suggest individuals can be "distanced, or separated, from…the community" (p. 58, Smith, 2012). This belief in separation from community and the focus on the individual expert has resulted in the implementation and normalization of extractive research processes that position the assumed distant/objective researcher as the sole creator and disseminator of knowledge (Said, 1978).

As research and its dissemination about the "othered" in the form of the written word became an engine of European imperialism, colonial power relations emerged that wrongfully established/reified White/university-based researchers as powerful and more knowledgeable, rational, and superior to those who were researched (e.g., indigenous peoples). This dichotomous thinking between those who did the research (i.e., had the power to pose the questions and then author their answers) and those who were researched cemented and normalized power hoarding by Eurocentric researchers (Said, 1978). White supremacist practices and values continue to shape the scientific enterprise today (Buchanan et al., 2021).

Centering Youth Voice Through Youth Participatory Action Research: One Response to White Supremacy in Developmental Science

Research that aims to understand and promote youth voice may be used to counteract white supremacist practices and notions in developmental science. Youth voice is defined as youth sharing their lived experiences, opinions, and beliefs in different social contexts (e.g., home, school), and the translation of these perspectives into meaningful outcomes (Checkoway, 2011; Sprague Martinez et al., 2018). Youth voice, as a form of authenticity that honors individuals' lived experiences, is an act of resistance against systems of oppression (Ginwright, 2011). Latinx youth who engage in justice-oriented spaces that center their voices in social change efforts develop positive social identities, socioemotional, psychological, and academic outcomes (Cabrera et al., 2014; Cammarota & Romero, 2009; Sprague Martinez et al., 2020).

In the context of research and school-based practices that promote youth engagement, youth voice has been described as a continuum (see Mansfield, 2018; Wong et al., 2010), wherein the most superficial level includes youth being seen as data sources where researchers and other adults extract knowledge—or "data"—from youth on their lived experiences. The second level includes youth sharing their perceptions of their lives, but with limited opportunities to shape important decisions through governance and policies that affect their own and community members' outcomes. The third level involves including youth as co-researchers and collaborators with adults on important issues; however, adults primarily guide the research and practice agendas. The most impactful level of incorporating youth voice into research and practice involves adults supporting youth as co-researchers, which includes building their leadership capacity and agency, so youth may reflect on and identify social issues they care about and implement efforts that address these issues (Anyon et al., 2018; Desai, 2019).

Youth voice is often facilitated in the context of YPAR, which is a radical epistemological approach— not a research method—in which youth are trained and guided alongside adults to critically analyze the political, economic, and social forces that maintain current social hierarchies that negatively impact their communities and are supported to disrupt these issues (Ginwright & Cammarota, 2002;

Ozer, 2016). Rodríguez and Brown (2009) outline three principles that guide YPAR. First, YPAR is situated in local community contexts and is inquiry based. That is, youth engage in research that is relevant to their lives to ensure the knowledge they co-create is relevant to their lived experiences, issues they care about, and can be used to inform tangible solutions that promote healing and liberation in their communities. As youth explore and critique the history and structural causes of social issues their communities face, youth can uplift narratives about their communities' cultural assets and shift the intervention "gaze" toward oppressive systems. In this way, YPAR is a form of liberation psychology (Comas-Díaz & Torres Rivera, 2020) that advances youths' positive development, but also their healing and liberation from systems of oppression (Ginwright & Cammarota, 2002; Haskie-Mendoza et al., 2018).

Second, YPAR is participatory in that youth are collaborators in the research process, as opposed to research "subjects" or "participants," who share power in co-developing action research agendas (Ozer, 2016). Thus, YPAR aims to promote youths' voices at the highest level of the continuum of youth engagement in research and practice (Domínguez, 2021). Viewing and engaging youth as authentic collaborators in research acknowledges that they are experts of their lives. Given that youth are co-researchers in the research process, youth-adult partnerships are vital in the YPAR process (Richards-Schuster & Timmermans, 2017; Serido et al., 2011). For example, adults scaffold youths' learning by co creating opportunities that facilitate collective recognition of links among youths' experiences (Richards-Schuster & Timmermans, 2017). Adult co-researchers also bring other partial knowledge, such as methodological expertise or university IRB access. Finally, YPAR aims to dismantle oppression on individual and structural levels with youth. The research generated in collaboration with youth can and should be used as evidence to inform policy and interventions that promote youths' well-being and liberation (Aldana & Richards-Schuster, 2021; Ballonoff Suleiman et al., 2021).

Given the demands, needs, and power structures within youth development settings, there are challenges associated with YPAR (Kornbluh et al., 2016; Ozer, 2016; Rose et al., 2021, under review; Teixeira et al., 2021). For example, YPAR is not inherently anti-racist even though the approach has been commonly used to address racial issues youth of color face (Aldana & Richards-Schuster, 2021). It is necessary to recognize that critical YPAR projects, which aim to promote anti-oppression, gaze upward at systems that create oppression and not remain fixated solely on the seductive practice of "extracting" youths' experiences. Thus, equity in YPAR is not guaranteed: adult researchers must intentionally discuss and challenge power differentials with youth—a responsibility that needs to be shouldered by developmentalists (Teixeira et al., 2021). For instance, developmentalists may initiate activities that allow team members (e.g., youth, adult community partners, developmentalists) to identify the power in the room by voicing who has access to power and decision-making in certain aspects of the work (e.g., allocation of funds in grants) and whether the team can shift where power is allocated.

A Research Roadmap for Centering Latinx Youth Voice in Developmental Science

Consistent with calls that describe and encourage social justice research practices with racial and ethnic minorities and immigrant communities (Cammarota & Fine, 2008; Caraballo et al., 2017; Hernández et al., 2013; Rivas-Drake et al., 2016; Rowley & Camacho, 2015) and work that outlines how developmental science may inform YPAR (Ballonoff Suleiman et al., 2021), we offer considerations and recommendations on how developmentalists may engage in research that elevates Latinx youths' voices. Specifically, we propose a cyclical roadmap (see Fig. 9.1) that outlines strategies that elevate youth voice throughout the entire research process, which include (1) engagement in personal researcher reflexivity, (2) the development of relationships with Latinx youth and communities, (3) the development of conceptual frameworks and research questions, and measures with youth (i.e., Research Design), and (4) the communication and dissemination of research findings alongside youth (i.e., Research Communication).

Our proposed roadmap builds on previous discussions of the need to incorporate community psychology and social justice research values and practices in the developmental science research process, but are not explicitly grounded in the principles of YPAR (see Hughes & Seidman, 2002; Rivas-Drake et al., 2016). Building on these discussions, our roadmap describes how engagement in these practices are in direct response to white supremacist practices in developmental science, which

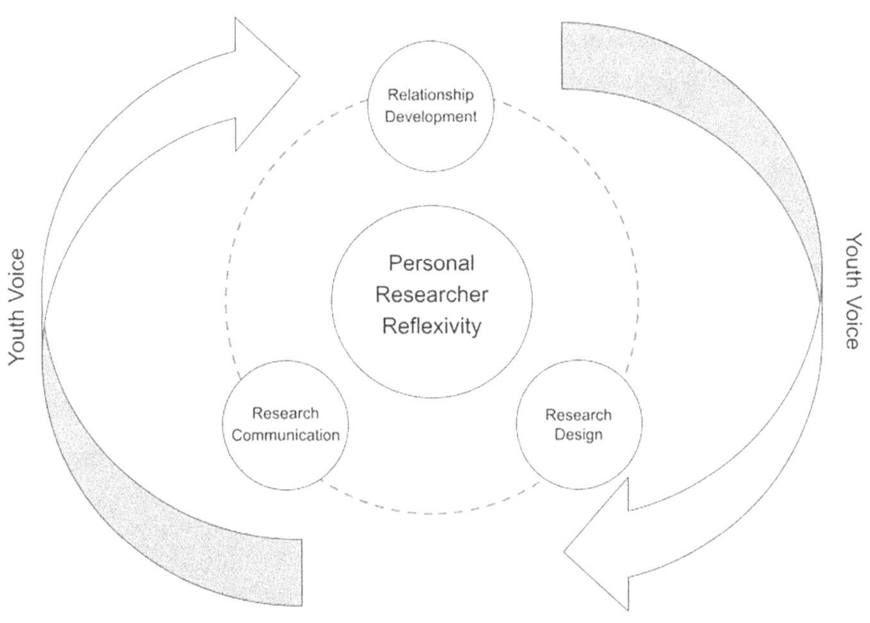

Fig. 9.1 A research roadmap for elevating Latinx youth voice in developmental science

increases the likelihood that ethical, mutually beneficial, liberatory, and healing research with Latinx youth is pursued. Involvement in these practices better allows developmental science to function as "just good developmental science" that has the potential to transform the academy (Lee et al., 2021; Rivas-Drake et al., 2016) and how adults in youth serving contexts understand and respond to the needs of Latinx youth.

Personal Researcher Reflexivity

Developmentalists must engage in continuous reflexivity around their positionality in the U.S.'s power hierarches (e.g., race- and gender-based hierarchies that privilege White people and cis-gender people), in academia, and in research with Latinx youth and their communities (Rivas-Drake et al., 2016; Hernández et al., 2013). Personal researcher reflexivity involves reflecting on and questioning one's values toward certain research questions and methodologies, one's relationships with youth, community members, and others involved in the work, the history of one's academic institution and its relation to local communities, and one's social location in the U.S.'s power hierarchies and how these positions shape the entire research process (Emirbayer & Desmond, 2012; Collins, 2002). Such reflection involves questioning and reconciling whether one's beliefs, values, and actions are congruent (Norton & Sliep, 2018; Sakamoto & Pitner, 2005). For example, one might recognize that they value working with Latinx youth and their communities in research settings, but that they do not have Latinx friends outside of work, indicating incongruence between one's values and behaviors.

The reflexivity process also involves reflecting on the socialization one received about quality and impactful research throughout one's academic training and the values one places on individualism and assumed work ethic—values that are informed by structural white supremacy (Lee et al., 2021; Okun, 2021). For example, developmentalists, similar to other academics, are in a system that values tangible outcomes in the form of written empirical articles, which is predicated on the assumption that one's individual work ethic is reflected in the number of articles published. In the context of YPAR, developmentalists are often encouraged to reflect on their positionality as adults to disrupt adultism in relationships with youth (Ballonoff Suleiman et al., 2021; Teixeira et al., 2021), but there is often minimal explicit discussion or training around how researchers engage in reflexive practices around their social positions in the U.S.'s white supremacist racial hierarchy and how their positionality shapes interactions with youth (Aldana & Richards-Schuster, 2021).

Developmentalists working with Latinx youth and their communities, in particular, may ask themselves the following questions in order to engage in reflexivity around their social connections with Latinx youth, how they view Latinx youth, their notions of Latinidad, and the liberation of all Latinx people:

1. What motivates me to do research with Latinx youth?
2. Who do I "picture" when I think of Latinx youth?
3. Do I identify as Latinx? From which ethnic group? How do I identify racially?
4. Did I grow up in a Latinx community?
5. Do I live in the same community as the youth and families with whom I work?
6. How do I explain the causes of the social issues Latinx youth face?
7. How do I conceive of sociopolitical success for Latinx youth? Are my notions rooted in solidarity, meritocracy, or respectability?
8. Who do I envision when I think of liberation of Latinx youth? Do I view the subjugation of Latinx youth as one dimensional or intersectional (i.e., queer, black, and indigenous)?

By answering and critically examining responses to the above questions, developmentalists can work to identify areas of growth within their social and research contexts. These reflexive questions may allow developmentalists to examine their own social identities and locations, as well as recognize whether they are working from outside the community or from within. Acknowledging any areas of disconnect, such as recognizing that certain social identities are not shared between developmentalists, youth, or their communities, may allow developmentalists to reflect on behavioral or structural changes that respond to these differences (e.g., having a diverse research team based on age, race, ethnicity, neighborhood residence).

Developmentalists may also engage in reflexivity around their views about whether Latinx youth should be co-creators in the research process and help determine how power will be negotiated in youth-adult relationships. Such questions individuals may ask themselves are:

1. Do I view Latinx youth as experts, or as credible sources, on their own lives?
2. Do I believe Latinx youth are capable of making systemic changes that affect their lives?
3. How do I view and respect the agency of Latinx youth? When youth offer solutions, do I respond by sharing what we can and cannot do? Or is my response more imaginative and related to how we can achieve our goals collectively?
4. How will I relinquish power or opportunities I am afforded by my PhD, racial, ethnic, or social class backgrounds while working with Latinx youth? For example, when invited to share my work with the public, do I ensure media creators interview youth so they can speak for themselves and gain coverage from the experience, depending on youths' interest in the experience or preferred method of disseminating their work?
5. How will YPAR advance anti-oppression, liberatory knowledge that does not reproduce deficit-based narratives about Latinx youth and render systems invisible?

Engaging in reflexivity around one's social identities (e.g., racial, ethnic, gender), positionality (e.g., citizenship status, body size, language use), and power as a developmentalist, and the systems that create one's experiences in society and the academy, should not be optional: this process applies to individuals who engage in quantitative, qualitative, and participatory methods (Lee et al., 2021). The documentation of reflexivity practices can be done through individual journal, audio or written memo recording, or in community with other developmentalists who engage in these practices (Merriam et al., 2001; Saldaña, 2017). The larger reflexivity practice is continuous and evolves throughout developmentalists' professional careers, as social roles and responsibilities shift. A continuous reflexive practice, however, does not need to be an arduous experience. Short, weekly reflections such as journaling, have been found to be beneficial and significantly impactful (Boden et al., 2006; Meyer & Willis, 2019). This process can elicit emotional responses from researchers, which may include anger, sadness, grief, or healing, particularly for scholars who are marginalized by systems of oppression (Lee et al., 2021). To address these emotions, developmentalists can decenter their experiences, examining them through the lens of others within the community. Such re-focusing techniques, as well as examining why emotions occur, are associated with more rewarding emotion processing experiences (Ayduk & Kross, 2010; Kross et al., 2005).

There are unique considerations for personal researcher reflexivity among developmentalists who do and do not identify as Latinx. For developmentalists who identify as Latinx and who also are from the same neighborhoods or schools as youth, relationship building with Latinx youth and communities can be exciting, rewarding, healing, and taxing (Haskie Mendoza et al., 2018). These scholars might be especially motivated to engage in youth voice practices, have insights into youths' experiences based on their own personal experiences, and may already be connecting with youth in informal ways. Because of these dynamics, the line between research and informal relationship development may be blurry, which may create relationships that feel taxing for both youth and Latinx developmentalists. Yet, Latinx developmentalists still occupy "outsider" positions while working with Latinx youth as they are adults, have advanced degrees, or may have other social background differences (see Chap. 5 in this book by Lorenzo-Blanco et al.; Rivas-Drake et al., 2016). Although there might be shared experiences between those who identify as Latinx, Latinx researchers are not the representative for all Latinx people. The constant reflection of one's intersectional Latinx identities alongside youth will better ensure that the root causes of social issues are understood and tackled from an intersectional perspective.

Developmentalists who do not identify as Latinx, but share other social identities and experiences as youth, may contribute to the elevation of youth voice but must remain critical of their positionality. For example, in working with queer Latinx youth who are undocumented, queer developmentalists of color who are or were undocumented in the U.S. may share similar experiences as youth due to being othered for their racial, sexual orientation, or documentation status, granting them some insight into youths' experiences. Yet, these developmentalists' experiences may remain different as they do not identify as Latinx or youth.

Given that most of academic research is done in teams, it is imperative that developmentalists' reflexivity extend to their own research groups. In the field of psychology, White faculty continue to be overwhelmingly represented among tenured and tenure-track faculty, making it likely that a White primary investigator (PI) will lead a team that has members who identify as people of color. Conversations surrounding power, identity, and addressing harm must become normative within the research team. Within the first few lab meetings, we suggest having explicit conversations about how power is defined and identifying how power dynamics may manifest within the space and the research process. For example, power may be discussed to determine whose perspective will be prioritized during disagreement in qualitative data analysis (e.g., will the PI have final say or will undergraduate research assistants decide?). Team members should also be integrated throughout all stages of the research process, such as in grant writing, protocol development, and interactions with the community. By integrating team members throughout the process, there is potential to cultivate new partnerships, identify research procedures that may perpetuate or reduce harm, and ensure all members are trained to engage with the community in culturally responsive ways. However, it is also important that faculty do not use their team members of color (e.g., Latinx youth, students, community collaborators) as sounding boards or representatives of their communities. Overall, developmentalists' ability to engage in reflexivity practices sets the stage for all aspects of the research process with Latinx youth, starting with the development of relationships.

Relationship Development

Consistent with perspectives on community-engaged research practices with youth and communities marginalized by systems of oppression (Haskie-Mendoza et al., 2018; Lee et al., 2021; Rivas-Drake et al., 2016; Rowley & Camacho, 2015), the development of authentic and genuine relationships with Latinx youth is a key component of the research process, which involves considering: (1) how relationships are initiated, (2) who comprises the research team, (3) where relationship building takes place, (4) and in what format and when. The formation of relationships *is* a part of the research process, as opposed to "contributing" to the process, as reciprocal and authentic relationships with youth sets the foundation for the success and impact of research in the lives of Latinx youth and communities (Rivas-Drake et al., 2016; Teixeira et al., 2021). The development of sustained and authentic relationships with Latinx youth and their communities directly challenges white supremacy cultural practices and notions in developmental science, particularly practices of paternalism, individualism, and power hoarding.

Developmentalists should be intentional about *how relationships with youth are initiated*. Oftentimes, the forces of academia push developmentalists to begin formulating visions for projects and collaborations without the insights of youth and their communities, which is a manifestation of paternalism as it implies that researchers' professional qualifications rather than youths' lived experiences should primarily guide decision-making about research processes. This sense of paternalism centers adults' agendas (e.g., disseminating publications in mainstream journals) and voices in meetings and collaborations with youth, which maintains power differences in adult-youth partnerships (Bettencourt, 2020). To counter the forces of paternalism, we encourage developmentalists to learn about work that is already being implemented by youth about their lived experiences, such as youth groups in communities and schools or youth serving community organizations that serve or are led by Latinx youth and community members. Connecting and supporting existing work led by youth is aligned with anti-racist and feminist practices that honor visible and less visible labor (Fine, 2018). Integrating into existing efforts that support Latinx youth, rather than creating new groups or programs, challenges perspectives that Latinx youth should be "given opportunities" or "invited" into research spaces—approaches that replicate power differentials between developmentalists and youth. Collaborating with youth as co-investigators better ensures that youth have actual say in the design and implementation of research studies as opposed to incorporating "youth perspectives" in research questions or design decisions that have already been made by developmentalists (Ozer, 2016; Teixeira et al., 2021). Co-creating research visions with youth recognizes the deeply symbiotic connection between youth, their communities, and developmentalists, which better ensures that power in decision making throughout the research process is shared among group members.

Developing authentic relationships with youth sets the stage for the development of co-led research agendas that are responsive to the needs of Latinx youth and communities. Building authentic partnerships with youth researchers provides a counter strategy to the dominant emphasis on individualism in research design (e.g., creating a project motivated by one's career goals/expectations or research grants). Given that youth are not restrained by the demands of academia, they are likely to be creative in the questions and solutions they propose to address social issues, and may draw on their racial, ethnic, and cultural resources in devising solutions for social change (Bañales et al., 2020; Cammarota & Romero, 2009; Pinetta et al., 2020). For instance, Latinx youth who endorse the value of familismo—a Latinx cultural value that emphasizes obligation, family support, obedience, and filial piety (Stein et al., 2014), may share research opportunities with their families and other community members in ways that resonate with members, such as in languages individuals speak or with less jargon, increasing the reach of the research. Latinx youth high in familismo may be especially creative in developing structural solutions that advance the lives of their families and community members.

As potential youth or community partners are identified, we encourage developmentalists and youth to discuss *who should comprise the research team* (e.g., other youth experts, leaders, or co-conspirators). Consistent with a team science approach (Buchanan et al., 2021; Caughy et al., Chap. 14 this book), developmentalists and youth may initiate collaborations with other academics, youth, undergraduate students, doctoral students, and community members who already engage in YPAR or other youth voice strategies with Latinx youth. These relationships allow for the sharing of knowledge on the intricacies of YPAR and resources (e.g., internal grants) that further sustain current youth efforts, as opposed to reinventing "the research wheel."

Latinx youth may consider their family members (e.g., grandparents, parents, siblings, cousins, aunts, or uncles) or other members of the community (e.g., religious figures, youth organizers, teachers) as vital to understanding the breadth and depth of social issues that impact their communities and may want to include these individuals on the research team. Honoring that Latinx youth are often embedded in family and communal systems that are grounded in cultural values of interconnection, familism, and communalism between members (Baker & Brookins, 2014; Stein et al., 2014) better ensures the rigor, reach, and relevance of the work (Ozer, 2016). Latinx and other youth identified community members have a keen sense of the most pressing issues facing their communities, scope of social issues, who the issues affect, and who should be involved in devising solutions to the issues. For example, for Latinx youth who are immigrants in emerging Latinx communities, or communities that have small but growing Latinx populations, their needs might involve increasing Spanish speaking services in schools, medical, and service centers (Booth et al., 2021). Yet, for Latinx youth who are 6th generation, who do not speak Spanish, and live in larger cities, their needs might involve accessing more culturally responsive resources that allow them to reconnect with Spanish or other languages indigenous to their communities. Communal research spaces that are intergenerational (e.g., research teams with youth, developmentalists, and other community members) and diverse along other axes of social identity better ensures that diverse lived experiences are reflected in research with youth and communities marginalized by oppression (Rivas-Drake et al., 2016; Sánchez Carmen et al., 2015), increasing the likelihood that research agendas are relevant and impactful to multiple members of youths' communities.

Given that whiteness and classism impose boundaries on community spaces (Bonds & Inwood, 2016), there should be explicit discussions between developmentalists and youth about *where relationship building and other YPAR activities take place* so youth can show up as themselves in safe and authentic ways. Latinx youth who are from low-income communities, undocumented in the U.S., or who have been involved in the juvenile (in)justice system experience surveillance by adults in schools and in the streets, which threatens their psychological and physical safety (Haskie-Mendoza et al., 2018; Yu et al., 2021). Thus, developmentalists should consider what locations or settings team meetings and the broader YPAR process occur. Depending on youths' social identities and positionality, YPAR activities may take place in public libraries, parks, restaurants, coffee shops, in houses of worship, at school, online, or as walking meetings throughout neighborhoods, if individuals are able (see Kornbluh et al., 2016; Ozer, 2016). As decisions are made on the settings where relationship building and other YPAR processes take place, it will be important for youth and developmentalists to consider whether other Latinx youth and people will be in the space, and whether other aspects of the environment (e.g., music by and for Latinx people) validate the lived experiences of Latinx youth (Yu et al., 2021). In the first author's work with Latinx youth who are primarily first- and second- generation immigrants, she and her team use WhatsApp to connect with youth during and after YPAR activities—a free phone app youth already use to connect with their family members across Latin America. For youth who are unable to use video platforms (e.g., Zoom) due to living in intergenerational homes with limited space or restraints on WiFi connections, the use of WhatsApp allows the team to keep an open line of communication with youth that is more accessible and engaging for youth.

Developmentalists should also consider *what formats and when relationship building* with Latinx youth takes place. Youth may want to speak in Spanish, English, Spanglish, an Indigenous language, or Portuguese, depending on their comfort and use of these languages, particularly in certain locations in the U.S. (Booth et al., 2021; Flores-González, 2017). For example, Latinx youth may prefer to speak Spanish but may not to do so in public spaces, given hostilities against Latinx people who speak Spanish in public or at school (Cammarota & Aguilera, 2012; Cammarota & Romero, 2009). Researchers should engage which languages Latinx youth speak, and how youth choose to use them, as they develop relationships with youth. The ability to have meetings with youth in the language and style of youths' choosing will ensure youths' voices are heard (literally). If developmentalists do not speak or use a language in ways youth prefer, we encourage researchers to develop teams that are inclusive of the languages youth speak and the social identities youth occupy, including researchers who identify as Latinx, those from similar ethnic groups, first or later generation immigrants, etc.

Finally, as developmentalists consider how to communicate with youth, there must be a consideration of *when relationship building opportunities occur*. Latinx youth from low-income backgrounds and first-generation immigrants often have limited transportation options, jobs or caregiving responsibilities after school (e.g., watching siblings), making it difficult for them to stay after school to attend activities (Ceballo et al., 2021; Haskie-Mendoza et al., 2018). Explicit discussions between developmentalists and youth about what times work for meetings, centering youths' schedules, is needed. In our experience working with Latinx youth who were primarily first- or second-generation immigrants with work and caregiving responsibilities, youth preferred to meet early Saturday or Sunday morning before work and when school was not in session.

The Development of Conceptual Frameworks, Research Questions, and Measures

Latinx youths' voice can be elevated to inform the development of conceptual frameworks, research questions, and measures that are grounded in youths' cultural, racial, ethnic lived experiences or other ways of knowing that have been marginalized by the academy (Baker & Brookins, 2014; Teixeira et al., 2021). Developmentalists may take a deductive-inductive approach, or an etic-emic approach (Fereday & Muir-Cochrane, 2006; Saldaña, 2017), to investigate whether current conceptual frameworks on Latinx youths' social experiences are relevant to their lives and to develop new frameworks and research questions that more accurately capture youths' experiences. Deductive-inductive approaches are well equipped to capture nuances in Latinx youths' complex social experiences, such as how youth define, conceive, and experience racial, ethnic, and cultural processes (Baker & Brookins, 2014; Cammarota & Romero, 2009). Centering Latinx youths' voices in the development of conceptual frameworks, research questions, and measures disrupts the white supremacist practices of assumed objectivity and dichotomous thinking.

The deductive, or "top-down" approach, involves relying on established or developing conceptual frameworks on Latinx youths' psychological and behavioral development to guide potential research questions youth and researchers may investigate together. Youth can review conceptual frameworks on Latinx youths' psychological development in written or visual form, and may be asked whether the frameworks accurately capture their lived experiences. For example, Bañales and Rivas-Drake (2022) created a theoretical model for anti-racist identity and action (MARIA) among Latinx youth differentially positioned in the U.S.'s system of white supremacy. The MARIA framework articulates that an anti-racist identity among Latinx youth is comprised of multiple psychological and behavioral components, such as youths' recognition of "linked fate" with racially marginalized communities, or the belief that Latinx youths' liberation from white supremacy is intricately connected with the oppression and the liberation of other youth marginalized by white supremacy, anti-Indigeneity, and anti-Blackness. The model is grounded on the assumption that youths' intersectional social identities and experiences inform their notions of race, ethnicity, community, and anti-racism.

To determine whether the MARIA model reflects the anti-racist identity and actions among Latinx youth within the participating community, developmentalists may ask youth whether they agree with theoretical assertions made in the model. For example, youth and developmentalists may discuss whether youths' involvement in anti-racism actions are, indeed, motivated by their sense of linked fate with other youth marginalized by systems of oppression. There may also be an explicit discussion about whether the model accurately captures their motivations for anti-racism actions based on their unique social positions in the U.S., which is especially important given the social diversity within Latinx communities (Azmitia, 2021; Bañales & Rivas-Drake, 2022). Engaging in a deductive approach on the assessment of the MARIA framework or other conceptual models on Latinx youths' psychological and behavioral development has the potential to directly address assumptions made in theoretical frameworks, such as how race, ethnicity, culture, racism, and anti-racism are defined. Discussing the assumptions of conceptual models with youth also recognizes that adult developmentalists are entirely objective in the development of models—models are informed by researchers' values, beliefs, and experiences. Although discussing and critiquing conceptual frameworks with youth is participatory, it does not fully elevate youth voice, as youth are presented with knowledge that has already been synthesized by adults (Mansfield, 2018; Wong et al., 2010). This approach is also limiting as youth might be fixated on information they are presented rather than thinking outside the "conceptual framework box" about other psychological factors important for their anti-racism identity and actions.

The inductive, or "bottom-up," approach involves youth discussing their understandings of social phenomena openly and freely (Ballonoff Suleiman et al., 2021; Ozer, 2016). Participatory research approaches, such as the sharing of one's family's immigration history, ancestral family tree, or intergroup dialogue (Aldana et al., 2016; Bañales & Rivas-Drake, 2022), may reveal the complexity of youths' understanding of complex social phenomena (Aldana et al., 2019). Methods that involve youth processing and sharing their personal and ancestral narratives in their own words can be healing for youth and communities (Haskie-Mendoza et al., 2018; Lee et al., 2021).

In the context of the MARIA framework (Bañales & Rivas-Drake, 2022), youth and developmentalists may discuss how youth conceive of race, ethnicity, racism, and community, which likely undergird anti-racist identities and actions (Bañales et al., 2021). For example, these conversations might reveal that second generation and other later generation Latinx youth do not always distinguish between race and ethnicity in defining their identity as Latinx (Flores-González, 2017), whereas Afro-Latinx youth, regardless of generation status, might be more likely than non-Afro-Latinx youth to more clearly distinguish race and ethnicity, identifying racially as Black and ethnically as Latinx (Salas Pujols, 2020). These different notions of race and ethnicity among diverse Latinx youth may inform the development of conceptual frameworks and research questions that apply to how different Latinx youth develop anti-racist identities and behaviors. Group-specific conceptual models and research questions among different Latinx youth are especially important because Latinx youth will have different roles in challenging white supremacy and other forms of oppression, depending on their unique social positions in the U.S.'s power hierarchies (see Bañales & Rivas-Drake, 2022).

For example, youth-adult teams may recognize that Afro-Latinas are excluded in conceptions of "Latinx-ness," and therefore may question how Afro-Latinas resist and heal from anti-Black racism in Latinx communities (Chavez-Dueñas et al., 2019; Salas Pujols, 2020). Given that youths' development of beliefs, feelings, and behaviors that challenge racism is informed by social context (Aldana et al., 2019), youth and adult research teams have an opportunity to rely on youths' expertise on which environments (e.g., neighborhoods, classes in school, home environments, peer networks) activate or reduce youths' commitment to disrupting and healing from white supremacy. Centering youths' expertise on which social contexts are important in their anti-racist development may reveal that, for example, Latinx youth might feel confident in their ability to advocate for the citizenship of undocumented people in the U.S. with their friends who also share the belief that no person is "illegal," but may feel less confident talking with their family members who hold deeply held stereotypes about what "good" immigrants in the U.S. look like (e.g., immigrants who comes to the U.S. the "right" way).

Youth and developmentalists who reflect on and critique assumptions about youths' beliefs, attitudes, and behaviors have an opportunity to co-create measures that are grounded in these diverse perspectives. For example, informed by youths' notions of what engaged citizenship looks like, youth and developmentalists may co-develop measures that aim to capture the civic and political engagement of Latinx communities who are undocumented in the U.S.—a community that does not have access to traditional forms of political participation (e.g., voting) due to legal constraints. Youth may share these developed measures with other youth for feedback to ensure items resonate with youth, are translated in languages youth use, and that the length and format of measures are accessible, increasing the reliability and validity of measures (Aldana et al., 2019; Peña, 2007; Knight et al., 2009).

The Communication and Dissemination of Research Findings

Centering Latinx youth voice continues in the communication and dissemination of research findings with youth. We offer three points as considerations in how Latinx youth may decide to communicate and disseminate research findings from YPAR: (1) youths' intended target audience, (2) form of communication, and (3) impact of the work. This approach moves beyond perspectives that suggest that research findings on youth development should be *shared with* youth for feedback prior to publication and dissemination. Instead, youth decide how to share information and how to act on it, which increases transparency about research between youth and developmentalists, the impact of findings in the lives of youth and communities, and disrupts hegemonic structures that limit the impact of YPAR (Singh et al., 2018). Thus, centering Latinx youths' voices in the communication and dissemination of research findings disrupts the white supremacy culture characteristics of dichotomous thinking and worship of the written word within academic research dissemination expectations.

Youth may want to share their findings with people or in contexts that are not be valued by traditional university standards. For example, when the third author conducted a YPAR project with system-involved, girls of color, including Latinas, they wanted to communicate their findings directly to incarcerated girls and police officers. Their lived experiences of danger at the hands of state-sanctioned violence, both inside and outside of carceral spaces, drove their urgency to inform these two audiences about the dangers of internalizing hegemonically authored narratives of girls pushed into juvenile legal system involvement. However, multiple structural, institutional constraints framed as "protections for vulnerable youth" (e.g., liability-focused IRB review processes at university, city and state levels) meant that ultimately youth co-researchers could not enter carceral facilities and assess girls' safety in those settings. While it is important to safeguard youth from potential mistreatment from research, it is important to critique and challenge institutional policies and practices that set narrow parameters ("one-right-way" as a form of dichotomous thinking) for research engagement that perpetuate paternalistic power dynamics that dictate who is "qualified" (e.g., adult academic researcher) to make decisions about research dissemination.

Combined with a lack of funding to conduct the research, there is inherent adultism in framing systems-impacted girls as individuals in need of protection as opposed to critical researchers who have the lived knowledge to generate questions that can hold systems of oppression accountable. In this example, institutional adherence to one right way of doing research with "human subjects" served as a strong barrier in not allowing this project to happen. However, in extending and pushing the literature on critical YPAR, the third author and colleagues have advanced structural recommendations for university-based researchers to engage in so that we engage the privilege and flexibility of our positions to create needed systems change in universities to better support anti-oppression, participatory research (Rose et al., 2021, under review).

It is also necessary that developmentalists co-navigate with Latinx youth regarding the most effective forms of communication to disseminate findings in order to maximize the impact of the work. There is an expectation that developmentalists must document findings—using a very specialized form of writing (e.g., APA style, empirical study)— to share their findings with their academic communities in order to demonstrate impact on their field. Furthermore, the expectations for scholarly impact hinges on producing as many publications as possible (i.e., quantity over quality and worship of the written word). Although writing one's research is valuable and can serve as a form of activism, we encourage developmentalists to not worship the written word and explore other forms of research dissemination youth are excited to engage. Youth may want to share their work in podcasts, art, social media, documentaries, op-eds, or in youth-led community forums, as these forms of communication may reach youth more quickly and directly (Cammarota & Romero, 2009; Kidd et al., 2018). For example, a primarily Latinx fifth grade YPAR group collected stories from community members to the prompt, "tell us about a time you felt you did or did not have the power to change something in your community" (Kohfeldt et al., 2016). To display these collective stories, youth decided to display them through a mural at their school. Findings should be shared in accessible ways that aligned with youths' values, energy, and critical racial consciousness development (Aldana et al., 2019; Lee et al., 2021). Poetry writing can also be used as a method to disseminate research findings or experiences in YPAR among Latinx youth and can be used as a form of activism as youth create and define narratives around their social identities and experiences in their own words, which may shape policy decisions around their education (Cammarota & Romero, 2009).

Summary and Conclusions

Mainstream approaches in developmental science are often grounded in white supremacist values and practices, resulting in research that does not capture and honor the voices and lived experiences of youth (Aldana & Richards-Schuster, 2021). Developmental science may function as transformative research: An approach to conducting research that centers youths' successes, love, hardships, and complexities around their lived experiences and is used to advance the thriving of youth and their communities. Such an approach to research can be achieved by elevating youths' voices in the research process.

To counteract white supremacy culture in developmental science, the current chapter proposed a "research roadmap" grounded in the principles of YPAR and other critical methodologies and frameworks (Haskie-Mendoza et al., 2018; Ozer, 2016; Rivas-Drake et al., 2016) to guide research that centers Latinx youths' voices in developmental science. Our proposed research road map highlights that research that aims to elevate Latinx youths' voices should be done throughout the entire research process, starting with developmentalists' engagement in personal researcher reflexivity, the development of authentic, sustained, and mutually beneficial relationships between developmentalists and Latinx youth, the formation of conceptual frameworks, research questions, and measures, and the communication of research findings and dissemination of knowledge to key stakeholders. This roadmap offers considerations and tangible suggestions for developmentalists of all career stages (e.g., undergraduates and graduate students, early, mid, and late career faculty) who aim to elevate Latinx youths' voices in their research. Elevating Latinx youths' voices in developmental science ensures that research is reliable, valid, meaningful to and impactful in youths' lives.

Given the complex interplay between manifestations of white supremacy in society, the larger scientific enterprise, and in developmental science (Aldana & Richards-Schuster, 2021; Buchanan et al., 2021; Okun, 2021; Lee et al., 2021), developmentalists' ability to engage in research that elevates Latinx youths' voices is largely dictated by the reward structures of academia. The structure and demands of academia, and, as a result, developmental science, do not always support YPAR and other participatory action approaches that are thoughtful, slow, and co-lead with youth and their communities. For example, early career scholars are encouraged to "hit the ground running" with their research agendas, which is an approach that runs counter to the principles of slow and thoughtful community-engaged and participatory research practices. Thus, the activities that are rewarded as impactful scholarly activity in tenure and promotion should include the extent to which developmentalists engage in community-engaged and participatory research practices, such as developing research partnerships with youth and their communities (Teixeira et al., 2021). Developmentalists who do not engage in YPAR or youth approaches that center youth voice may serve as co-conspirators (see Love, 2019) by supporting efforts to change tenure and promotion guidelines in their department or organization to factor YPAR and community-based research practices in promotion decisions. To advance the careers of early career faculty who already engage in YPAR, more senior developmentalists should consider inviting early career individuals to serve as co-principal investigators or co-investigators, not solely as consultants, to honor their expertise, support their professional development, and the development of their research team.

Including the values of YPAR in tenure and promotion should encourage all developmentalists, not only women of color and other scholars marginalized by systems of oppression who often engage in community-engaged research, to engage in community-engaged research. However, developmentalists who are advantaged by white supremacy, particularly cis-heterosexual White men, should engage in YPAR in teams that include girls, femmes, and women of color who have meaningful leadership roles. In addition to the reward structure of tenure, developmentalists who engage in YPAR should be supported through the development of grants that support youth engaged research, grant writing, and startup funds that acknowledge that youth incentives for research is not the same as paying for the informal time youth give as they develop relationships with researchers (e.g., developmentalists paying for smoothies during initial relationship building meetings). In all, supporting the training and development of YPAR scholars is an investment in universities themselves (Teixeira et al., 2021).

To change the values of tenure and promotion, the training of what it means to be an engaged, critical, and impactful developmentalist must begin as early as undergraduate and doctoral training. Consistent with the personal researcher reflexivity exercises outlined above, developmentalists should encourage students to reflect on their assumptions on credible ways of knowing, whether they believe youth are experts on the lived experiences, and the responsibility of developmentalists to promote liberation for youth and communities marginalized by systems of oppression (Buchanan et al., 2021; Teixeira et al., 2021). These early educational experiences set the stage for potential lifelong engagement in YPAR and other community-engaged practices among emerging developmentalists.

In all, the current chapter provided a research roadmap to center Latinx youths' voices in developmental science with the aim to capture and promote unconditional love for Latinx youth and their communities through youth engaged research. We encourage future work that considers how the voices of specific Latinx youth (e.g., queer Latina girls) and other youth marginalized by systems of oppression may be elevated throughout the research process (see Anyiwo et al., 2021; Johnston-Goodstar, 2013; Maker Castro et al., 2021; for examples with Black, Indigenous, and immigrant-origin youth). Our vision for developmental science is that it becomes synonymous with community-engaged research practices that respect and honor youths' lived experiences and translate into tangible outcomes that promote healing and liberation among communities marginalized by white supremacy and other systems of oppression.

References

Aldana, A., & Richards-Schuster, K. (2021). Youth-led antiracism research: Making a case for participatory methods and creatives strategies in developmental science. *Journal of Adolescent Research, 36*(6), 654–685. https://doi.org/10.1177/07435584211043289 10.1177%2F07435584211043289

Aldana, A., Richards-Schuster, K., & Checkoway, B. (2016). Dialogic pedagogy for youth participatory action research: Facilitation of an intergroup empowerment program. *Social Work with Groups, 39*(4), 339–358. https://doi.org/10.1080/01609513.2015.1076370

Aldana, A., Bañales, J., & Richards-Schuster, K. (2019). Youth anti-racist engagement: Conceptualization, development, and validation of an anti-racism action scale. *Adolescent Research Review, 4*(4), 369–381. https://doi.org/10.1007/s40894-019-00113-1

Anyiwo, N., Richards-Schuster, K., & Jerald, M. C. (2021). Using critical media literacy and youth-led research to promote the sociopolitical development of Black youth: Strategies from Our Voices. *Applied Developmental Science, 25*(3), 201–216. https://doi.org/10.1080/10888691.2021.1906246

Anyon, Y., Bender, K., Kennedy, H., & Dechants, J. (2018). A systematic review of youth participatory action research (YPAR) in the United States: Methodologies, youth outcomes, and future directions. *Health Education & Behavior, 45*(6), 865–878. https://doi.org/10.1177/1090198118769357

Ayduk, Ö., & Kross, E. (2010). Analyzing negative experiences without ruminating: The role of self-distancing in enabling adaptive self-reflection. *Social and Personality Psychology Compass, 4*, 841–854. https://doi.org/10.1111/j.1751-9004.2010.00301.x

Azmitia, M. (2021). Latinx adolescents' assets, risks, and developmental pathways: A decade in review and looking ahead. *Journal of Research on Adolescence, 31*(4), 989–1005. https://doi.org/10.1111/jora.12686

Baker, M. A., & Brookins, C. C. (2014). Toward the development of a measure of sociopolitical consciousness: Listening to the voices of Salvadoran youth. *Journal of Community Psychology, 42*(8), 1015–1032. https://doi.org/10.1002/jcop.21668

Ballonoff Suleiman, A., Ballard, P. J., Hoyt, L. T., & Ozer, E. J. (2021). Applying a developmental lens to youth-led participatory action research: A critical examination and integration of existing evidence. *Youth and Society, 53*(1), 26–53. https://doi.org/gfxf4j

Bañales, J., & Rivas-Drake, D. (2022). Showing up: A theoretical model of anti-racist identity and action for Latinx youth. *Journal of Research on Adolescence, 32*(3), 999–1019. https://doi.org/10.1111/jora.12747

Bañales, J., Hoffman, A. J., Rivas-Drake, D., & Jagers, R. J. (2020). The development of ethnic-racial identity process and its relation to civic beliefs among Latinx and Black American adolescents. *Journal of Youth and Adolescence, 49*(12), 2495–2508. https://doi.org/gv2v

Bañales, J., Aldana, A., Richards-Schuster, K., & Merritt, A. (2021). Something you can see, hear, and feel: A descriptive, exploratory mixed-methods analysis of youths' articulations about racism. Journal of Adolescent Research. https://doi.org/10.1177/07435584211062117

Bettencourt, G. M. (2020). Embracing problems, processes, and contact zones: Using youth participatory action research to challenge adultism. *Action Research, 18*(2), 153–170. https://doi.org/10.1177/1476750318789475

Boden, C. J., Cook, D., Lasker-Scott, T., Moore, S., & Shelton, D. (2006). Five perspectives on reflective journaling. *Adult Learning, 17*(1–4), 11–15.

Bonds, A., & Inwood, J. (2016). Beyond white privilege: Geographies of white supremacy and settler colonialism. *Progress in Human Geography, 40*(6), 715–733. https://doi.org/f9btdn

Bonilla-Silva, E. (2015). The structure of racism in color-blind, "post-racial" America. *American Behavioral Scientist, 59*(11), 1358–1376. https://doi.org/10.1177/0002764215586826

Booth, J., Huerta, C., & Thomas, B. (2021). The role of bilingualism in Latino youth experiences of acculturation stress when living in an emerging Latino community. *Qualitative Social Work, 20*(4), 1059–1077. https://doi.org/10.1177/1473325020923012

Buchanan, N. T., Perez, M., Prinstein, M. J., & Thurston, I. (2021). Upending racism in psychological science: Strategies to change how our science is conducted, reported, reviewed & disseminated. *American Psychologist, 76*, 1097. https://doi.org/10.1037/amp000090510.31234/osf.io/6nk4x

Cabrera, N. L., Milem, J. F., Jaquette, O., & Marx, R. W. (2014). Missing the (student achievement) forest for all the (political) trees: Empiricism and the Mexican American studies controversy in Tucson. *American Educational Research Journal, 51*(6), 1084–1118. https://doi.org/10.3102/0002831214553705

Cammarota, J., & Aguilera, M. (2012). "By the time I get to Arizona": Race, language, and education in America's racist state. *Race Ethnicity and Education, 15*(4), 485–500. https://doi.org/10.1080/13613324.2012.674025

Cammarota, J., & Fine, M. (2008). *Revolutionizing education: Youth participatory action research in motion*. Routledge Taylor & Francis Group.

Cammarota, J., & Romero, A. F. (2009). A social justice epistemology and pedagogy for Latina/o students: Transforming public education with participatory action research. *New Directions for Youth Development, 2009*(123), 53–12. https://doi.org/10.1002/yd.314

Caraballo, L., Lozenski, B. D., Lyiscott, J. J., & Morrell, E. (2017). YPAR and critical epistemologies: Rethinking education research. *Review of Research in Education, 41*(1), 311–336. https://doi.org/10.3102/0091732X16686948

Ceballo, R., Cranford, J. A., Alers-Rojas, F., Jocson, R. M., & Kennedy, T. M. (2021). What happens after school? Linking Latino adolescents' activities and exposure to community violence. *Journal of Youth and Adolescence, 50*, 2007–2020. https://doi.org/gv2t

Chavez-Dueñas, N. Y., Adames, H. Y., Perez-Chavez, J. G., & Salas, S. P. (2019). Healing ethno-racial trauma in Latinx immigrant communities: Cultivating hope, resistance, and action. *American Psychologist, 74*(1), 49–62. https://doi.org/10.1037/amp0000289

Checkoway, B. (2011). What is youth participation? *Children and Youth Services Review, 33*(2), 340–345. https://doi.org/10.1016/j.childyouth.2010.09.017

Collins, P. H. (2002). *Black feminist thought: Knowledge, consciousness, and the politics of empowerment*. Routledge.

Comas-Díaz, L., & Torres Rivera, E. (Eds.). (2020). *Liberation psychology: Theory, method, practice, and social justice*. American Psychological Association. https://doi.org/10.1037/0000198-000

Desai, S. R. (2019). Youth participatory action research: The nuts and bolts as well as the roses and thorns. In K. K. Strunk & L. A. Locke (Eds.), *Research methods for social justice and equity in education*. Palgrave Macmillan.

Domínguez, A. D. (2021). Venceremos!: Challenging school barriers with Latinx youth participatory action research. *Journal of Latinos and Education*, 1–15. https://doi.org/10.1080/15348431.2021.1935258

Emirbayer, M., & Desmond, M. (2012). Race and reflexivity. *Ethnic and Racial Studies, 35*(4), 574–599. https://doi.org/10.1080/01419870.2011.606910

Fereday, J., & Muir-Cochrane, E. (2006). Demonstrating rigor using thematic analysis: A hybrid approach of inductive and deductive coding and theme development. *International Journal of Qualitative Methods, 5*(1), 80–92. https://doi.org/gcdvss

Fine, M. (2018). *Just research in contentious times: Widening the methodological imagination*. Teachers College Press.

Flores-González, N. (2017). *Citizens but not Americans: Race and belonging among Latino millennials*. NYU Press.

García Coll, C., Crnic, K., Lamberty, G., Wasik, B. H., Jenkins, R., Garcia, H. V., & McAdoo, H. P. (1996). An integrative model for the study of developmental competencies in minority children. *Child Development, 67*(5), 1891–1914. https://doi.org/10.1111/j.1467-8624.1996.tb01834.x

Ginwright, S. (2011). Hope, healing, and care. *Liberal Education, 97*(2), 34–39.

Ginwright, S., & Cammarota, J. (2002). New terrain in youth development: The promise of a social justice approach. *Social Justice, 29*(4), 82–95.

Haskie-Mendoza, S., Tinajero, L., Cervantes, A., Rodriguez, J., & Serrata, J. (2018). Conducting youth participatory action research (YPAR) through a healing-informed approach with system-involved Latinas. *Journal of Family Violence, 33*(8), 605–612. https://doi.org/10.1007/s10896-018-9996-x

Hernández, M. G., Nguyen, J., Casanova, S., Suárez-Orozco, C., & Saetermoe, C. L. (2013). Doing no harm and getting it right: Guidelines for ethical research with immigrant communities. *New Directions for Child and Adolescent Development, 2013*(41), 43–60. https://doi.org/10.1002/cad.20042

Hughes, D. L., & Seidman, E. (2002). In pursuit of a culturally anchored methodology. In T. A. Revenson & A. R. D'Augelli (Eds.), *Ecological research to promote social change: Methodological advances from community psychology* (pp. 243–255). Kluwer Academic/ Plenum Publishers.

Johnston-Goodstar, K. (2013). Indigenous youth participatory action research: Re-visioning social justice for social work with indigenous youths. *National Association of Social Workers, 58*(4), 314–320. https://doi.org/10.1093/sw/swt036

Jost, J. T., & Hunyady, O. (2005). Antecedents and consequences of system-justifying ideologies. *Current Directions in Psychological Science, 14*(5), 260–265. https://doi.org/10.1111/j.0963-7214.2005.00377.x

Kiang, L., Espino-Pérez, K., & Stein, G. L. (2020). Discrimination, skin color satisfaction, and adjustment among Latinx American youth. *Journal of Youth and Adolescence, 49*(10), 2047–2059. https://doi.org/10.1007/s10964-020-01244-8

Kidd, S., Davidson, L., Frederick, T., & Kral, M. J. (2018). Reflecting on participatory, action-oriented research methods in community psychology: Progress, problems, and paths forward. *American Journal of Community Psychology, 61*(1–2), 76–87. https://doi.org/10.1002/ajcp.12214

Knight, G. P., Roosa, M. W., & Umaña-Taylor, A. J. (2009). *Studying ethnic minority and economically disadvantaged populations: Methodological challenges and best practices*. American Psychological Association.

Kohfeldt, D., Bowen, A. R., & Langhout, R. D. (2016). They think kids are stupid: YPAR and confrontations with institutionalized power as contexts for children's identity work. *Revista PuertoRriqueña de Psicología, 27*(2), 276–291.

Kornbluh, M., Neal, J. W., & Ozer, E. J. (2016). Scaling-up youth-led social justice efforts through an online school-based social network. *American Journal of Community Psychology, 57*(3–4), 266–279. https://doi.org/10.1002/ajcp.12042

Kross, E., Ayduk, O., & Mischel, W. (2005). When asking "why" does not hurt distinguishing. Rumination from reflective processing of negative emotions. *Psychological Science, 16*(9), 709–715. https://doi.org/10.1111/j.1467-9280.2005.01600.x

Lee, B. A., Ogunfemi, N., Neville, H. A., & Tettegah, S. (2021). Resistance and restoration: Healing research methodologies for the global majority. *Cultural Diversity and Ethnic Minority Psychology*. https://doi.org/10.1037/cdp0000394

Love, B. L. (2019). *We want to do more than survive: Abolitionist teaching and the pursuit of educational freedom*. Beacon Press.

Maker Castro, E., López Hernández, G., Karras-Jean Gilles, J., Novoa, A., & Suárez-Orozco, C. (2021). Everyone collaborated and came together: The civic promise (and pitfalls) of yPAR for immigrant-origin students in an era of deportation. *Cultural Diversity and Ethnic Minority Psychology., 28*, 427. https://doi.org/10.1037/cdp0000429

Mansfield, K. C. (2018). Listening to student voice: Toward a more holistic approach to school leadership. *Journal of Ethical Educational Leadership, 1*, 10–27.

Merriam, S. B., Johnson-Bailey, J., Lee, M. Y., Kee, Y., Ntseane, G., & Muhamad, M. (2001). Power and positionality: Negotiating insider/outsider status within and across cultures. *International Journal of Lifelong Education, 20*(5), 405–416.

Meyer, K., & Willis, R. (2019). Looking back to move forward: The value of reflexive journaling for novice researchers. *Journal of Gerontological Social Work, 62*(5), 578–585. https://doi.org/10.1080/01634372.2018.1559906

Norton, L., & Sliep, Y. (2018). A critical reflexive model: Working with life stories in health promotion education. *South African Journal of Higher Education, 32*(3), 45–63. https://doi.org/10.20853/32-3-2523

Okun, T. (1999/2021). *White supremacy culture*. https://www.whitesupremacyculture.info/

Ozer, E. J. (2016). Youth-led participatory action research: Developmental and equity perspectives. *Advances in Child Development and Behavior, 50*(2016), 189–207. https://doi.org/10.1016/bs.acdb.2015.11.006

Peña, E. D. (2007). Lost in translation: Methodological considerations in cross-cultural research. *Child Development, 78*(4), 1255–1264. https://doi.org/10.1111/j.1467-8624.2007.01064.x

Pinetta, B. J., Martinez, B., Cross, F. L., & Rivas-Drake, D. (2020). Inherently political? Associations of parent ethnic – Racial socialization and sociopolitical discussions with Latinx youths ' emergent civic engagement. *American Journal of Community Psychology, 66*, 1–12. https://doi.org/10.1002/ajcp.12435

Ponterotto, J. G. (2005). Qualitative research in counseling psychology: A primer on research paradigms and philosophy of science. *Journal of Counseling Psychology, 52*(2), 126–136. https://doi.org/10.1037/0022-0167.52.2.126

Richards-Schuster, K., & Timmermans, R. (2017). Conceptualizing the role of adults within youth-adult partnerships: An example from practice. *Children and Youth Services Review, 81*, 284–292. https://doi.org/10.1016/j.childyouth.2017.07.023

Rios, V. M., & Vigil, J. D. (2017). *Human targets: Schools, police, and the criminalization of Latino youth*. University of Chicago Press.

Rivas-Drake, D., Camacho, T. C., & Guillaume, C. (2016). Just good developmental science: Trust, identity, and responsibility in ethnic minority recruitment and retention. *Advances in Child Development and Behavior, 50*, 161–188. https://doi.org/gv2q

Rodríguez, L. F., & Brown, T. M. (2009). From voice to agency: Guiding principles for participatory action research with youth. *New Directions for Youth Development, 123*, 19–34. https://doi.org/10.1002/yd.312

Rodriguez, M. C., & Morrobel, D. (2004). A review of Latino youth development research and a call for an asset orientation. *Hispanic Journal of Behavioral Sciences, 26*(2), 107–127. https://doi.org/10.1177/0739986304264268

Rose, R., Singh, S., Berezin, M., & Javdani, S. (2021). *Roses have thorns for a reason: The promises and perils of critical youth participatory research with girls of color* (under review).

Rowley, S. J., & Camacho, T. C. (2015). Increasing diversity in cognitive developmental research: Issues and solutions. *Journal of Cognition and Development, 16*(5), 683–692. https://doi.org/10.1080/15248372.2014.976224

Said, E. A. (1978). *Orientalism*. Pantheon Books.

Sakamoto, I., & Pitner, R. O. (2005). Use of critical consciousness in anti-oppressive social work practice: Disentangling power dynamics at personal and structural levels. *British Journal of Social Work, 35*(4), 435–452. https://doi.org/10.1093/bjsw/bch190

Salas Pujols, J. (2020). 'It's about the way I'm treated': Afro-Latina black identity development in the third space. *Youth & Society, 54*(4), 593–610. https://doi.org/gv2r

Saldaña, J. (2017). *The coding manual for qualitative researchers* (4th ed.). Sage.

Sánchez Carmen, S. A., Domínguez, M., Greene, A. C., Mendoza, E., Fine, M., Neville, H. A., & Gutiérrez, K. D. (2015). Revisiting the collective in critical consciousness: Diverse sociopolitical wisdoms and ontological healing in sociopolitical development. *Urban Review, 47*(5), 824–846. https://doi.org/10.1007/s11256-015-0338-5

Serido, J., Borden, L. M., & Perkins, D. F. (2011). Moving beyond youth voice. *Youth & Society, 43*(1), 44–63. https://doi.org/10.1177/0044118x09351280

Singh, S., Granski, M., & Victoria, M. d. P., & Javdani, S. (2018). The praxis of decoloniality in researcher training and community-based data collection. *American Journal of Community Psychology, 62*(3–4), 385–395. https://doi.org/10.1002/ajcp.12294

Smith, L. T. (2012). *Decolonizing methodologies: Research and indigenous peoples* (2nd ed.). Zed Books. (Original work published 1999).

Spencer, M. B. (2006). Revisiting the 1990 special issue on minority children: An editorial perspective 15 years later. *Child Development, 77*(5), 1149–1154. https://doi.org/10.1111/j.1467-8624.2006.00925.x

Spencer, M. B. (2021). Acknowledging bias and pursuing protections to support anti-racist developmental science: Critical contributions of phenomenological variant of ecological systems theory. *Journal of Adolescent Research, 36*(6), 569–583. https://doi.org/10.1177/07435584211045129

Sprague Martinez, L., Richards-Schuster, K., Teixeira, S., & Augsberger, A. (2018). The power of prevention and youth voice: A strategy for social work to ensure youths' healthy development. *Social Work, 63*(2), 135–143. https://doi.org/gdctbz

Sprague Martinez, L., Tang Yan, C., McClay, C., Varga, S., & Zaff, J. F. (2020). Adult reflection on engaging youth of color in research and action: A case study from five US cities. *Journal of Adolescent Research, 35*(6), 699–727. https://doi.org/10.1177/0743558420906086

Stein, G. L., Cupito, A. M., Mendez, J. L., Prandoni, J., Huq, N., & Westerberg, D. (2014). Familism through a developmental lens. *Journal of Latina/o Psychology, 2*(4), 224–250. https://doi.org/10.1037/lat0000025

Suárez-Orozco, C., Yoshikawa, H., Teranishi, R., & Suárez-Orozco, M. (2011). Growing up in the shadows: The developmental implications of unauthorized status. *Harvard Educational Review, 81*(3), 438–473.

Teixeira, S., Augsberger, A., Richards-Schuster, K., & Sprague Martinez, L. (2021). Participatory research approaches with youth: Ethics, engagement, and meaningful action. *American Journal of Community Psychology, 68*, 142. https://doi.org/10.1002/ajcp.12501

Umaña-Taylor, A. J., Diversi, M., & Fine, M. A. (2002). Ethnic identity and self-esteem of Latino adolescents: Distinctions among the Latino populations. *Journal of Adolescent Research, 17*(3), 303–327. https://doi.org/10.1177/0743558402173005

Valencia, B. G., Caporale, J., & Romero, A. J. (2020). I had (my dreams) on hold, I had to…: Mexican descent youth discuss factors that lead to leaving high school. *Urban Education*. https://doi.org/10.1177/0042085920909869

Wong, N. T., Zimmerman, M. A., & Parker, E. A. (2010). A typology of youth participation and empowerment for child and adolescent health promotion. *American Journal of Community Psychology, 46*(1–2), 100–114. https://doi.org/10.1007/s10464-010-9330-0

Yu, M. V. B., Liu, Y., Soto-Lara, S., Puente, K., Carranza, P., Pantano, A., & Simpkins, S. D. (2021). Culturally responsive practices: Insights from a high-quality math afterschool program serving underprivileged Latinx youth. *American Journal of Community Psychology, 68*, 323. http://dx.doi.org.proxy.cc.uic.edu/10.1002/ajcp.12518

Chapter 10
Action, but Make It Critical: The Measurement and Developmental Processes of Critical Action for Black and Latinx Youth

Channing J. Mathews, Josefina Bañales, N. Keita Christophe, Alexis S. Briggs, and Elan C. Hope

Introduction

The COVID-19 and racism pandemics of 2020 underscore that the United States (U.S.) continues to be plagued with white supremacy, anti-Blackness, and xenophobia (Sobo et al., 2020). Black people continue to be murdered by police and armed civilians, Black and Brown children remain in cages at the U.S.-Mexico border, Asian people are murdered and assaulted because they are blamed for the spread of COVID-19, and Indigenous communities continue to live on reservations with no running water (Society on Research in Child Development, 2020). Racism infringes on the positive development of people of color (Causadias & Umaña-Taylor, 2018; Coll et al., 1996; Seaton et al., 2018), yet communities of color have resisted and continue to resist racism. Young people of color resist through *critical action*, behavior that disrupt racism and other systems of oppression (Aldana et al., 2019; Diemer et al., 2017; Mathews et al., 2019).

In this chapter, we provide novel insights for critical action research among Black and Latinx youth. Marginalized groups (i.e., Asian-Americans, Indigenous people, queer people of color) have varied sociopolitical experiences. We focus on Black and Latinx youth given their similarities in sociopolitical movements and

C. J. Mathews (✉)
Department of Psychology, University of Virginia, Charlottesville, VA, USA
e-mail: cjmathews@virginia.edu

J. Bañales
Department of Psychology, University of Illinois, Chicago, Chicago, IL, USA

N. K. Christophe
Department of Psychology, McGill University, Montreal, QC, Canada

A. S. Briggs · E. C. Hope
Department of Psychology, North Carolina State University, Raleigh, USA

© The Author(s), under exclusive license to Springer Nature Switzerland AG 2023
D. P. Witherspoon, G. L. Stein (eds.), *Diversity and Developmental Science*,
https://doi.org/10.1007/978-3-031-23163-6_10

fights for civil rights and the more robust literature on these populations (Ortiz, 2018). First, we describe what critical action is and how critical action is typically measured. We then describe barriers and facilitators of critical action for Black and Latinx youth and conclude with recommendations for research on youth critical action.

Conceptualizing Critical Action Among Black and Latinx Youth

Critical action is behavior to combat oppression and promote social justice (Aldana et al., 2019; Diemer et al., 2017). We refer to critical action broadly, to encompass activism (e.g., protesting), political participation (e.g., voting), and community involvement (e.g., volunteering), which youth use to seek social justice. In this conceptualization, we specifically consider youth's intent to disrupt the sociopolitical status quo as a defining feature of critical action. While activism is typically associated with social justice for marginalized communities, political and community engagement can also be used to dismantle systems of oppression (Hope & Spencer, 2017; Watts & Hipolito-Delgado, 2015). For example, in the 2020 Presidential election, 68% of youth reported that voting was a way to disrupt racism and violence against people of color (Lundberg et al., 2021). Conversely, youth are involved in activism that restrict the rights of marginalized people (e.g., Second Amendment rights movement; Hayat et al., 2020). Regarding community engagement, youth translate documents between their native language and English to help family and community members navigate U.S. political systems (Jensen, 2008; Stepick et al., 2008). This critical action ensures all members of the community, regardless of first language, have access to resources. Given these nuances, we consider critical action broadly to include any behavior that Black and Latinx youth use to dismantle systems of oppression.

The civic engagement literature has not typically attended to the social justice goals of youth critical action (Watts & Flanagan, 2007). This oversight is not trivial and promotes a deficit-based approach. For instance, researchers often cite that Black and Latinx youth score lower than White youth on general civic knowledge including U.S. democracy, civic life, and the role of citizens (National Center for Education Statistics, 2018). These traditional metrics do not encompass the racialized ways that Black and Latinx youth experience and interact with political systems (Kerrison et al., 2018). To fully understand critical action in Black and Latinx youth as a developmental asset, we must consider how oppression and motivation to achieve liberation shape the breadth of activities that youth participate in (Mathews et al., 2019).

Black and Latinx youth are undeniably affected by institutional (e.g., residential segregation, access to healthcare), cultural (e.g., media representation of Black Americans as violent and deviant), and interpersonal (e.g., microaggressions; bias) racism (Williams & Mohammed, 2009). Historical and enduring structural racism limits the political power of marginalized communities and function as barriers to civic engagement among Black and Latinx people (Diemer, 2012; Hope & Bañales, 2019). Equally, experiencing and understanding racism may motivate youth to use critical action to challenge systems of oppression (Hope et al., 2020a, b; Pinedo et al., 2021). We must expand what behaviors "count" as critical action and consider the goals of youth critical action. Such a conceptualization of critical action foregrounds the innovative ways Black and Latinx youth overcome structural barriers to harness political power and voice in social issues that affect their communities.

Ways to Categorize Critical Action

There are several ways to categorize and define critical action. In a broad engagement framework (Ekman & Amnå, 2012), critical action can be civic participation where there is interest in politics with the goal to address social and political issues. Critical action within this framework can also be political participation, which includes formal political participation such as voting and contacting political representatives, extra-parliamentary legal activism such as boycotts and involvement in social movements, and extra-parliamentary illegal activism which includes civil disobedience or politically motivated violence. Finally, critical action could include active non-participation that is motivated by dissatisfaction with mainstream politics.

Other frameworks of critical action focus on specific forms of oppression. For example, Aldana and colleagues (2019) conceptualize critical action against racism across three dimensions: interpersonal, communal, and political change. Interpersonal critical action is behavior that is between individuals within their local social context, like challenging a friend or family member who makes a racially biased comment or joke. Critical action can also be communal, which includes participation in collective efforts like youth organizing groups. Finally, critical action is political change action, which aligns with political participation and extra-parliamentary activism as outlined by Ekman and Amnå (2012). Political change action includes participation in protests and engagement with public officials. While some overlap may exist between broad and specific frameworks of critical action, specific frameworks expand the multidimensionality of types of actions taken to challenge unique forms of systemic oppression.

Other scholars conceptualize critical action according to perceived or relative risk, such that action can be low- or high-risk (Corning & Myers, 2002; Hope et al., 2019). Low-risk critical actions are relatively safe and present less long-term or consequential risk for youth. Low-risk critical action includes confronting a comment or joke rooted in oppression or wearing a t-shirt in support of oppressed communities or social justice causes. Critical action can also be high-risk, where there is greater likelihood for long-term or more severe consequences. High-risk critical action includes attending protests where arrest is likely or participating in a demonstration that could result in bodily harm.

Finally, Watts and colleagues (2015) consider critical action across two dimensions: agents of critical action and targets of critical action. In this framework agents of critical action can be an individual or collective group. This framework also considers the target or intended outcome of the critical action. Targets of critical action can be internal, which focus on individual beliefs and attitudes that might inform intergroup interactions or policy opinions. Targets of critical action can also be external, which include structural and institutional policies and practices. For instance, an individual may seek change through dialogue with a close friend (internal action) with the goal of changing and oppressive local policy (external action). Collectives can also work toward internal change through group activities and dialogues and toward external change through mobilization and direct actions (e.g., protests, community organizing). This is similar to interpersonal, communal, and political change actions (Aldana et al., 2019) and low-risk activism (Hope et al., 2019).

There are several unifying themes across these various conceptualizations of critical action. First, critical action occurs within and beyond traditional political engagement, inclusive of interpersonal interactions and investment in one's community. Additionally, critical action can be individually and in community with others. Finally, critical action can target social justice change at the individual level and via institutional policies and the larger structures of society. Though critical action has been conceptualized broadly, critical action measurement has not fully captured the multidimensional and intersectional experiences of Black and Latinx youth. This discrepancy presents an opportunity to reimagine developmentally appropriate measures that align with how Black and Latinx youth challenge systemic oppression relevant to their social contexts. Moreover, we can create measures that acknowledge the structural barriers to the participation and consider types of disengagement within the scope of critical action.

Measuring Critical Action Among Black and Latinx Youth

Critical action, by definition, is action performed to combat oppression and seek liberation (Diemer et al., 2017), yet developmental scientists cannot fully understand critical action, its causes, and its consequences without attending to the continued impact that structural racism and other forms of oppression have on Black and Latinx youth (Hope & Spencer, 2017). In addition to alignment with the specific conceptualizations described above, critical action measures should include items that are grounded in relevant ethnic-racial experiences, assess intersectional forms of oppression, account for the varied behaviors youth may use, and measure the intention, frequency, and risk of critical action. In doing so, measures will be more contextually and developmentally appropriate, and research will more accurately reflect the lived experiences of Black and Latinx youth.

Existing Measures of Critical Action

Critical action is measured in three different ways (see Table 10.1). The first set of measures consider critical action, generally, across traditional and extra-parliamentary political activities. These measures include the critical action subscale of the Critical Consciousness Scale (CCS; Diemer et al., 2017) the political action subscale of the Youth Involvement Inventory (YII; Pancer et al., 2007), and the Critical Consciousness Inventory (CCI; Thomas et al., 2014). A second set of measures focuses on critical action that addresses a specific form of oppression, typically racial oppression. These include the Anti-Racism Action Scale (ARAS; Aldana et al., 2019) and the Measure of Adolescent Critical Consciousness (MACC, McWhirter & McWhirter, 2016). A third group of scales examines critical action specific to a particular ethnic-racial group, most often used with Black or African American samples. These measures include the Black Community Activism Orientation Scale (BCAOS; Hope et al., 2019), the Involvement in African American Activism Scale (IAAS; Szymanski, 2012), and the Multidimensional Measure of Black Activism Scale (MMBA; Thomas, 2001).

Sample Characteristics

Measures of critical action are often developed with and administered to youth of color, with many measures developed with Black or African American youth. For measures developed with ethnically diverse samples, Black and Latinx youth are most widely represented, with lesser representation of Asian-American/Pacific Islander youth. Few measures include Indigenous youth as a part of the sample (see Aldana et al., 2019; Diemer et al., 2017 for examples). These critical action scales validated across different ethnic-racial groups are often broad and may be useful to understand how youth of color challenge social inequality, generally. These broader scales may miss the nuance of types of critical action that are specific to cultural norms within an ethnic-racial group as well as differences in how systems of oppression shape the lived experiences of different youth of color. Some behaviors may appear similar, but the cultural and justice-related motivation of such behaviors may not be. For instance, Black and Latinx youth might organize food distribution to address food desserts in their community, while other youth participation may be rooted solely in altruism, rather than the underlying injustice. Thus, it is important to consider differences in experiences of social inequality across contexts and how these differences may inform youth critical action among racially and ethnically diverse youth.

Critical action subscales of critical consciousness measures assess participation in activism and politics, regardless of the respondents' connection with specific forms of oppression. Such broad assessments may suggest that all persons have an equivalent understanding and exposure to each type of inequity. Yet, given the

Table 10.1 Overview of studies included in measurement review

Scale name and reference	Age	Race/Ethnicity	Social issue assessed	Type of action (Community, Political, Interpersonal)	Risk assessment	Frequency assessment	Scoring
Anti-Racism Action Scale (ARAS); Aldana et al. (2019)	13–19	Black/African American, White/European American, Latino/Hispanic, Asian/Pacific Islander, Multiracial, Native American, Arab/Middle Eastern	Anti-racism action	Interpersonal, Political, and Community	None	Previous 2 months	0 = No 1 = Yes
Critical Consciousness Scale (CCS); Critical Action: Sociopolitical Participation Subscale; Diemer et al. (2017, 2020)	13–19	Black/African American, Multiracial, Native American, Asian/Asian American	General social inequality	Political and Community	None	Annual	1 (Never did this) – 5 (Did this at least once a week)
Black Community Activism Orientation Scale (BCAOS); Hope et al. (2019)	14–29	Black/African American	Anti-black racism action; involvement in black issues	Community, Political, and Interpersonal	Yes	N/A	1 (Extremely likely) – 5 (Extremely unlikely)
The Measure of Adolescent Critical Consciousness (MACC); Critical Behavior Subscale McWhirter and McWhirter (2016)	14–19	Latinx (majority Mexican American)	Anti-Racism action	Political	None	N/A	1 (Strongly agree) – 4 (Strongly disagree
Youth Involvement Inventory (YII); Pancer et al. (2007)	16–20	Canadian Students (Ontario) described as ethnically diverse	Political, Community, and Service action	Community, Political, and Interpersonal	None	Annual	0 (Never did this) – 4 (Did this a lot)

Scale name and reference	Age	Race/Ethnicity	Social issue assessed	Type of action (Community, Political, Interpersonal)	Risk assessment	Frequency assessment	Scoring
Involvement in African American Activism Scale (IAAAS) Szymanski (2012)	18–79	Black/African American	Anti-Racism action; involvement in black issues	Community and Political	None	N/A	1 (Very untrue of me) – 7 (Very true of me)
The Critical Consciousness Inventory (CCI); Thomas et al. (2014)	18–25	Back/African American, White, Asian/Pacific Islander, Multiracial	Promoting equality and challenging oppression	Interpersonal	None	N/A	Guttman Scaling
The Multidimensional Measure of Black Activism (MMBA); Thomas (2001)	18+	Black/African American and Black Biracial	Anti-Black Racism; Black community involvement	Political, Community, and Interpersonal	None	N/A	1 (Very likely) – 4 (Not at all)

Note. Four of the five critical action items included in Diemer et al.'s (2020) Short Critical Consciousness Scale (ShoCCS) also appear in the critical action: sociopolitical participation subscale of the Short Critical Consciousness Scale (CCS-S; Rapa et al., 2020). Both the ShoCCS and CCS-S are shortened versions of Diemer et al.'s (2017) Critical Consciousness Scale (CCS) multitude of identities individuals hold, it is likely that youth have varied experiences including direct personal experience, vicarious experiences, or no experience at all with specific forms of oppression (Godfrey & Burson, 2018). As critical consciousness scales are most often developed with diverse ethnic groups, variation across identities and experiences is likely masked.

A majority of racially specific critical action measures focus on the experiences of Black populations. Black American communities have a historical legacy of civil rights activism against racial injustice, which is reflected in current measurement. For example, three scales that focus specifically on Black youth (BCAOS, IAAS, MMBA) all include items that focus on protest, boycotting, and involvement in Black sociopolitical issues (Hope et al., 2019; Szymanski, 2012; Thomas, 2001). Each scale differs in how they assess the types or dimensions of critical action. The IAAS scale assumes the unidimensionality of Black activism. The BCAOS and MMBA are multidimensional and assess risk and intention, respectively. The multidimensional approach allows for heterogeneity and within-group variation in the types of critical action that youth might use. This consideration highlights the diversity of experiences and perspectives that individuals use to challenge social inequities.

There is a notable absence of scales that focus specifically on Latinx communities, which could capture activism that is especially relevant to issues facing this population. One exception is with the MACC, which was developed with Latinx high school students (McWhirter & McWhirter, 2016). While the MACC was developed with Latinx youth, the items that assess critical action are not specific to the unique experiences of Latinx communities in the U.S. Measures that contextualize the experiences of Latinx populations are integral to understanding the critical action behaviors of Latinx groups. Further, Latinx communities vary widely across ethnicity and race and a myriad of issues may be differentially salient among Latinx groups. Masking these differences within critical consciousness measures homogenizes the experiences of heterogeneous ethnic-racial groups, which can be misleading if certain phenomena do not function similarly within each group.

For example, immigration, citizenship, and property rights are often central social issues that prompt Latinx activism in the U.S. (Schmitz et al., 2020; Terriquez, 2017). Terriquez and colleagues (2018) highlighted how Latinx youth activists during the Development, Relief, and Education for Aliens Minors Act (DREAM) pushed the movement to consider the ways that coming out as undocumented was connected to coming out as LGBTQ+. Such overlapping identities were cited as inspirations for various types of activism during the DREAMERS movement. Yet none of the widely used critical action scales to date feature items that are specific to immigration or to intersectional aspects of activism, and thus do not fully capture the range of actions that occur across Latinx youth who are immigrants or are from immigrant families. Future measures should consider how activism may differ across and within ethnic-racial groups, with specific attention to how multiple forms of marginalization may lead to different forms and frequency of activism across individuals. For example, Terriquez (2015) found that Latinx undocumented youth who also identified as queer were more civically engaged than their straight peers.

Bringing a more intersectional lens to measurement may shed light on how multiple forms of oppression may lead to (or detract from) higher motivation for critical action across different social issues relevant to Black and Latinx youth.

Addressing Social Issues with Multiple Forms of Critical Action

Beyond adding a lens intersectionality and items specific to the unique experiences faced by Black and Latinx youth, critical action research must assess different forms of critical action. Scholars have recently begun to incorporate this level of specificity into their measurement of critical action. For instance, the ARAS (Aldana et al., 2019) focuses on anti-racism critical action across several dimensions such as interpersonal action (e.g., talked with friends about issues related to race, ethnicity, discrimination, and/or segregation), community action (e.g., joined a club working on issues related to race, ethnicity, discrimination and/or segregation), and political action (e.g., called, written, emailed an elected official). Such focus on a specific type of oppression allows for deeper exploration on the multidimensionality of actions individuals can take to challenge social inequities. Both the IAAS and BCAOS assess anti-racist action but differ from the ARAS in two important ways. First, the former scales were created and validated with Black populations. Second, these scales broadly assess different ways that Black people enact anti-racism and advocate for Black communities. This includes promoting awareness of Black issues or joining groups that emphasize Black pride (Hope et al., 2019; Szymanski, 2012).

Online critical action is not explicitly captured in existing measures (Anyiwo et al., 2020; Bañales et al., 2019). Internet and social media use are associated with critical action (Xenos et al., 2014) and involvement on these platforms may be considered forms of critical action. Social media provides access to information, digital resources, and a community of civically engaged peers outside of physically bound social circles. Access to information and connections via social media can increase awareness, political opinion sharing, and participation in in-person demonstrations (Boulianne et al., 2020; Valenzuela, 2013). Youth may also use social media to overcome time, location, and access-based barriers to in-person critical action. By incorporating social media into measures of critical action, scholars can gain insight into the creative ways that youth overcome these participation barriers. It will be important for researchers to consider how actions that are typically in-person (i.e., attending national or community organization meetings) translate to a virtual context (particularly in light of the COVID-19 pandemic). This may require adapting existing scales to assess which actions are conducted online, in-person, or in both contexts to parse out potentially informative and predictive differences in frequency across these spaces. For instance, there might be different factors that determine whether and how frequently youth "challenged or checked a friend who uses a racial slur or makes a racial joke" (Aldana et al., 2019) in-person versus on social media.

Measuring Motivation, Frequency, and Risk in Critical Action

Critical action scales address issues related to promoting social justice and challenging any form of oppression, with varying degrees of specificity (Diemer et al., 2015). For instance, the critical action subscale of the CCS (Diemer et al., 2017) asks whether participants have participated in a protest or worked on a political campaign. A fundamental assumption embedded in this type of measurement is that these actions are social justice oriented and that the activities that youth participate in are seeking to address structural inequality. The benefit of broad measurement is that you can assess youth participation across a variety of issues that might be relevant across various communities and in reference to various social issues. The downside is that it is difficult to determine whether these actions are *critical.* How well these measures can distinguish between youth who are engaged because they are invested in social change, versus youth who act to fulfill a requirement (e.g., build a strong resume for college), is an open question. There may be types of critical action, beyond what is seen as traditional activism, that communities leverage to resist and survive oppressive conditions (e.g., mutual aid funds; community mentoring). Capturing the motivational aspect of critical action sheds light on the factors that encourage critical action across the lifespan, despite the barriers that exist to such participation in Black and Latinx communities.

Many critical action measures assess frequency of critical action, which may be conflated with an assessment of motivation (i.e., higher frequency as stronger intentions to challenge social oppression). Higher frequency of participation may imply stronger commitment to social justice, but this association has not been thoroughly examined in critical action research. Indeed, in one study of minoritized adolescents (42.2% Black, 25.9% Latinx) critical motivation and critical action measured by Diemer et al.'s (2020) scale were not significantly correlated with one other (Kiang et al., 2021). This finding may call into question whether greater motivation consistently translates into more critical action. Given the importance of peers during adolescence, it may be that youth engage in frequent critical actions as a form of peer acceptance, rather than having a critical view of the purpose of such actions or the motivation to rectify social inequities. Researchers have not assessed which factors are associated with motivations for critical action and how such motivations may differ across populations. Recent IRT analyses of the critical action subscale of the CCS and the Short Critical Consciousness Scale (ShoCCS) suggests that critical action items only capture those who have higher motivations to act rather than those who are average or lower in their action motivation (Diemer et al., 2020). Future measures should consider the various motivations that might lead youth to critical action given that youth likely vary in their motivation to act given various social and structural obstacles that may impede their participation. Further, understanding motivation for critical action allows for a broad range of types of engagement, and even purposeful disengagement.

Facilitators of and Barriers to Critical Action

Ethnic-racial identity and socialization, being the target of racism, intergroup dialogue, and citizenship likely play a role in critical action (Anyiwo et al., 2018; Mathews et al., 2019). Below, we outline ways these sociocultural factors may facilitate or impede youth critical action.

Ethnic-Racial Identity and Socialization

Critical action among Black and Latinx youth may depend on ethnic-racial identity—how one's race or ethnicity fits into their sense of self (Umaña-Taylor et al., 2014). While ethnic-racial identity may drive critical action through various causal pathways (see Mathews et al., 2019 for examples), youths' understanding of the values, histories, and traditions of their ethnic-racial group may serve as an impetus for critical action. Critical action may also be influenced by ethnic-racial socialization messages, or messages parents communicate to children about the meaning of race and ethnicity (Hughes et al., 2006). Socialization messages, particularly preparation for bias, which are geared toward helping youth anticipate and cope with racism (Priest et al., 2014) may help socialize youth to respond to and cope with oppression through critical action.

While these factors have broadly been implicated in predicting critical action among Black and Latinx youth (Anyiwo et al., 2018; Mathews et al., 2019), it is important to understand and measure how factors such as ethnic-racial identity and socialization may be differentially predictive of critical action depending on types of action, risk level of the action, and motivation behind the action being measured. For instance, in a study using the ARAS to examine how discrimination, ethnic-racial centrality, preparation for bias socialization, and their interactions were associated with interpersonal, communal, and political forms of critical action among Black college students, Christophe et al. (2022) found that each form of action had a different set of significant predictors. If the authors used sum scores of the ARAS, these differences would be masked. Similarly, Hope et al. (2020a, b) found using the BCAOS that Black adolescents and young adults endorsed varying levels of low- and high-risk orientation toward critical action depending on their age, psychological anticipation of racism, and profile of ethnic-racial identity.

Exposure to Racism and Discrimination

Experiences of racism and racial discrimination are central drivers of critical action among youth of color. Researchers have found that more exposure to interpersonal racial discrimination (Hope et al., 2018; Szymanski, 2012; White-Johnson, 2012) and cultural racism (Hope et al., 2019) is related to more critical action among Black youth. Experiences with institutional racism are sometimes related to more critical action (Hope & Jagers, 2014; Hope et al., 2019). Similarly, among Latinx youth, more experiences of interpersonal racism are related to more critical action (Ballard, 2015; Hope et al., 2016). More exposure to discrimination and unequal treatment may empower Black and Latinx youth to engage in critical action as a way to dismantle these unequal and oppressive sociopolitical systems. Racism has a clear impact on Black and Latinx youth's lived experiences and Black and Latinx youth may respond to this racial oppression with critical action, which may include intergroup discussions of shared experiences. By considering the role of racism and discrimination in relation to critical action, researchers could have more accurate insights into how such factors shape Black and Latinx critical action trajectories and experiences.

Intergroup Dialogue Between Youth of Color

Intergroup dialogue is a one example of a culturally responsive practice that may stimulate critical action among Black and Latinx youth (Aldana et al., 2016, Checkoway & Aldana, 2013). Intergroup dialogue involves providing youth with different social identities opportunities to explore, critique, and challenge systems of oppression, and the ways marginalization informs youths' social identities, intergroup relations, and the structure in which their lives are embedded (see Dessel & Rogge, 2008 for review). Interracial dialogue is a form of intergroup dialogue that involves bringing youth of color and white youth together to engage in conversations about race, racism, and collective action against racism (Richards-Schuster & Aldana, 2013). Within these spaces, youth discuss their racial identity, racial stereotypes, and racial prejudices about the other racial group with members of their own racial group. Next, youth have a combined group discussion on the topics discussed in the separate racial affinity groups, with specific discussion about cross-racial coalition building to dismantle anti-Blackness, xenophobia, and other intersectional forms of oppression that negatively affect both communities. Interracial dialogues between Black and Latinx youth have the potential to increase interracial solidarity and cross-racial critical action, given the emphasis on building group-based identity and trust across groups that involves addressing racial and social issues that negatively impact other ethnic-racial groups.

Citizenship Status

Undocumented Black and Latinx youth in the U.S. might engage in critical action differently than youth who are U.S. citizens. Undocumented youth face barriers in access to civic and political institutions. For example, undocumented people cannot vote and face risks when interfacing with the government (e.g., completing the U.S. Census). Along with risk, there is likely variability in the extent to which undocumented Black and Latinx youth are motivated to engage in critical action *because of* their undocumented status. While some undocumented individuals might limit critical action for fear of deportation (Abrego, 2011), others may find strength in engaging in critical action that advances the well-being of undocumented communities (DeAngelo et al., 2016; Perez et al., 2010; Suárez-Orozco et al., 2015). Undocumented Black and Latinx youth's critical action might depend on the level of support they receive (Gonzales, 2016; Katsiaficas et al., 2019), which may stem from conversations with parents about family documentation status, parental ethnic-racial socialization, and involvement in community groups that support critical action. However, the risks associated with revealing one's undocumented status, particularly with respect to protecting the identities and safety of friends or family, may limit youth engagement (Enriquez & Saguy, 2016).

From the initial evidence presented in these above sections, there is a complex interplay between factors such as ethnic-racial identity, ethnic-racial socialization, coalition building, and citizenship status in predicting critical action, and how these predictive factors may vary based on action type, motivation, and risk. These associations may also differ based on the racial-ethnic background and sociopolitical history of each group, further underscoring the need for within-group studies. By attending to specificity in critical action measurement, scholars are better equipped to analyze the barriers and facilitators of action in Black and Latinx youth.

Developmental Considerations for Critical Action

Next, given the overwhelming focus on critical action in adolescence, we describe considerations for measuring critical action during childhood and the transition to adulthood.

Critical Action in Childhood

Research on critical action is less prevalent among child samples as compared to adolescent samples. Children cannot typically engage in critical action independently, thus we must expand our measurement and conceptualization of critical action to include behaviors that are both reasonable and feasible in childhood. One area that scholars have considered is prosocial behavior. Prosocial behaviors include helping behavior (e.g., helped a classmate with homework), sharing (e.g., shared school supplies with a classmate), concern for others, and volunteering (Lozada et al., 2017; Syvertsen et al., 2015). Children spend much of their time in classrooms, playgrounds, and neighborhoods where they learn civic principles and processes. As children engage in prosocial behaviors, they may do so to reduce unequal opportunities that they begin to recognize as unfair (Hope, 2016). For instance, a child can help a neighbor clean up their property as part of recognition of limited government services in their community. Capturing nascent critical action in childhood has implications for implementing developmentally appropriate critical action opportunities during early encounters with oppression.

Critical Action in the Transition to Adulthood

Two important contexts for critical action in adulthood are college and career. Black and Latinx college students use critical action to seek racial justice on college campuses and in the community at-large (Linder et al., 2019). For example, the activism of undocumented Latinx college students has brought national attention to the passage and implementation of the Development, Relief, and Education for Alien Minors (DREAM) Act (DeAngelo et al., 2016; Katsiaficas et al., 2019). Similarly, Black college student activism has amplified the Black Lives Matter Movement and challenged police brutality (Mwangi et al., 2018; White, 2016). The quantitative measurement of such activism and its impact is burgeoning as scholars employ multimethod approaches (e.g., media content analyses) to capture these historic moments as they unfold (Zuckerman et al., 2019).

Black and Latinx students continue engagement beyond college. Among a national sample of adults, Black adults had the highest levels of community engagement and had the fastest increase in political engagement from ages 18–24 when compared to Asian, Latinx, and White adults (Wray-Lake et al., 2020). Black adults also demonstrated the highest level of political engagement at age 30 (Wray-Lake et al., 2020). Latinx adults who felt a strong sense of common fate with other Latinx folks were connected to communities with undocumented populations, and who had previously engaged in protests were most supportive of immigration activism throughout their adulthood (Wallace & Zepeda-Millán, 2020). Given that the transition to adulthood involves balancing career, family, and other life responsibilities, Black and Latinx adults might shift their critical action to meet other responsibilities. Critical action might also extend into career choices as a means to stay connected to seeking social justice change for their communities. Recent work suggests that youth who are actively involved in critical action are more likely to choose majors and careers that integrate their interests in social justice (Connor, 2011; Nicholas et al., 2019). Given these mixed findings, future measures of critical action must consider the longitudinal impacts of sustained critical action among Black and Latinx communities, and the factors that facilitate or slow such engagement as youth transition to adulthood.

Recommendations for Research and Practice

Given the current advances in research on youth critical action, we have three specific recommendations to better understand the causes and correlates of youth critical action. Overall, scholars should choose critical action measures that align with the research question and sample population and are developmentally and contextually relevant.

Recommendation #1: Effective Measurement of Critical Action

Measures of critical action should be developmentally appropriate across childhood, adolescence, and emerging adulthood. There are several ways to improve measurement of civic action. Importantly, measures of critical action should be domain specific and inclusive of a variety of culturally relevant and developmentally appropriate types of action. Researchers have operated under the assumption that critical action in one domain is predicted by and predictive of similar factors of critical action in another domain. However, youth may be inclined toward critical action in one domain (e.g., anti-racism) but not equally inclined to pursue critical action in another domain (e.g., environmentalism). There have been some discrepant findings regarding factors that predict critical action for Black and Latinx youth. These discrepancies may be resolved through measures that consider specific domains and broadly define what types of behaviors constitute critical action. This may be especially true as researchers find that sociocultural factors such as racial identity, gender socialization, or racism may be differentially related to critical action across various dimensions (Hope et al., 2019). Measures of critical action should also consider motivation and risk. Low and high-risk activism occur with different prevalence and are predicted differently (Hope et al., 2019). Finally, measures of critical action should explicitly consider time. Many measures of critical action ask youth to reflect retrospectively on actions (Aldana et al., 2019; Diemer et al., 2017) or intentions for future actions (Hope et al., 2019). Measurement should take into account smaller increments of time for retrospective reports, including daily, weekly, or monthly reports. Further, measurement must account for actions that are inherently less frequent (e.g., voting in a Presidential election) to explain how engagement may differ in a given time period.

Recommendation #2: Critical Action Before and Beyond Adolescence

The vast majority of research on critical action has focused on adolescence (Diemer et al., 2020; Heberle et al., 2020). Scholars suggest that critical action can begin in childhood and continue well into adulthood (Heberle et al., 2020). In childhood research, it is important to understand the early experiences of critical action (e.g., prosocial behavior) and how those early experiences shape critical action into adolescence. Researchers should also use longitudinal methods to examine how sociocultural factors and critical action in adolescence translate into critical action in and beyond emerging adulthood. This is important as researchers find that early political involvement predicts lifelong political engagement (Wray-Lake et al., 2020; Wray-Lake & Abrams, 2020). More research is needed to understand if these same trends extend from traditional politics into critical action that prioritizes liberation.

Recommendation #3: Critical Action in Practice

We also recommend that research on youth critical action includes youth. Actions that are deemed critical to researchers may not be deemed critical by youth. Thus, youth should be given opportunities to co-create research to define critical action and how youth understand and experience barriers and facilitators to critical action (Aldana et al., 2019). This will include qualitative approaches where youth interview each other to define critical action according to current youth trends and perspectives. This also includes participatory methods (e.g., photovoice and youth participatory action research) that explicitly connect critical action research to opportunities for youth critical action. Youths' voices should be centered in how critical action is defined (Aldana et al., 2019). For example, behaviors that are deemed critical by scholars may not be deemed critical by youth. Thus, youth should be given opportunities to co-create research and organize spaces that focus on understanding and increasing their critical action. As youth develop over time, these participatory methods are also especially well-suited to capture changes in what actions 'count' as critical; it is possible that developmental changes and/or changes in one's understanding of ethnic-racial inequalities may cause an action youth were performing all along to become progressively more critical in nature. Ultimately, by including youth in the research process, scholars may come to a more complete understanding of the nature of critical action, as well as the barriers, facilitators, antecedents, and consequences of critical action.

Conclusion

This chapter reviewed the current conceptualization and measurement of critical action and offered suggestions on how to advance research on the construct. For a more comprehensive understanding of youth critical action, we encourage future research to ground critical action measures in groups' varied ethnic-racial experiences (particularly examining other marginalized group histories, such as in Asian American and Indigenous groups), account for the varied behaviors youth may use, and measure intention, frequency, and risk of critical action. We also recommend that scholars consider sociocultural facilitators and barriers of critical action with the aim of understanding the social structures and psychological factors that promote, or impede, youths' access to opportunities to challenge social injustice. In all, research on youth critical action should be grounded in their sociocultural experiences, which may differ across childhood, adolescence, and adulthood.

References

Abrego, L. J. (2011). Legal consciousness of undocumented Latinos: Fear and stigma as barriers to claims-making for first- and 1.5-generation immigrants. *Law & Society Review, 45*(2), 337–370. https://doi.org/cpgt9d

Aldana, A., Richards-Schuster, K., & Checkoway, B. (2016). Dialogic pedagogy for youth participatory action research: Facilitation of an intergroup empowerment program. *Social Work with Groups, 39*, 339–358. https://doi.org/gb83

Aldana, A., Bañales, J., & Richards-Schuster, K. (2019). Youth anti-racist engagement: Conceptualization, development, and validation of an anti-racism action scale. *Adolescent Research Review, 4*(4), 369–381. https://doi.org/gb82

Anyiwo, N., Bañales, J., Rowley, S. J., Watkins, D. C., & Richards-Schuster, K. (2018). Sociocultural influences on the sociopolitical development of African American youth. *Child Development Perspectives, 12*(3), 165–170. https://doi.org/f5rm

Anyiwo, N., Palmer, G. J., Garret, J. M., Starck, J. G., & Hope, E. C. (2020). Racial and political resistance: An examination of the sociopolitical action of racially marginalized youth. *Current Opinion in Psychology, 35*, 86–91. https://doi.org/f9c9

Ballard, P. J. (2015). Longitudinal links between discrimination and civic development among Latino and Asian adolescents. *Journal of Research on Adolescence, 26*(4), 723–737. https://doi.org/f9fcj2

Bañales, J., Aldana, A., Richards-Schuster, K., Flanagan, C. A., Diemer, M. A., & Rowley, S. J. (2019). Youth anti-racism action: Contributions of youth perceptions of school racial messages and critical consciousness. *Journal of Community Psychology, 49*(8), 3079–3100. https://doi.org/fkbf

Boulianne, S., Koc-Michalska, K., & Bimber, B. (2020). Mobilizing media: Comparing TV and social media effects on protest mobilization. *Information, Communication & Society, 23*(5), 642–664. https://doi.org/ggkt4w

Causadias, J. M., & Umaña-Taylor, A. J. (2018). Reframing marginalization and youth development: Introduction to the special issue. *American Psychologist, 73*(6), 707–712. https://doi.org/gd6g7s

Checkoway, B., & Aldana, A. (2013). Four forms of youth civic engagement for diverse democracy. *Children and Youth Services Review, 35*(11), 1894–1899. https://doi.org/f5gwms

Christophe, N. K., Martin Romero, M. Y., Hope, E. C., & Stein, G. L. (2022). Critical civic engagement in Black college students: Interplay between discrimination, racial centrality, and preparation for bias. *American Journal of Orthopsychiatry, 92*, 144.

Coll, C. G., Lamberty, G., Jenkins, R., McAdoo, H. P., Crnic, K., Wasik, B. H., & García, H. V. (1996). An integrative model for the study of developmental competencies in minority children. *Child Development, 67*(5), 1891–1914. https://doi.org/cbjhg7

Connor, J. O. (2011). Youth organizers as young adults: Their commitments and contributions. *Journal of Research and Adolescence, 21*(4), 923–942. https://doi.org/cgb2cn

Corning, A. F., & Myers, D. J. (2002). Individual orientation toward engagement in social action. *Political Psychology, 23*, 703–729. https://doi.org/cqrz2s

DeAngelo, L., Schuster, M. T., & Stebleton, M. J. (2016). California DREAMers: Activism, identity, and empowerment among undocumented college students. *Journal of Diversity in Higher Education, 9*(3), 216–230. https://doi.org/f84x7g

Dessel, A., & Rogge, M. E. (2008). Evaluation of intergroup dialogue: A review of the empirical literature. *Conflict Resolution Quarterly, 26*(2), 199. https://doi.org/d9rr5s

Diemer, M. A. (2012). Fostering marginalized youths' political participation: Longitudinal roles of parental political socialization and youth sociopolitical development. *American Journal of Community Psychology, 50*(1–2), 246–256. https://doi.org/fx28sp

Diemer, M. A., McWhirter, E. H., Ozer, E. J., & Rapa, L. J. (2015). Advances in the conceptualization and measurement of critical consciousness. *The Urban Review, 47*(5), 809–823. https://doi.org/gf82x6

Diemer, M. A., Rapa, L. J., Park, C. J., & Perry, J. C. (2017). Development and validation of the critical consciousness scale. *Youth & Society, 49*(4), 461–483. https://doi.org/gf82x9

Diemer, M. A., Pinedo, A., Bañales, J., Mathews, C. M., Frisby, M. B., Harris, E. M., & McAlister, S. (2020). (Re)centering action in critical consciousness. *Child Development Perspectives.* Advanced Online Publication. https://doi.org/gh7kq6

Ekman, J., & Amnå, E. (2012). Political participation and civic engagement: Towards a new typology. *Human Affairs, 22*(3), 283–300. https://doi.org/f23gvr

Enriquez, L. E., & Saguy, A. C. (2016). Coming out of the shadows: Harnessing a cultural schema to advance the undocumented immigrant youth movement. *American Journal of Cultural Sociology, 4*(1), 107–130. https://doi.org/gbz7

Godfrey, E. B., & Burson, E. (2018). Interrogating the intersections: How intersectional perspectives can inform developmental scholarship on critical consciousness. *Science, 161*, 17–38. https://doi.org/ggdnfq

Gonzales, R. G. (2016). *Lives in limbo: Undocumented and coming of age in America.* University of California Press.

Hayat, N., Kiesa, A., Lundberg, K., & Medina, A. (2020). *Young women's political engagement in elections and beyond.* Center for Information & Research on Civic Learning and Engagement. https://circle.tufts.edu/latest-research/young-womens-political-engagement-elections-and-beyond

Heberle, A. E., Rapa, L. J., & Farago, F. (2020). Critical consciousness in children and adolescents: A systematic review, critical assessment, and recommendations for future research. *Psychological Bulletin, 146*(6), 525. https://doi.org/gb86

Hope, E. C. (2016). Preparing to participate: The role of youth social responsibility and political efficacy on civic engagement for Black early adolescents. *Child Indicators Research, 9*(3), 609–630. https://doi.org/f8wxgh

Hope, E. C., & Bañales, J. (2019). Black early adolescent critical reflection of inequitable sociopolitical conditions: A qualitative investigation. *Journal of Adolescent Research, 34*(2), 167–200. https://doi.org/gfgzkg

Hope, E. C., & Jagers, R. J. (2014). The role of sociopolitical attitudes and civic education in the civic engagement of black youth. *Journal of Research on Adolescence, 24*(3), 460–470. https://doi.org/10.1111/jora.12117

Hope, E. C., & Spencer, M. B. (2017). Civic engagement as an adaptive coping response to conditions of inequality: An application of Phenomenological Variant of Ecological Systems Theory (PVEST). In N. Cabrera & B. Leyendecker (Eds.), *Handbook on positive development of minority children and youth* (pp. 421–435). Springer Science Business Media. https://doi.org/gb88

Hope, E., Keels, M., & Durkee, M. (2016). Participation in Black Lives Matter and deferred action for childhood arrivals: Modern activism among Black and Latino college students. *Journal of Diversity in Higher Education, 9*, 203–215. https://doi.org/f84zd5

Hope, E. C., Velez, G., Offidani-Bertrand, C., Keels, M., & Durkee, M. I. (2018). Political activism and mental health among Black and Latinx college students. *Cultural Diversity & Ethnic Minority Psychology, 24*(1), 26–39. https://doi.org/gcwmgf

Hope, E. C., Gugwor, R., Riddick, K. N., & Pender, K. N. (2019). Engaged against the machine: Institutional and cultural racial discrimination and racial identity as predictors of activism orientation among Black youth. *American Journal of Community Psychology, 63*(1–2), 61–72. https://doi.org/gftk4q

Hope, E. C., Cryer-Coupet, Q. R., & Stokes, M. N. (2020a). Race-related stress, racial identity, and activism among young Black men: A person-centered approach. *Developmental Psychology, 56*(8), 1484–1495. https://doi.org/gb87

Hope, E. C., Smith, C. D., Cryer-Coupet, Q. R., & Briggs, A. S. (2020b). Relations between racial stress and critical consciousness for Black adolescents. *Journal of Applied Developmental Psychology, 70*, 101184. https://doi.org/gjkshp

Hughes, D., Rodriguez, J., Smith, E. P., Johnson, D. J., Stevenson, H. C., & Spicer, P. (2006). Parents' ethnic-racial socialization practices: A review of research and directions for future study. *Developmental Psychology, 42*(5), 747–770. https://doi.org/bdqxq6

Jensen, L. A. (2008). Immigrants' cultural identities as sources of civic engagement. *Applied Development Science, 12*(2), 74–83. https://doi.org/cgfv82

Katsiaficas, D., Volpe, V., Raza, S. S., & Garcia, Y. (2019). The role of campus support, undocumented identity, and deferred action for childhood arrivals on civic engagement for Latinx undocumented undergraduates. *Child Development, 90*(3), 790–807. https://doi.org/gjkzst

Kerrison, E. M., Cobbina, J., & Bender, K. (2018). "Your pants won't save you" why black youth challenge race-based police surveillance and the demands of Black respectability politics. *Race and Justice, 8*(1), 7–26. https://doi.org/ggk3mj

Kiang, L., Christophe, N. K., & Stein, G. L. (2021). Differentiating pathways between ethnic-racial identity and critical consciousness. *Journal of Youth and Adolescence, 50*, 1–15. https://doi.org/10.1007/s10964-021-01453-9

Linder, C., Quaye, S. J., Lange, A. C., Roberts, R. E., Lacy, M. C., & Okello, W. K. (2019). "A student should have the privilege of just being a student": Student activism as labor. *The Review of Higher Education, 42*(5), 37–62. https://doi.org/gb89

Lozada, F. T., Jagers, R. J., Smith, C. D., Bañales, J., & Hope, E. C. (2017). Prosocial behaviors of Black adolescent boys: An application of a sociopolitical development theory. *Journal of Black Psychology, 43*(5), 493–516. https://doi.org/gbjrhv

Lundberg, K., Kiesa, A., & Medina, A. (2021). *The 2020 election is over, but young people believe in continued engagement.* Center for Information & Research on Civic Learning and Engagement. https://circle.tufts.edu/latest-research/2020-election-over-young-people-believe-continued-engagement

Mathews, C. J., Medina, M. A., Bañales, J., Pinetta, B. J., Marchand, A. D., Agi, A. C., Miller, S. M., Hoffman, A. J., Diemer, M. A., & Rivas-Drake, D. (2019). Mapping the intersections of adolescents' ethnic-racial identity and critical consciousness. *Adolescent Research Review, 5*, 363–379. https://doi.org/f5rq

McWhirter, E. H., & McWhirter, B. T. (2016). Critical consciousness and vocational development among Latina/o high school youth: Initial development and testing of a measure. *Journal of Career Assessment, 24*(3), 543–558. https://doi.org/f8xvkd

Mwangi, C. A. G., Thelamour, B., Ezeofor, I., & Carpenter, A. (2018). "Black elephant in the room": Black students contextualizing campus racial climate within US racial climate. *Journal of College Student Development, 59*(4), 456–474. https://doi.org/gf9mzz

National Center for Education Statistics. (2018). *NAEP report card: Civics* (National Assessment of Educational Progress (NAEP): Civics Assessments). U.S. Department of Education, Institute of Education Sciences. https://www.nationsreportcard.gov/civics

Nicholas, C., Eastman-Mueller, H., & Barbich, N. (2019). Empowering change agents: Youth organizing groups as sites for sociopolitical development. *American Journal of Community Psychology, 63*(1–2), 46–60. https://doi.org/gfvmx9

Ortiz, P. (2018). *An African American and Latinx history of the United States* (Vol. 4).

Pancer, S. M., Pratt, M., Hunsberger, B., & Alisat, S. (2007). Community and political involvement in adolescence: What distinguishes the activists from the uninvolved? *Journal of Community Psychology, 35*(6), 741–759. https://doi.org/dgh6tw

Perez, W., Espinoza, R., Ramos, K., Coronado, H., & Cortes, R. (2010). Civic engagement patterns of undocumented Mexican students. *Journal of Hispanic Higher Education, 9*(3), 245–265. https://doi.org/10.1177/1538192710371007

Pinedo, A., Durkee, M. I., Diemer, M., & Hope, E. C. (2021). Disentangling racial discrimination and critical action among Black and Latinx college students: What role do peers play? *Cultural Diversity & Ethnic Minority Psychology, 27*(3), 546–557. https://doi.org/gb9b

Priest, N., Walton, J., White, F., Kowal, E., Baker, A., & Paradies, Y. (2014). Understanding the complexities of ethnic-racial socialization processes for both minority and majority groups: A 30-year systematic review. *International Journal of Intercultural Relations, 43*, 139–155. https://doi.org/gb9c

Rapa, L. J., Bolding, C. W., & Jamil, F. M. (2020). Development and initial validation of the short critical consciousness scale (CCS-S). *Journal of Applied Developmental Psychology, 70*, 101164. https://doi.org/10.1016/j.appdev.2020.101164

Richards-Schuster, K., & Aldana, A. (2013). Learning to speak out about racism: Youths' insights on participation in an intergroup dialogues program. *Social Work With Groups, 36*(4), 332–348. https://doi.org/gb9d

Schmitz, R. M., Coley, J. S., Thomas, C., & Ramirez, A. (2020). The cyber power of marginalized identities: Intersectional strategies of online LGBTQ+ Latinx activism. *Feminist Media Studies*, 1–20. https://doi.org/gb9f

Seaton, E. K., Gee, G. C., Neblett, E., & Spanierman, L. (2018). New directions for racial discrimination research as inspired by the integrative model. *American Psychologist, 73*(6), 768–780. https://doi.org/gd7msd

Sobo, E. J., Lambert, H., & Heath, C. D. (2020). More than a teachable moment: Black lives matter. *Anthropology & Medicine, 27*(3), 243–248. https://doi.org/gg5jhp

Society on Research in Child Development. (2020). *Addressing inequities in education during the COVID-19 Pandemic: How education policy and schools can support historically and currently marginalized children and youth*. Society for Research in Child Development. https://www.srcd.org/sites/default/files/resources/FINAL_AddressingInequalitiesVolume-092020.pdf

Stepick, A., Stepick, C. D., & Labissiere, Y. (2008). South Florida's immigrant youth and civic engagement: Major engagement: Minor differences. *Applied Development Science, 12*(2), 57–65. https://doi.org/b79gkz

Suárez-Orozco, C., Katsiaficas, D., Birchall, O., Alcantar, C. M., Hernandez, E., Garcia, Y., Michikyan, M., Cerda, J., & Teranishi, R. T. (2015). Undocumented undergraduates on college campuses: Understanding their challenges and assets and what it takes to make an undocufriendly campus. *Harvard Educational Review, 85*(3), 427–463. https://doi.org/gb9p

Syvertsen, A. K., Wray-Lake, L., & Metzger, A. (2015). *Youth civic and character measures toolkit*. Search Institute.

Szymanski, D. M. (2012). Racist events and individual coping styles as predictors of African American activism. *Journal of Black Psychology, 38*(3), 342–367. https://doi.org/ccx55h

Terriquez, V. (2015). Intersectional mobilization, social movement spillover, and queer youth leadership in the immigrant rights movement. *Social Problems, 62*(3), 343–362. https://doi.org/f7mz7b

Terriquez, V. (2017). Legal status, civic organizations, and political participation among Latino young adults. *The Sociological Quarterly, 58*(2), 315–336. https://doi.org/gbkc

Terriquez, V., Brenes, T., & Lopez, A. (2018). Intersectionality as a multipurpose collective action frame: The case of the undocumented youth movement. *Ethnicities, 18*(2), 260–276. https://doi.org/gdfqgf

Thomas, O. (2001). *Collective action, sociopolitical attitudes, and social change preferences: A multidimensional measure of Black activism*. (Doctoral Dissertation, Michigan State University), ProQuest Dissertations Publishing.

Thomas, A. J., Barrie, R., Brunner, J., Clawson, A., Hewitt, A., Jeremie-Brink, G., & Rowe-Johnson, M. (2014). Assessing critical consciousness in youth and young adults. *Journal of Research on Adolescence, 24*(3), 485–496. https://doi.org/ggsphs

Umaña-Taylor, A. J., Quintana, S. M., Lee, R. M., Cross, W. E., Rivas-Drake, D., Schwartz, S. J., Syed, M., Yip, T., Seaton, E., & Ethnic and Racial Identity in the 21st Century Study Group. (2014). Ethnic and racial identity during adolescence and into young adulthood: An integrated conceptualization. *Child Development, 85*(1), 21–39. http://doi.org/f5srww

Valenzuela, S. (2013). Unpacking the use of social media for protest behavior: The roles of information, opinion expression, and activism. *American Behavioral Scientist, 57*(7), 920–942. https://doi.org/f425zn

Wallace, S. J., & Zepeda-Millán, C. (2020). Do Latinos still support immigrant rights activism? Examining Latino attitudes a decade after the 2006 protest wave. *Journal of Ethnic and Migration Studies, 46*(4), 770–790. https://doi.org/gb9g

Watts, R. J., & Flanagan, C. (2007). Pushing the envelope on youth civic engagement: A developmental and liberation psychology perspective. *Journal of Community Psychology, 35*(6), 779–792. https://doi.org/dg33xm

Watts, R. J., & Hipolito-Delgado, C. P. (2015). Thinking ourselves to liberation?: Advancing sociopolitical action in critical consciousness. *The Urban Review, 47*(5), 847–867. https://doi.org/gb9j

White, K. (2016). Black lives on campuses matter: Reflecting on the rise of the new Black student movement. *Soundings, 63*(63), 86–97. https://doi.org/gbj9

White-Johnson, R. L. (2012). Prosocial involvement among African American young adults: Considering racial discrimination and racial identity. *Journal of Black Psychology, 38*(3), 313–341. https://doi.org/bzh6dt

Williams, D. R., & Mohammed, S. A. (2009). Discrimination and racial disparities in health: Evidence and needed research. *Journal of Behavioral Medicine, 32*(1), 20–47. https://doi.org/ch3dsc

Wray-Lake, L., & Abrams, L. S. (2020). Pathways to civic engagement among urban youth of color. *Monographs of the Society for Research in Child Development, 85*(2), 7–154. https://doi.org/ghqq47

Wray-Lake, L., Arruda, E. H., & Schulenberg, J. E. (2020). Civic development across the transition to adulthood in a national US sample: Variations by race/ethnicity, parent education, and gender. *Developmental Psychology, 56*(10), 1948. https://doi.org/f5rv

Xenos, M., Vromen, A., & Loader, B. D. (2014). The great equalizer? Patterns of social media use and youth political engagement in three advanced democracies. *Information, Communication & Society, 17*(2), 151–167. https://doi.org/gc7mzf

Zuckerman, E., Matias, J. N., Bhargava, R., Bermejo, F., & Ko, A. (2019). Whose death matters? A quantitative analysis of media attention to deaths of Black Americans in police confrontations, 2013–2016. *International Journal of Communication, 13*, 4751–4777. https://hdl.handle.net/1721.1/123475

Chapter 11
Toward Developmental Science That Meets the Challenges of 2044: Afrofuturist Development Theory, Design, and Praxis

Brendesha M. Tynes, Matthew Coopilton, Joshua Schuschke, and Ashley Stewart

> THERE ARE BLACK PEOPLE IN THE FUTURE. –Alisha B. Wormsley

> Little did we know at the beginning of this term that Afrofuturism would become now, as we've been forced to adapt to new uses of technology- not just to complete the school term, but just to go about our lives. –Tanarive Due

> We believe it is our right and responsibility to write ourselves into the future. –Walidah Imarisha and adrienne maree brown

Artist Alisha Wormsley's Afrofuturist interdisciplinary body of work includes a billboard that reads, "There are Black People in the Future." When the billboard was placed in a recently gentrified Philadelphia neighborhood, developers removed the text. Wormsley's website (https://alishabwormsley.com/) notes the removal transformed the sentence into a movement. The artist encouraged others to use these words for the betterment of the world around them. We use them in our efforts to transform developmental science. Specifically, we believe that the possibilities of a liberated Black future, where Black people not only exist, but thrive, are essential to healthy socio-emotional, cultural, and academic development (Love, 2019). This chapter outlines Afrofuturist Development, a theory, design lens, and praxis for Black child, adolescent and emerging adult thrival.

B. M. Tynes (✉) · M. Coopilton
University of Southern California, Los Angeles, CA, USA
e-mail: btynes@rossier.usc.edu

J. Schuschke
Johns Hopkins University, Baltimore, MD, USA

A. Stewart
Georgia State University, Atlanta, GA, USA

© The Author(s), under exclusive license to Springer Nature Switzerland AG 2023
D. P. Witherspoon, G. L. Stein (eds.), *Diversity and Developmental Science*,
https://doi.org/10.1007/978-3-031-23163-6_11

Though we are moving toward a time where the US will be majority-minority, this demographic shift does not automatically mean positive outcomes for all people of color. All of the authors have at one point lived in California, a state that is 39.4% Latine,[1] 34.7% white, 15.1% Asian, 14.6% multi-racial, and 5.4% African American (McGhee et al., 2021). Yet, we have witnessed rampant anti-Blackness in the state, either through our research or personally or both. The Education Trust-West's (2015) report "Black Minds Matter: Supporting the Educational Success of Black Children in California" outlines how Black K-12 students have the worst performance on just about every indicator from school suspensions to graduation rates. Our research supports these findings of anti-Black teacher bias through observations and interviews. To provide an example, we conducted research in a magnet high school that was majority Latine and African American. In this majority-minority school, teachers often engaged in anti-Black racism. One of the computer science instructors had very clear ideas about who could do computer science well and who could not. Demonstrating lowered expectations of Black students, the instructor described Black students as "not able to do anything," "knuckleheads," and "slower" than other students (Schuschke et al., 2023). We have numerous accounts like this one, suggesting Black children have vastly different school experiences, sometimes even in the same classrooms, compared to their white, Asian, and Latine counterparts at "diverse" schools.

This teacher's writing Black students off in computer science is a continuation of a decades-long process of writing Black people out of the future. In describing the digital boom of the 1990s, Nelson (2002) calls the promise of a raceless, disembodied future a "founding fiction of the digital age" (p. 1). She further notes that the digital divide narrative along with the myth of a raceless utopia were frames of reference for thinking about race and gender in the digital age. "In these frameworks the technologically enabled future is by its very nature unmoored from the past and from people of color" (Nelson, 2002, p. 6). In this case, Black people are disappeared from the future. Nelson's argument is that the "myth of black disingenuity with technology" (borrowing from historian Evelynn Hammonds) ignores Black people's role in modernization and their contributions to technological development (p. 6).

The tech industry appears to also buy into this myth. Despite the fact that Black people are innovators across a wide range of technologies and platforms, the industry has displayed an unwillingness to accept Black workers or promote Black ideas. Google and Facebook both report that Black tech workers only make up 2% of their companies' workforces. Meanwhile, Microsoft and Apple have similar abysmal numbers at 3% and 6%, respectively (Lago, 2020). This lack of representation has created hostile work environments for those Black individuals who are there (Barton, 2021), and a loss of opportunity to benefit from Black people's deep wells of creativity and innovation for those who are not. The historical and contemporary pervasiveness of racism in the tech industry, alongside the inability to foster possibilities within schools, has systemically limited many Black people from being able to reach their full potential.

---

[1] We use Latine as a gender neutral form of the word Latino.

These forms of systemic racism also extend to developmental science. Scholars note the multidisciplinary field has recurring conceptual flaws in which Black populations are labeled as "pathological" or "nonnormative" and that there is a lack of cultural competence among researchers (Swanson et al., 2003). Psychology, a field from which many developmental scientists hail, has a history fraught with attempts to prove Black inferiority under the guise of science (Guthrie, 2004). Profound inequities also persist in grants where white researchers are 1.7 times more likely than African American/Black researchers to be funded (Taffe & Gilpin, 2021). This compromises the quality of the science that is produced, as Black scholars are more likely to have the training needed to study these populations.

Despite the wide range of problems, developmental science and education research have adequately addressed how race and racism impact aspects of development (e.g., García Coll et al., 1996). They have also done groundbreaking work on the importance of attending to culture in learning and development (e.g., Boykin, 1994; Lee, 2007). What has not been accounted for is systemic efforts to disappear and erase Black people from the future. This is important, particularly considering, as Tanarive Due (Bruce, 2020) points out, the (Afro)future is now. Developmental science needs theories that incorporate Black people's experiences of anti-Black racism, while at the same time leveraging their technological prowess and culture. Doing so has the potential to create a profoundly different future for Black people and the world. We argue Afrofuturism can provide a lens that addresses this gap when integrated with developmental and educational frameworks, African American psychology, and liberation psychology. This chapter describes Afrofuturism and Afrofuturist Development theory, design, and praxis, which integrates key tenets, findings, and frameworks from extant literature. We then outline principles of Afrofuturist Development theory and provide a case study that shows the theory's utility as a design lens and praxis.

What Is Afrofuturism?

Though scholars have noted that Afrofuturism has existed earlier in history, the term was not coined until the 1990s when critic Mark Dery (1994) called it "speculative fiction that treats African American themes and addresses African American concerns in the context of twentieth century techno-culture- and more generally, African American signification that appropriates images of technology and a prosthetically enhanced future..." (p. 180). Founders like Alondra Nelson advanced the field in reaction to popular mythology around Black people and technology in the 1990s. It was in this context that Nelson started the online community called Afrofuturism in the fall of 1998. She defines Afrofuturism as "African American voices with other stories to tell about culture, technology and things to come" (Nelson, 2002, p. 9). Other early definitions show how the field emerged as an intervention on this attempted erasure, disappearing and painting of blackness as primitive (Eshun, 2003; Nelson, 2002). Kodwo Eshun (2003) describes Afrofuturism as "a program for recovering the histories of counter-futures created in a century hostile to Afrodiasporic projection…a tool kit developed for and by Afrodiasporic intellectuals," and he predicted that "the imperative to code, adopt, adapt, translate, misread, rework and revision these concepts...is likely to persist in decades to come" (p. 301).

Reynaldo Anderson and Charles Jones (2016) note how twentieth-century Afrofuturism was primarily concerned with techno-culture, the digital divide, technology, music, and literature of the West. They expand early definitions, calling Afrofuturism a Pan-African movement and proposing Afrofuturism 2.0. They define this version as "an early twenty-first century technogenesis of Black identity reflecting counter histories, hacking and or appropriating the influence of network software database logic, cultural analytics, deep remixability, neurosciences, enhancement and augmentation, gender fluidity, posthuman possibility, the speculative sphere, with transdisciplinary applications" (Anderson & Jones, 2016, p. x). It has five dimensions: metaphysics; aesthetics, including art, music and literature; theoretical and applied science; social sciences; and programmatic spaces like exhibitions, community organizations, and online forums. Each of these dimensions are interwoven in our conceptions of Afrofuturism as it applies to developmental science.

In her review of Afrofuturism scholarship, Toliver (n.d.) notes that there is no consensus on what Afrofuturism is. The review also outlines themes in scholarship which include a focus on combating dystopian realities—for example, imagining worlds free of racism, a way of reclaiming and recovering lost histories, as well as addressing diverse issues such as the environment, religion, and gender. Despite the range of topics covered in Afrofuturism scholarship, some scholars argue there is room for expansion. For example, Lavender III (2019) calls for more of a focus on the fantastic or native science that may include spirit work. Similarly, Susana Morris' work argues an Afrofuturist Feminism (Morris, 2012, 2016) whose defining features include the creation of parallel feminist universes, remixing of dominant futurist discourses, and Black women as agents of change (Morris, 2016). She notes that it is a critical epistemology reflecting the shared tenets of Afrofuturism and Black feminist thought, which can be used to understand Black speculative fiction (Morris, 2012). Like Morris and other scholars, we propose to extend Afrofuturism to developmental and education science.

Afrofuturist Development Theory, Design, and Praxis

Afrofuturist Development is a theory of Black thrival that centers the full humanity of Black children, adolescents, and emerging adults, drawing from developmental science, learning/educational science, Black psychology, liberation psychology, and Afrofuturism. Afrofuturist Development explains how optimal physical, psychological, emotional, cognitive, spiritual, academic, civic, and social outcomes are attained by allowing Black people to be their glorious selves, by meeting individuals' developmental needs and nurturing/leveraging assets, including the wells of spirituality and limitless creativity Black young people possess, within and across contexts and developmental stages, by providing systems of equitable educational and work opportunities, by subverting and transcending anti-Blackness, and by a reorientation toward the past, present, future, and spacetime. Central to the theory is the idea that Black thrival can be realized when key community members, researchers, teachers, and clinicians are trained in Black people's histories, cultures, and development as well as critical race theory across fields (e.g., medicine, education, mental health, politics).

Afrofuturist Development theory as design and praxis positions Black children, adolescents, and emerging adults as creators of their present and futures, as people who have individual and collective power to resist anti-Black racism. At the same time, when putting the theory into practice, researchers and practitioners should recognize that anti-Black racism is not the totality of a Black person's being (Quashie, 2021). Below, we outline the theory's ten core principles, which are distinct, but closely related, along with the theories, experiences, and research that inform them. We also provide examples of design and possible applications.

1. Afrofuturist learning and development is multi-level, unfolds across developmental stages, and is the product of spiraling, racialized interactions between an individual, technology, and their environments.
2. Contexts in which children and young people learn, play, and work center their full humanity, as well as foster Black aliveness and innovations of speculative futures void of oppressions.
3. Black history knowledge, literacies, and historical consciousness are key building blocks for a liberated present and future; they also protect children and young people from negative outcomes that may result from risks in their environment.
4. Technological ingenuity and power are a part of every Black person's heritage and therefore if not already possessed or expressed, an individual is assumed to be highly capable of developing these strengths.
5. Critical consciousness and action are competencies taught in developmentally appropriate ways to support Black people as they navigate and resist oppressive media, technology, institutions and social practices.
6. Community members, teachers, employers, researchers, and clinicians must have training in Black children and young people's histories, technocultures, and development for optimal outcomes.
7. Systems of equity that resist anti-Black racism, create a sense of belonging, and provide Afrofuturist developmental learning and community possibilities are key drivers of optimal development.
8. Shared and unique developmental needs are met and developmental assets (e.g., imagination and creativity) leveraged and nurtured across contexts.
9. Toolkits for future physical and mental well-being are tailored to an individual's needs in a given setting.
10. Black children and young people, often powered by their Afrofuturistic selves, imagine, build, and reinvent liberated, pro-Black futures and technologies.

1. Afrofuturist Learning and Development Is Multi-Level, Unfolds Across Developmental Stages, and Is the Product of Spiraling, Racialized Interactions Between an Individual, Technology, and Their Environments For the first principle, we draw from extensions of Bronfenbrenner's Ecological Systems theory (1979). Expanding on the five core systems initially proposed by Bronfenbrenner, Johnson and Puplampu (2008) put forth the *techno-subsystem* as a dimension of the microsystem and include children's interactions with technological tools such as computers and phones used for communication, recreation, and information, as well as interpersonal interactions occurring via technological mediums. Johnson (2010) later extends the notion of the techno-subsystem by introducing the techno-microsystem to provide insights into the mechanisms by which technology and Internet use might contribute to an individual's development. Contrary to early theorists whose depictions of these processes have been two-dimensional, Johnson suggests that learning outcomes such as physical, social, cognitive, and emotional development are the products of continual "spiraling" interactions between individual characteristics and the use of technologies. These technologies can be used for the purposes of communication, recreation, and information (i.e., the techno-subsystem) across the child's immediate social contexts including home and school. Johnson's argument goes beyond the notion of bidirectional relationships between an individual and their environment to suggesting that all of these environments are interacting with each other and the individual simultaneously when technology is involved. It is these spiraling, simultaneous interactions that shape the development of youth.

Afrofuturist Development argues these spiraling interactions between the individual, technology, and their environment are inherently racialized. Black children and young people need tools to counter, resist, critique, navigate, cope, and create despite the racism and anti-Blackness built into technologies. Their family, communities, and schools play integral roles in this process. Family strengths such as adaptability of family roles and parent racial socialization, for example, can dramatically shape young peoples' online, schooling, and work experiences (Bentley et al., 2008; Tynes et al., 2016).

Principle one, as with each of the following principles of Afrofuturist development theory, can be applied across a range of contexts in a number of ways. Developmental scientists might apply this principle by considering research questions and designs that recognize the dynamic nature of how multiple system levels and contexts combine to impact outcomes, keeping in mind the strengths and challenges of individuals across contexts. The novel coronavirus pandemic which significantly impacted every system level, and the international shift from in-person to online learning in 2020 provides a timely case. In these settings, students were engaging with school through the use of technology while in their homes. As principle one suggests, these "spiraling" interactions of academic and home settings with technology uniquely shaped students' outcomes in profound ways.

An Afrofuturist developmental perspective takes this consideration a bit further by considering the racialization of students' experiences. In this case, Black students were disproportionately impacted by the coronavirus both directly in cases where rates of positive testing, hospitalizations, and death were higher in Black communities, but also indirectly, in cases where Black parents are more likely to be essential workers (Center for Disease Control, 2020) which has implications for how students experience distance learning. Black parents also took on the responsibility of educating their children when school districts failed; some designed curricula or provided experiences that helped students learn about Black history and culture (Edwards & Edwards, 2020). Adapting an Afrofuturist developmental perspective means examining how outcomes may shift multiple times based on events or policies in different contexts. It also means reimagining what manifestations learning and success can take, while engaging with Black students when making these decisions. Research designs should more expansively and holistically capture the learning that is taking place and the developmental competencies being achieved as a result of the ways these systems are shifting and interacting with the developing child or young person.

2. Contexts in Which Children and Young People Learn, Play, and Work Center Their Full Humanity, as Well as Foster Black Aliveness and Innovations of Speculative Futures Void of Oppressions

For principle two, we build on other developmental models that argue that Black people's positionality in terms of race, and their cultures—including their language and spirituality—should be centered. Over the last three and a half decades, developmental scientists, particularly those from racially minoritized backgrounds, have taken seriously the role of race in the development of Black children and youth. One of the most foundational models to guide the field in its efforts to center the racialized experiences of children and youth of color is García Coll and colleagues' (1996) integrative model for the study of developmental competencies in minority children. It asserts that an emphasis on the race-related experiences of youth in a white-dominated society is critical for research on their development. More recently, the integrative model has also been applied when examining young people's online experiences in digital settings (Tynes et al., 2019).

We extend this model to Afrofuturist Development and underscore the need to focus specifically on Black people because of their unique experiences with anti-Blackness in the US and the ways that they continue to create and innovate for the purposes of joy and liberation. We also argue, in line with many other scholars, that Black cultural practices, including their spirituality, for centuries have and continue to fuel their survival and thrival through community engagement, resource sharing, and fostering hope (Brown & Brown, 2003; Cone, 2010; Harding, 1997; Hope et al., 2019; Mattis et al., 2019; Neville et al., 2009). In a framework that explores the role of urban Black spirituality in the development of Black young people, Mattis and colleagues (2019) attend to the fact that Black people engage with faith and spirituality, not only in physical spaces, but in digital spaces as well. Similarly, other cultural practices and characteristics such as music and celebrations have been established for promoting positive development for Black youth (Evans, 2020; Hunter et al., 2016; Van Steenis, 2020). Developmental models that seek to represent and understand Black people in their full humanity, should have cultural and spiritual dimensions figured prominently.

Our idea of full humanity takes into account the fact that all (Black) people, all genders, sexual orientations, (dis)abilities, religions, and SES backgrounds are made, as Chanda Prescod-Weinstein (2021) notes, of the same stardust and deserve to live in their fullness. It is also informed by Na'im Akbar's (1984) conceptualization of the core African personality as a spiritual or divine substance that "reflects a deep inner sense of self that reaches back before contact with Europeans and unites the African person with everything in the universe" (Kambon & Bowen-Reid, 2009: p. 63). Obasi and Smith (2009) further note that being human is to be actively engaged with community; this is perhaps most well represented in the African philosophy of Ubuntu, a way of living that would affirm the statement "I am because we are." They also highlight how Akan metaphysics would define the person as both spiritual and physical with an interchange between the soul (which originates from the Supreme being), spirit (the sunsum or building block of all living things in the universe), and the body. In addition, a core part of our understanding of full humanity is what Kevin Quashie (2021) theorizes as "aliveness": "a quality of being, a term of habitat, a manner and aesthestic, a feeling-or many of them, circuits in an atmosphere…we are totality: we are and are of the universe; we are and are of a black world…" (p. 14). We understand aliveness in the Black world to be a way of being or imagined reality where Black people are human without qualification, where anti-Blackness is not the sum of who they are, and where they are worthy as they are. It means Black children, adolescents, and emerging adults are allowed to love, heal, develop, play, pursue their interests, fail, try again, grieve, fight for their rights, build their futures, or just be.

Applying principle two of Afrofuturist development theory would include a consideration of the full humanity of Black people and the role technology may play in expressing or exploring that humanity. When developing interventions, programs, and curricula, scholars/educators might consider intersectionality along with their rich cultures, including the expansive spiritual lives of Black young people and the innovative use of technology for the purposes of liberation and world-building in the face of systemic oppression. Scholars interested in educating through a lens of Black liberation highlight the importance of considering the role of spirituality and the Black church, in particular in the lives of Black children and young people (Coles & Stanley, 2021). For example, Emdin's (2016) work highlights how theological pedagogies might be leveraged to cultivate shared dialogical practices using call and response, a common practice unique to Black church culture. Educators would do well to learn from Black churches as sites of Black culture, political socialization, artistry, and spirituality if they want to enact Afrofuturist pedagogies that promote learning and development in Black young people.

As Black people deal with the pandemic, climate change, and systemic racism, it is also important to understand how they are embracing African spirituality for community and liberation (Burton, 2020). Perhaps the most popular example is Beyoncé representing herself as the Yoruba goddess Oshun in recent years (Grady, 2020). Oshun is the goddess of sweet waters, love, femininity, sensuality, and wealth. In the song, Black Parade, which was first released in June of 2020 after the death of

George Floyd, she invokes Oshun, Yemoja (another Yoruba goddess), and the ancestors. The lyrics (Beyoncé., 2020) read:

> I'm going back to the south
> I'm going back, back, back, back, back
> Where my roots ain't watered down
> Growing, growing like a Baobab tree
> Of life on fertile ground
> Ancestors put me on game
> Ankh charm on gold chains, with my Oshun energy…
> Baby sister reppin' Yemoja (Yemoja)
> Trust me, they gon' need an army
> Rubber bullets bouncing off me…
> Waist beads from Yoruba…
> Pandemic fly on the runway in my hazmat
> Children runnin' through the house and my art all Black
> Ancestors on the wall let the ghosts chit chat…

The song suggests she is not only connecting to the goddesses and ancestors, but that she, her family, and Black people can receive protection from them when protesting and during pandemic life. Developmental scientists could explore the role African spirituality might play in socio-political development, how it inspires creativity, or how it informs ways of coping through challenging times.

3. Black History Knowledge, Literacies, and Historical Consciousness Are Key Building Blocks for a Liberated Present and Future; They Also Protect Children and Young People from Negative Outcomes That May Result from Risks in Their Environment

Principle three is informed by research that highlights the importance of Black history knowledge and literacies. For optimal development, Black children and young people require a reorientation toward the past that counters the whitewashed curricula that they are exposed to in US schools and across the diaspora. King (2019) argued that Black history is generally taught through a white epistemic lens; therefore, Black history should be reframed through the "epistemologies, gazes, and imaginations" of Black people. Using Black histories and frameworks to enhance education serves the explicit goal of elevating Black students' intellectual prowess. In her book *Cultivating Genius*, Muhammad (2020) highlights how an Historically Responsive Literacy (HRL) framework encourages Black student development in four distinct ways: (a) Identity development: understanding the self, others, and their accompanying histories; (b) Skill development: gaining proficiency and experiencing joy when it comes to literacy across content areas; (c) Intellectual development: Knowledge of content and concepts with an expansive worldview; and (d) Criticality: Understanding power, injustice, and how to resist and dismantle systems. Not only do Black histories serve as protective factors for students navigating racist school environments (Chapman-Hilliard & Adams-Bass, 2016; King, 2019; Spencer, 1995), these frameworks promote healthy racial identity formations and provide an entry point into theorizing Black futures.

Black history knowledge and HRL are essential for the praxis of Afrofuturist development as any creation or imagining of a future must begin with the past. The first author's experience in the African History Club, led by her teacher Richard James, provides the first of two examples that show its importance across age groups. She was part of the club from grades 3–5 and during that time became a writer and editor for the club's newsletter, learned about a range of historical events and figures, and went on her first trip without any of her family where she learned about the history of Philadelphia. She was essentially able to practice her future as she went on to become the vice president of her senior class, president of the National Honor Society, and editor-in-chief of the Cass Technical High School yearbook. She later majored in history as an undergrad at Columbia University and then taught history in Detroit.

Our second example shows how educators using Black history are able to teach concepts centering Black figures, moments, and phenomenon in speculative contexts. For example, Love (2004) uses critical race theory's counter-storytelling to feature various Black historical figures in an imagined community meeting to critique the application of the landmark Brown v Board of Education ruling. From an Afrofuturist lens, educators who deploy these readings and teaching strategies are able to provide lessons that provide a road map to the future. This praxis serves two purposes: (a) to teach adolescents about Black individuals and events outside of their historical moments to provide contextually relevant content knowledge; and (b) encourage the development of Black speculative thought among students that tie the past, present, and future together.

4. Technological Ingenuity and Power Are a Part of Every Black Person's Heritage and Therefore If Not Already Possessed or Expressed, an Individual Is Assumed to Be Highly Capable of Developing These Strengths

For principle four, we use Afrofuturism to counter the myth of Black technological disingenuity and posit that Black people actually have technological ingenuity that is both a part of their heritage and developed in contemporary contexts. Our definitions of technology span a range of traditional and non-traditional. In his book titled "Afrofuturism Rising: The Literary Prehistory of a Movement," Lavender III (2019) argues that early versions of Afrofuturism can be seen in work that is centuries old. He notes that "Afrofuturism provides a new way to both literally and figuratively decode the dreams of Black freedom in a country built on slave labor" (Lavender III, 2019, p. 25). He goes on to argue that technologies are not Western or necessarily mechanical and that spiritual technologies such as the clandestine coding of slave songs and writings of people like Fredrick Douglass, Olaudah Equiano, and Phillis Wheatley represent the embryonic stage of Afrofuturism. In addition, he suggests that practical knowledge that helps Black people solve problems in their community or that allows them to be freed physically or psychologically from bondage is a freedom (or spiritual) technology (Lavender III, 2019). This may include ritual practices, dance, art, and other cultural traditions. Afrofuturism, he notes, provides a lens to understand how spiritual technologies work: the material and non-physical worlds function in complementary ways to produce practical ends, including freedom.

Ron Eglash's research on ethnomathematics has shown evidence of computational and technological thinking such as recursion, fractals, cybernetics, and coding in African and Afro-diasporic cultural practices; these have influenced the development of computational technologies globally—for example, the binary logic that all contemporary digital devices use may have originated in Bamana divination practices in West Africa (e.g., Eglash, 1995, 2007). Other examples include the fact that West African peoples such as the Dogon had vast knowledge of the universe including the Sirius star system, which predates modern technology by millennia.

We note that many children and young people grow up in families that wield the power of these cultural technologies to transcend oppressive forces in the US. The Founder of BLKCreatives, Melissa Kimble, posted a tweet that became a part of Twitter's Black Lives Matter campaign in 2020. It read, "this world does not move without Black creativity" (Kimble, 2020). This is literally the case in the US, where for centuries Black labor, inventions, scientific contributions, and cultural productions have driven our economy and American culture. One recent example is the popular social media platform, TikTok, which is wildly popular because Black creatives create content. The platform has been accused of racist practices by Black creators who have their content appropriated by whites who take it and are able to amass fortunes (Pruitt-Young, 2021). Recently they staged a strike and journalists and users noted the creativity on the platform virtually came to a halt (Pruitt-Young, 2021). Other examples include the numerous Black inventors during and after slavery, many of whom did not get credit for their inventions because of racist US patent laws (Johnson, 2017). Black children and young people should understand this legacy of innovation and see their technological ingenuity, creativity and innovation as part of essence of who they are.

For principle four we highlight the fact that the technological ingenuity and innovation Black people possess can be seen from a very young age to adulthood. To illustrate how this principle might be put into practice, we provide examples from community outreach from the Center for Empowered Learning and Development with Technology, Black Lives Matter and the broader Movement for Black Lives, and Verzuz online music events created during the pandemic. First, the first and third author conducted app development bootcamps with Black students in south Los Angeles where K-12 students were able to design apps that solved problems in their community (Tynes & Schuschke, 2023). Their designs were based on their interests and were incredibly innovative, from an Afro-Latina 7th grader who made a more user-friendly WebMD without seeing the site to the 11-year-old who came up with an idea for an African history app because she was not learning about herself in southern California schools. We found every student exuded technological innovation that, if developed over time, could completely transform their communities and the tech industry. Thus far, schools do not assess or reward them for these strengths. Afrofuturist Development would radically transform the nature of schooling such that they would be further developed and valued.

Second, we have seen the profound impact of Black technological ingenuity and power with the creation and rise of the Black Lives Matter Movement, founded by Patrisse Cullors, Opal Tometi, and Alicia Garza. The movement began with the use of the hashtag #BlackLivesMatter after George Zimmerman was acquitted after killing African American teen Trayvon Martin. It grew to become arguably the largest movement in US history in the wake of the murder of George Floyd in 2020 (Buchannon et al., 2020). We note that many Black teens led protests across the country (Bennett, 2020) and in our own work, teens and emerging adults used the Internet to organize and educate others about Black history, systemic racism, and police brutality. We have data from Instagram in May and June of 2020 that show young people creating informative stories which resembled high school and college lessons (Tynes, 2022). Third and similar to these powerful examples, in March 2020, at a time when uncertainty and distress were heightened due to the Covid-19 pandemic and social unrest related to the death of Ahmaud Arbery, Black people across generations gathered in community using digital technologies to celebrate Black music and culture. Notably, *Verzuz* digital live music battles, in which some of music's most celebrated Black artists played twenty of their biggest songs, garnered millions of views on Instagram's live streaming platform, as well as millions of live Tweets during the battles (https://verzuztv.com/). The artistry, celebration, laughter, and community-building on display in these digital experiences allows for families and educators, alike, to engage in conversations with their youth about Black music history and the role of music and art in Black liberation, coping, and thrival.

5. Critical Consciousness and Action Are Competencies Taught in Developmentally Appropriate Ways to Support Black People as They Navigate and Resist Oppressive Media, Technology, Institutions, and Social Practices

Principle five draws on Black feminist thought, critical race theory, and liberation psychology to suggest that critical consciousness and action should be embedded across contexts in developmentally appropriate ways. The concept of critical consciousness comes from Black liberation psychologist Frantz Fanon (1967), as interpreted by Brazilian critical literacy educator Paulo Freire (2018); it refers to the development of capacities to critically analyze the contradictions that drive society (e.g., class struggles and struggles between white supremacist systems and anti-racist/decolonial movements), and the ability to use this knowledge to take action to liberate oneself and one's community from oppression. Consciousness raising has also been a key part of Black feminist praxis, and Black feminists emphasize the ability to critique patriarchy and misogynoir, and to take action against it; critical consciousness involves recognizing that Black women face oppression at the intersections of anti-Blackness, patriarchy, and heterosexism, and that they have developed a range of strategies to liberate themselves from this oppression (Collins, 2002; Crenshaw, 2017; Taylor, 2017). As mentioned above, we are building on Susana Morris' efforts to merge these Black feminist practices with Afrofuturism (Morris, 2012, 2016).

Critical consciousness development for Black children and young people should include an explicit focus on systemic racism (see Chap. 10 for a further discussion of critical consciousness and methodology). Legal scholars developed critical race theory (CRT) as a lens to understand the ways racism is embedded in United States' law (Bell, 1992; Williams, 1991). Ladson-Billings and Tate (1995) argue that using CRT in education helps to underscore the fact that race and racism shape educational policies and inequalities. Scholars have since extended CRT's analysis to examine the specificity of antiblackness, via BlackCrit (Dumas & ross, 2016). The application of CRT to the field of education has allowed for a broad systems analysis of educational laws and policies and has faced recurring and renewed resistance from political opponents. These critical analyses of race extend into the online learning space as well; where Tynes et al. (2021) put forth critical race digital literacy (CRDL) as a way for youth to understand race-related media and technology. The development of critical consciousness through a systemic analysis of technologies is an inherent feature of Afrofuturist development.

An Afrofuturist Developmental lens leverages the benefits of critical consciousness by highlighting the ways that racism is embedded in technology and how culturally specific practices with media can serve liberatory purposes for Black youth. For instance, Tynes et al. (2021) found that youth, in general, have a wide range of ability levels in detecting racialized mis/disinformation from algorithms, to websites, to social media posts. For Black youth in particular, teachers can begin developing their critical consciousness and CRDL early by providing lessons about algorithmic erasure and bias against Black women, encouraging sound critique of news sites reporting of Black social movements, and vetting information from influential figures and organizations. From an Afrofuturist perspective, critical consciousness is not simply a negation of existing oppression, it also involves imagining other ways of living and thriving that could replace it; critical literacies researchers have begun to identify a need for learning spaces that foster this kind of imagination (Storm & Jones, 2021). We define critical speculative imagination as the capacity to conjure, enact, and rehearse future worlds free from oppression, and we consider it to be a part of Afrofuturist Development that can be taught, developed, and examined by developmental scientists and educators.

6. Community Members, Teachers, Employers, Researchers, and Clinicians Must Have Training in Black Children and Young People's Histories, Technocultures, and Development for Optimal Outcomes

For principle six, we envision a future where teachers, clinicians, researchers, employers and all who serve Black people would be required to have training in their histories, cultures, and development. They should be well versed in critical perspectives in their respective fields. They should also be culturally competent, anti-racist, and have a deep understanding of how to apply their knowledge to the design of learning and development contexts. For example, culturally responsive approaches to technology education have emerged from the work of culturally relevant/responsive/sustaining scholars (Gay, 2013; Ladson-Billings, 1995; Paris, 2012). Scott et al.'s (2015) culturally responsive computing (CRC) emphasizes an intersectional and strengths-based approach to technology integration in classrooms. According to Scott et al. (2015), CRC is guided by five tenets:

1. All students are capable of digital innovation.
2. The learning context supports transformational use of technology.
3. Learning about one's self along various intersecting sociocultural lines allows for technical innovation.
4. Technology should be a vehicle by which students reflect and demonstrate understanding of their intersectional identities.
5. Barometers for technological success should consider who creates, for whom, and to what ends rather than who endures socially and culturally irrelevant curriculum. (p. 420–421).

CRC's cultural and intersectional approach to teaching with technology highlights student innovation, particularly among Black girls, through culturally specific practices.

For educators using an Afrofuturistic Development lens, there are many commonalities with CRC and other equity-minded approaches to education. Afrofuturism, however, accentuates the possibilities for not only creating content, but empowering Black students to construct their own futures through identifying with their desired personal and communal goals. Educators and developmental scientists will need training in how to design experiences for young people that not only include transformational uses of technology but they must be spirit-feeding. In the app development bootcamps previously mentioned, students' spirits soared as they imagined and practiced their roles as designers, brainstormed ideas, and showed off their creations in their presentations. They even burst with excitement when we took music and dance breaks where they taught the first author the latest dance moves and shared their favorite songs. One of the young app designers mentioned "this is like school, only fun." To be capable of creating spirit-feeding experiences, an individual or team would need years of training in their content area, pedagogical and technological skills, and knowledge of Black cultures that inspires a deep love for and awe of the students. This training is necessary across settings if we are serious about Black people's thrival.

7. Systems of Equity That Resist Anti-Black Racism, Create a Sense of Belonging, and Provide Afrofuturist Developmental Learning and Community Possibilities Are Key Drivers of Optimal Development

For principle seven, instead of reproducing systems of oppression, Afrofuturist learning and development creates systems of equity across contexts and developmental stages. These systems may be evident in a range of contexts, from micro-interactions in the microsystem all the way through to laws and policies that shape Black people's lived experiences. A core aspect of this principle is creating multiple opportunities for success and not only transformational, but spirit-feeding technology experiences in educational settings and beyond. We draw on a range of evidence-based theories, models and practices for this principal, including the Talent Quest Model of school reform (Boykin & Ellison, 2009). Initially proposed by the former Center for Research on the Education of Students Placed At-Risk, the Talent Quest Model (TQM; Boykin, 2000) is a comprehensive paradigm for school reform that argues the traditional function of schooling to classify, sort, and weed out should be shifted to maximizing every student's full potential. Six themes make up the TQM operating principles (2009): (1) overdetermination of success, or creating multiple pathways to enhance student outcomes; (2) optimizing classroom transaction, including building on cultural, community, and family assets; (3) multiple barometers for success, such as enhancement of neighborhood and family involvement; (4) co-construction or the idea that social and cultural factors impact TQM implementation; (5) evidence-based activities; and (6) continuous improvement. This model requires a rethinking of traditional schooling and the creation of systems of equity that are designed for Black student success not their failure. Overdetermining student success and building on cultural, community, and family assets may create safe spaces for students and create an environment where their need for belonging might be met.

In applying principle seven, researchers should try to understand how schools and programs might transform into systems of equity as well as identify potential barriers to reaching this goal. For example, to understand potential threats to creating a sense of belonging, a vital component of a system of equity, Gray and colleagues (2018) proposed an ecological framework and note several research strategies. These include: (1) recognizing policies and practices that may hinder school belonging efforts; (2) identify potential boundaries, whether conceptual or operational, to opportunity structures; (3) develop tools that assess belongingness; and (4) situate studies of belongingness within historical context. An Afrofuturist lens would also add how technology might be used to create or hinder a sense of belonging and broader systems of equity so Black students thrive.

8. Shared and Unique Developmental Needs Are Met and Developmental Assets (e.g., Imagination and Creativity) Leveraged and Nurtured Across Contexts

Principle eight highlights the fact that Black thrival outcomes are more likely when developmental needs are met and assets are leveraged and nurtured. Needs and assets vary across developmental stages. They also vary across groups and the most widely recognized studies of assets measure shared, sometimes Eurocentric types of assets (e.g., Search Institute, 2018). The Search Institute's (2018) study of 40 developmental assets among 121,157 US adolescents in grades 6–12 showed the more assets participants have the more likely they exhibit thriving indicators, including doing well in school, maintaining good health, helping others, overcoming adversity, and exhibiting leadership. The study included internal and external assets and found of the 40 assets, African Americans possess an average of 20. This number is not comparable across groups, and we argue this is because the assets framework does not capture the vast array of assets that are unique to Black adolescents such as racial literacy, technological ingenuity, and creativity. To assess creative activities among participants, for example, researchers asked about having three or more lessons in music, art, etc. This can lead to bias in developmental science along with a lack of construct equivalence because it would not measure the brilliant self- or family-taught children and young people or those who do not have access to formal lessons. It is also biased for not including other assets African Americans possess, including their forms of communalism, defined as interdependence and an inclination to share and complete tasks for the good of the group (Tyler et al., 2006), religious support (Gooden & McMahon, 2016), work ethic (Hill, 2003), and verve- "an especial receptiveness to and orientation toward high levels of sensate and physical stimuli" (Tyler et al., 2006, p. 367). There is also what African Americans call "Making a Way Out of No Way"; this is the African American spirit of persisting even in the face of the bleakest circumstances (see the description of the exhibit at the National Museum of African American History & Culture website). It is a type of resourcefulness, resilience, and will that stand in the face of centuries of white terror and a people who daily dream up and practice new ways to oppress Black people. These and countless other unique assets among Black children and young people should be leveraged and cultivated.

To apply principle eight we point to researchers who have designed technology programs with young people's assets as a central feature. Gaskins (2021) provides an example of this in her work on what she calls Afrofuturism 3.0 and the resulting techno-vernacular creative (TVC) production, including do-it-yourself technoculture and combining cultural and religious traditions (Gaskins, 2016). There are three main modes of techno-vernacular creativity: reappropriation or reclaiming cultural artifacts in an effort to subvert hegemonic social systems, improvisation-performing, creating, problem solving in such a way that new practices and artwork are invented- and remixing or "rearranging materials to produce something new" (Gaskins, 2021, p. 11). Her work highlights how educators who center these modes of TVC learning can support students in becoming key drivers of innovation (Gaskins, 2021). She further argues that culturally relevant making should be a part of the core curriculum (design and computational action). Similar to Gaskins work, Holbert and colleagues (2020) created Remixing Wakanda, a project for Black girls, that allowed them to design and produce futuristic artifacts that built on their family histories and critiqued existing inequity. This aligns with Winchester's III (2019) Afrofuturist design framework and his assertion that Afrofuturism allows for deeper reflection and ideation on design concepts as well as more equitable inclusive design.

9. Toolkits for Future Physical and Mental Well-being Are Tailored to an Individual's Needs in a Given Setting

The ninth principle highlights the importance of an approach to supporting the physical and mental well-being of young people that not only considers, but centers, their unique needs. For Black young people, specifically, this can include providing supports and facilitating skills that allow them to critically evaluate and make sense of their lived experiences in an antiblack world, as well as imagine and feel efficacious in building future worlds in which their thriving is centered. With the rise in suicide rates for Black adolescents over the last decade (Bridge et al., 2018), it is critical that Black youth are in environments that help them to build what Rheeda Walker (2020) calls "psychological fortitude." Psychological fortitude refers to an individual's ability to think, work, be in relationships, and ultimately pursue one's life goals and dreams. As it relates to Afrofuturist development, psychological fortitude can also be applied to young people's abilities to imagine liberated futures for themselves or their capacity to be innovative and creative in their use of technologies as tools for their own learning and development.

A learning and development toolkit grounded in Afrofuturist foundations can include skills in using digital technologies and media, as well as the unique cultural assets of Black people, to resist oppressive systems through the prioritization of joy, rest, and imagination. Parents, educators, clinicians, and others interacting or sharing space with Black young people should be equipped to assist in the development of such toolkits. These systems and actors within these systems play a role in promoting joy, and the overall adjustment of Black students (Tichavakunda, 2021). Black young people's position as a historically marginalized and subjugated community in the US context requires that institutions leverage cultural assets to promote optimal well-being and expand the imaginations of young people's future selves and future worlds.

10. Black Children and Young People, Often Powered by Their Afrofuturistic Selves, Imagine, Build, and Reinvent Liberated, Pro-Black Futures, and Technologies

The tenth and final principle highlights the development of young peoples' futures that are rooted in possibilities for Black liberation. It takes as fundamental children and young people's right to explore the universe (Prescod-Weinstein, 2021) both without and within. This principle also positions Black people as leaders in the creation of pro-Black futures across settings, communities, and nations. Drawing from literature on future and possible selves (Markus & Nurius, 1986) and concepts from Afrofuturism (Anderson & Jones, 2016; Gaskins, 2016), the futures Black people create begin with envisioning possible realities and identities for themselves. Central to the final principle is Black people's (re)orientation to spacetime. A primary tenet of Black Quantum Futurism (Phillips, 2015), which is informed by quantum physics, futurist traditions and Black/African traditions of consciousness, time and space, is the idea that African descended people should be made aware of their power to see, create or choose a desired future. Phillips argues that Black people can manipulate spacetime into a desired future to bring about the future's reality. The realities they create can be a broad range of imagined experiences and visions that come to fruition. This may include setting a goal to create a "spirit technology" (Wildman & Stockly, 2021) that creates transcendent experiences free from oppression, envisioning a desired outcome, and finding the resources needed to meet the goal despite any obstacles.

Individuals can also create possible identities or Afrofuturistic selves, which often drive the futures they mold into being. Afrofuturistic selves are multidimensional self-concepts that center (a) the individual within the context of Black history; (b) Black community, organization, and movement involvement; (c) redefining notions of success; and (d) the iterative design of an (Afro)future self that may be immediate or long-term, five seconds away or ten years (Schuschke & Tynes, in preparation). The development of Black youth's Afrofuturistic selves allows them to situate themselves within a larger diasporic community and political history. Through interactions with media and technology, Black creativity and imagination allow adolescents to build toward educational, career, and community goals that are not only focused on individual success, but movement toward a future that is not shaped by racism or white normativity and neoliberalism. Thus, the Afrofuturistic self represents one component of Afrofuturist Development among Black young people and provides a culturally specific identity construct for researchers and practitioners to explore.

Social media platforms and the unique ways that youth engage their affordances allow them to project multiple identities, including their future selves. For example, Schuschke and Tynes (in preparation) highlight the agentic use of Instagram and meme culture by Darius, a Black male high school student. Findings revealed that Darius, an aspiring child psychologist, used memes to encourage online discussions about mental and socio-emotional health. In this example, Darius is able to bend spacetime by using social media as a tool to perform practices that reflect his desired future identity. The implications for praxis in developing these Afrofuturistic selves are for educators to allow for Black youth identity exploration in online environments. Teachers, counselors, and mentoring adults should encourage cultural creativity and facilitate youth-directed digital practices. Specifically, this form of praxis can be achieved by connecting youth to online networks related to their aspirations and by allowing adolescents to become immersed in those communities and spaces. From there, students can take hold of identities and practices in ways they deem relevant and vital to their future success.

A Case Study of Afrofuturist Development: Designing an Abolitionist Video Game and a Critical Game Jam

Afrofuturist Development theory can support research, design, and praxis, and we provide a case study of how it might do so below. The second author (Matthew Coopilton) and colleagues from the University of Southern California Center for Empowered Learning and Development with Technology have launched a participatory design research pilot study of critical game literacies learning. We define this as the capacity to play, analyze, modify, and create games in ways that challenge systemic oppression; we are asking how learning these capacities might nurture critical speculative imagination, part of the critical consciousness component of Afrofuturist Development.

The first part of this project is *Kai UnEarthed*, a video game about young people learning, growing, and healing in an unpoliced future. A video trailer, text-based prototype, and screenshots are available on the game's website: kaiunearthed.com. Coopilton and their colleagues have designed the game for use in abolitionist learning spaces (Love, 2019) and movements for the abolition of police and prisons.

Emerging from participatory design activities in high school classrooms and abolitionist movements, *Kai UnEarthed* rehearses (Boal, 2014) liberated pro-Black futures, inspired by speculative theories in the (post)humanities, including Afrofuturism. It builds on emerging efforts to design Afrofuturist games, such as Brooks and Kosminsky's (n.d.) *Afrorithms from the Future* game.

The second part of the project is a critical game jam that Coopilton led in July 2021; it was a week-long series of workshops informed by Afrofuturist Development. Eight new participants joined our existing team of five participant-researcher-designers, and together we imagined pro-Black and queer games through design prompts, discussions, presentations, and hands-on prototyping activities. The week included a playtest of *Kai UnEarthed* as well as another abolitionist game that participants are developing. Participants were eight young adults (M = 22.3, range = 21–26, SD = 1.69). Six identify as Black, one as Latino (Brazilian), and one as Mexican and white; four identify as non-binary, and four as men. The participatory nature of this project is similar to Youth Participatory Action Research (see Chap. 9 for details); we created game-worlds that model real actions we could take to liberate our communities.

We documented and analyzed this process of participatory design research (Bang & Vossoughi, 2016) in order to iteratively improve our games and learning activities and to refine the theories embodied in them (Barab & Squire, 2004; Crippen & Brown, 2018). We conducted interviews, video recordings, observations, and pre- and post-assessments, and we gathered design documents that participants created. We are in the process of conducting a thematic analysis (Boyatzis, 1998) of these data sources. Full findings will be reported in future publications, but here we highlight how two principles of the Afrofuturist Development theory shaped the learning experience: principle ten-Afrofuturistic selves and principle five-critical consciousness. We note also that the designers and researchers all had the requisite training outlined in principle six.

Principle #10 -Afrofuturistic Selves (Re-imagining the Role of Game Designer)

In the post-workshop video recordings, several participants described how the game jam helped them to imagine possible near futures for themselves, including new vocational roles. During the workshops, Olivia Peace, one of the co-designers of *Kai UnEarthed*, reflected on her own process of overcoming perfectionism while working on the project through multiple iterations of prototypes. Nathann highlighted how Olivia's role modeling of this process helped him to imagine transitioning into possible roles as a professional game designer: "I've learned that it's invaluable for getting into your desired field and job—to make things. Overcoming imposter syndrome and perfectionism, and the importance of collaboration." This shows that Afrofuturistic development is not about attempting to approximate a perfect ideal society or self, it is about creating and sharing prototypes of possible futures.

Several participants described how the game jam helped them think about ways they might approach their work with a more critical lens. They talked about how they learned from the critical character creation workshop that professional video game artist Michelle Ma facilitated. For example, Gabriel said

> I see myself in the game industry and I would like to become a 3D modeler or something that falls in the scope of the arts, and that's one of the reasons that I really, really liked Michelle's workshop because I will take that for life, and how she taught us to critically think about the character that you just made, that made me think "why are you making a man, or why haven't you made a disabled person."

This quote shows the ways participants imagined their future selves is consistent with the notion of Afrofuturistic selves (Schuschke et al., 2023), a core part of Afrofuturist Development theory. They imagined critically, beyond existing constraints of the game industry, intertwining their vocational paths with larger community struggles for liberation. They imagined a future self who questioned patriarchy and ableism as well as how they might build alternatives to oppressive representations in commercial games.

There was also a meshing of the past, present, and future into a unified experience of time which resonates with Phillips' (2015) notion of Black Quantum Futurism, which Brendesha Tynes (first author) shared to open the game jam: "We are here to bend spacetime this week. I wanted to remind you of your power to do so." Participants reflected on this idea throughout the week. For example, Nathann said:

> We're trying to imagine futures like it's a separate thing, separate from us and where we are now, and then being like, able to condense that down and bend spacetime so that we can imagine what it'd be like if we had to make a game right now. It's sort of like picking and plucking those things…we would imagine in that perfect or ideal scenario, and then seeing how far we can actually get in this moment.

Principle #5- Fostering Critical Consciousness Among Participants

Several participants discussed how the critical game jam helped them imagine possible futures for their people (e.g., Black people, queer people), and actions they might take to participate in movements that are building and fighting for these futures today. In their post-assessment, Cam wrote that: "The workshops have provided me a better layer for being able to think through games that can challenge racism in a creative and entertaining way." Light Bird said that *Kai UnEarthed* is an example of the kind of design praxis, "informed by global justice and intersectional feminism," that they want to develop in their own work.

Several participants expressed a desire to create games set entirely in worlds inspired by African and Latin American societies. Gabriel argued that "there are really poor countries because of capitalism, because capitalism exports poverty. If there is technology in the north of the world it is because there is poverty in the south of the world. We need to change that." His framing challenges the association of technology with imperialist notions of progress, civilization, and whiteness, echoing some of the core themes of Afrofuturism (Anderson & Jones, 2016; Nelson, 2002). He imagined a game based on the Quilombos, the communities of enslaved people in Brazil who escaped and built autonomous camps at war with the plantations (similar to the maroons in the Caribbean):

> It got me thinking in a game like *Banished*. What if we could run and escape through a portal and build a new settlement. But in a new world, maybe escape to the future, maybe build some kind of Wakanda. Then send back expeditions and save our brothers without violence. In *Banished* you're only focused on building your society for balance, without militarization. In that game could have a building like a museum where they teach about the history of black and latinxs people. Even the buildings could be aesthetically different, inspired in afrofuturism.

This design idea resonates with the work of political scientist Neil Roberts (2017), who imagines freedom as marronage between the past and the future, as well as the work of educator and psychologist Asa Hilliard (1995) who imagined the collective development of Black people as an extension of maroon practices into the present and future. It is an example of the Black History Knowledge principle of Afrofuturist Development combined with principle five.

In the game jam, participants prototyped experiences rooted in freedom dreams (Kelley, 2002) that are spacetime-bending, sharing them with each other in the current moment. This compatibility of iterative design practices, critical consciousness, and Afrofuturist temporality demonstrates the potential of Afrofuturist Development.

Conclusion

This chapter outlines how Afrofuturist developmental uses of technology that are not only transformational but spirit-feeding, well-being-enhancing, and spacetime-bending can promote Black thrival. We see proposing the theory as Imarisha and brown (2015) note as a responsibility to write Black children, adolescents, and emerging adults into the future. We believe adhering to principles outlined here makes a small contribution to ensure this is the case. At minimum, we see the theory as a way to enhance rigor of developmental science and educational research. We also hope that it inspires transformations across settings that serve Black people.

The Critical Game Jam is one case study of what Afrofuturist Development looks like in practice. It is one of many emerging learning environments that embody the principles of this theory in various ways. The theory can be used to help shape the design of these environments, to study how people learn and develop within them, and to inform future designs through an iterative process of design research (Bang & Vossoughi, 2016; Barab & Squire, 2004). Afrofuturist Development can also be used to inform research on existing developmental contexts and practices in schools, homes, neighborhoods, health services, media, technology, etc. It can serve as a set of concepts that caregivers, nurturers, educators, practitioners, and mentors in these settings can use as guides while they support young Black people who are navigating increasingly complex technologically mediated lives. It can help us all to recognize the potential that young Black people hold. Given the necessary developmental support, they are capable of building a future tomorrow and in 2044, where Black people not only exist—they thrive.

References

Akbar, N. I. (1984). Africentric social sciences for human liberation. *Journal of Black Studies, 14*(4), 395–414. https://doi.org/10.1177/2F002193478401400401

Anderson, R., & Jones, C. (Eds.). (2016). *Afrofuturism 2.0: The rise of astro-blackness*. Lexington Books.

Bang, M., & Vossoughi, S. (2016). Participatory design research and educational justice: Studying learning and relations within social change making. *Cognition and Instruction, 34*(3), 173–193. https://doi.org/10.1080/07370008.2016.1181879

Barab, S., & Squire, K. (2004). Design-based research: Putting a stake in the ground. *Journal of the Learning Sciences, 13*(1), 1–14. https://doi.org/10.1207/s15327809jls1301_1

Barton, L. (2021, March 4). What it's like to be a Black man in tech. *Harvard Business Review*, https://hbr.org/2021/03/what-its-like-to-be-a-black-man-in-tech

Bell, D. (1992). *Faces at the bottom of the well: The permanence of racism*. Basic Books.

Bennett, J. (2020, June 26). These teen girls are fighting for a more just future. *New York Times*. https://www.nytimes.com/2020/06/26/style/teen-girls-black-lives-matter-activism.html

Bentley, K. L., Adams, V. N., & Stevenson, H. C. (2008). Racial socialization. In H. A. Neville, B. M. Tynes, & S. O. Utsey (Eds.), *Handbook of African American psychology* (pp. 155–268). Sage Publications.

Beyoncé. (2020). Black parade [Song]. *On The Lion King: The Gift*. Parkwood: Columbia. https://music.apple.com/us/album/black-parade-single/1519555113

Boal, A. (2014). *Theatre of the oppressed* (pp. 97–104). Routledge.

Boyatzis, R. E. (1998). *Transforming qualitative information: Thematic analysis and code development*. Sage.

Boykin, A. W. (1994). Afrocultural expression and its implications for schooling. In E. R. Hollins, J. E. King, & W. C. Hayman (Eds.), *Teaching diverse populations: Formulating a knowledge base* (pp. 50–51). State University of New York Press.

Boykin, A. W. (2000). The talent development model of schooling: Placing students at promise for academic success. *Journal of Education for Students Placed at Risk (JESPAR), 5*(1–2), 3–25. https://doi.org/10.1080/10824669.2000.9671377

Boykin, A. W., & Ellison, C. M. (2009). The talent quest model and the educating of African American children. In H. A. Neville, B. M. Tynes, & S. O. Utsey (Eds.), *Handbook of African American psychology* (pp. 237–252). Sage Publications.

Bridge, J. A., Horowitz, L. M., Fontanella, C. A., Sheftall, A. H., Greenhouse, J., Kelleher, K. J., & Campo, J. V. (2018). Age-related racial disparity in suicide rates among US youths from 2001 through 2015. *JAMA Pediatrics, 172*(7), 697–699. https://doi.org/10.1001/2Fjamapediatrics.2018.0399

Bronfenbrenner, U. (1979). *The ecology of human development*. Harvard University Press.

Brooks, L., & Kosminsky, E. (n.d.). *Afrorithms from the future*. https://www.afrorithmsfromthefuture.org/

Brown, R. K., & Brown, R. E. (2003). Faith and works: Church-based social capital resources and African American political activism. *Social Forces, 82*(2), 617–641. https://doi.org/10.1353/sof.2004.0005

Bruce, D. (2020). Afrofuturism: From the past to the living present. *UCLA Magazine*. https://newsroom.ucla.edu/magazine/afrofuturism

Buchannon, L., Bui, Q., & Patel, J. (2020, July 3). Black Lives Matter may be the largest movement in U.S. history. *New York Times*. https://www.nytimes.com/interactive/2020/07/03/us/george-floyd-protests-crowd-size.html

Burton, N. (2020, July 31). How some Black Americans are finding solace in African spirituality. *Vox*. https://www.vox.com/2020/7/31/21346686/orisha-yoruba-african-spirituality-covid

Center for Disease Control. (2020). *Risk of exposure to Covid-19*. https://www.cdc.gov/coronavirus/2019-ncov/community/health-equity/racial-ethnic-disparities/increased-risk-exposure.html

Chapman-Hilliard, C., & Adams-Bass, V. (2016). A conceptual framework for utilizing Black history knowledge as a path to psychological liberation for black youth. *Journal of Black Psychology, 42*(6), 479–507. https://doi.org/10.1177/2F0095798415597840

Coles, J. A., & Stanley, D. (2021). Black liberation in teacher education:(Re) envisioning educator preparation to defend Black life and possibility. *Northwest Journal of Teacher Education, 16*(2), 1–24. https://doi.org/10.15760/nwjte.2021.16.2.6

Collins, P. H. (2002). *Black feminist thought: Knowledge, consciousness, and the politics of empowerment*. Routledge.

Cone, J. (2010). *A Black theology of liberation*. Orbis Books.

Crenshaw, K. W. (2017). *On intersectionality: Essential writings*. The New Press.

Crippen, K., & Brown, J. (2018). Design-based research. In B. Frey (Ed.), *The SAGE encyclopedia of educational research, measurement, and evaluation* (Vol. 1–4, pp. 490–493). Sage Publications. https://doi.org/10.4135/9781506326139

Dery, M. (1994). Black to the future: Interviews with Samuel R. Delany, Greg Tate, and Tricia Rose. In *Flame wars* (pp. 179–222). Duke University Press.

Dumas, M. J., & Ross, K. M. (2016). "Be real black for me" imagining BlackCrit in education. *Urban Education, 51*(4), 415–442. https://doi.org/10.1177/2F0042085916628611

Education Trust-West. (2015). *Black minds matter: Supporting the educational success of Black children in California*. https://west.edtrust.org/resource/black-minds-matter-supporting-the-educational-success-of-black-children-in-california/

Edwards, E., & Edwards, C. (2020, August 17th). *Raising Afro-Latino children in the era of COVID & Black Lives Matter*. Successful Black Parenting. https://successfulblackparenting.com/2020/08/17/vidas-negras-raising-afro-latino-children-in-the-era-of-covid-black-lives-matter/

Eglash, R. (1995). African influences in cybernetics. In C. H. Gray (Ed.), *The cyborg handbook* (pp. 17–27). Routledge.

Eglash, R. [TED Global 2007]. (2007, June). *Ron Eglash: The fractals at the heart of African designs*. Ted Talks. https://www.ted.com/talks/ron_eglash_the_fractals_at_the_heart_of_african_designs?utm_campaign=tedspread&utm_medium=referral&utm_source=tedcomshare

Emdin, C. (2016). *For White folks who teach in the hood... and the rest of y'all too: Reality pedagogy and urban education*. Beacon Press.

Eshun, K. (2003). Further considerations of Afrofuturism. *CR: The New Centennial Review, 3*(2), 287–302. https://doi.org/10.1353/ncr.2003.0021

Evans, J. (2020). Connecting Black youth to critical media literacy through hip hop making in the music classroom. *Journal of Popular Music Education, 4*(3), 277–293. https://doi.org/hp9g

Fanon, F. (1967). *Black skin, white masks*. Grove Press.

Freire, P. (2018). *Pedagogy of the oppressed*. Bloomsbury Publishing USA.

García Coll, C., Crnic, K., Lamberty, G., Wasik, B. H., Jenkins, R., Garcia, H. V., & McAdoo, H. P. (1996). An integrative model for the study of developmental competencies in minority children. *Child Development, 67*(5), 1891–1914. https://doi.org/10.2307/1131600

Gaskins, N. (2016). Afrofuturism on Web 3.0. In R. Anderson & C. Jones (Eds.), *Afrofuturism 2.0: The rise of astro-blackness* (pp. 27–44). Lexington Books.

Gaskins, N. R. (2021). *Techno-vernacular creativity and innovation: Culturally relevant making inside and outside of the classroom*. MIT Press.

Gay, G. (2013). Teaching to and through cultural diversity. *Curriculum Inquiry, 43*(1), 48–70. https://doi.org/f4mb9k

Gooden, A. S., & McMahon, S. D. (2016). Thriving among African-American adolescents: Religiosity, religious support, and communalism. *American Journal of Community Psychology, 57*(1–2), 118–128. https://doi.org/10.1002/ajcp.12026

Grady, C. (2020, July 31). Meet the African goddess at the center of Beyoncé's Black is King. *Vox*. https://www.vox.com/culture/2020/7/31/21349403/beyonce-black-is-king-oshun-osun-yoruba-goddess

Gray, D. L., Hope, E. C., & Matthews, J. S. (2018). Black and belonging at school: A case for interpersonal, instructional, and institutional opportunity structures. *Educational Psychologist, 53*(2), 97–113. https://doi.org/fwtp

Guthrie, R. V. (2004). *Even the rat was white: A historical view of psychology*. Pearson Education.

Harding, V. (1997). Religion and resistance among antebellum slaves: 1800–1860. In T. E. Fulop & A. J. Raboteau (Eds.), *African American religion: Interpretive essays in history and culture* (pp. 107–130). Routledge.

Hill, R. B. (2003). *The strengths of Black families*. University Press of America.

Hilliard, A. G. (1995). *The maroon within us: Selected essays on African American community socialization*. Black Classic Press.

Holbert, N., Dando, M., & Correa, I. (2020). Afrofuturism as critical constructionist design: Building futures from the past and present. *Learning, Media and Technology, 45*(4), 328–344. https://doi.org/10.1080/17439884.2020.1754237

Hope, M. O., Lee, D. B., Hsieh, H. F., Hurd, N. M., Sparks, H. L., & Zimmerman, M. A. (2019). Violence exposure and sexual risk behaviors for African American adolescent girls: The protective role of natural mentorship and organizational religious involvement. *American Journal of Community Psychology, 64*(1–2), 242–255. https://doi.org/10.1002/ajcp.12341

Hunter, M. A., Pattillo, M., Robinson, Z. F., & Taylor, K. Y. (2016). Black placemaking: Celebration, play, and poetry. *Theory, Culture & Society, 33*(7–8), 31–56. https://doi.org/gjtm6v

Imarisha, W., & Brown, A. (2015). *Octavia's brood: Science fiction stories from social justice movements*. AK Press.

Johnson, G. (2010). Internet use and child development: The techno-microsystem. *Australian Journal of Educational and Developmental Psychology (AJEDP), 10*, 32–43.

Johnson, S. (2017). America's always had black inventors–even when the patent system explicitly excluded them. *The Conversation*.

Johnson, G. M., & Puplampu, K. P. (2008). Internet use during childhood and the ecological techno-subsystem. *Canadian Journal of Learning and Technology, 34*(1), n1.

Kambon, K., & Bowen-Reid, T. (2009). Africentric theories of African American personality: Basic constructs and assessment. In H. A. Neville, B. M. Tynes, & S. O. Utsey (Eds.), *Handbook of African American psychology* (pp. 61–73). Sage Publications.

Kelley, R. D. (2002). *Freedom dreams: The Black radical imagination*. Beacon Press.

Kimble, M. (2020, June 17th). *This world does not move without Black creativity.* [Tweet]. Twitter. https://twitter.com/melissa_kimble/status/1273318376052490242

King, L. J. (2019). Interpreting Black history: Toward a Black history framework for teacher education. *Urban Education, 54*(3), 368–396. https://doi.org/gmmj9r

Ladson-Billings, G. (1995). Toward a theory of culturally relevant pedagogy. *American Educational Research Journal, 32*(3), 465–491. https://doi.org/dx3gcj

Ladson-Billings, G., & Tate, W. F. (1995). Toward a critical race theory of education. *Teachers College Record, 97*(1), 47–68. https://doi.org/gpkkrs

Lago, C. (2020, November 18th). *Black representation in tech: What the figures don't tell us.* TechMonitor. https://techmonitor.ai/leadership/workforce/black-representation-tech-what-figures-dont-tell

Lavender, I., III. (2019). *Afrofuturism rising: The literary prehistory of a movement*. The Ohio State University Press.

Lee, C. D. (2007). *Culture, literacy, and learning: Taking bloom in the midst of the whirlwind*. Teachers College Press.

Love, B. J. (2004). Brown plus 50 counter-storytelling: A critical race theory analysis of the "majoritarian achievement gap" story. *Equity & Excellence in Education, 37*(3), 227–246.

Love, B. L. (2019). *We want to do more than survive: Abolitionist teaching and the pursuit of educational freedom*. Beacon Press.

Markus, H., & Nurius, P. (1986). Possible selves. *American Psychologist, 41*(9), 954–969.

Mattis, J. S., Palmer, G. J., & Hope, M. O. (2019). Where our bright star is cast: Religiosity, spirituality, and positive Black development in urban landscapes. *Religions, 10*(12), 1–24. https://doi.org/hp9h

McGhee, E., Paluch, J., & Hsieh, V. (2021, August 17th). *New census data sheds light on California's changes*. Public Policy Institute of California. https://www.ppic.org/blog/new-census-data-sheds-light-on-californias-changes/

Morris, S. M. (2012). Black girls are from the future: Afrofuturist feminism in Octavia E. Butler's "Fledgling". *Women's Studies Quarterly, 40*(3/4), 146–166.

Morris, S. M. (2016). More than human: Black feminisms of the future in Jewelle Gomez's The Gilda Stories. *The Black Scholar, 46*(2), 33–45. https://doi.org/10.2307/23333483

Muhammad, G. (2020). *Cultivating genius: An equity framework for culturally and historically responsive literacy*. Scholastic.

Nelson, A. (2002). Introduction: Future texts. *Social Text, 20*(2), 1–15.

Neville, H. A., Tynes, B. M., & Utsey, S. O. (Eds.). (2009). *Handbook of African-American psychology*. Sage Publications.

Obasi, E. M., & Smith, A. J. (2009). African psychology, or Sahku Sheti: An application of the art of spiritual liberation and illumination of African people. In H. A. Neville, B. M. Tynes, & S. O. Utsey (Eds.), *Handbook of African-American psychology* (pp. 47–61). Sage Publications.

Paris, D. (2012). Culturally sustaining pedagogy: A needed change in stance, terminology, and practice. *Educational Researcher, 41*(3), 93–97. https://doi.org/gcv663

Phillips, R. (2015). *Black quantum futurism*. The Afrofuturist Affair/House of Future Sciences Books.

Prescod-Weinstein, C. (2021). *The disordered cosmos: A journey into dark matter, spacetime, and dreams deferred*. Bold Type Books.

Pruitt-Young, S. (2021, July 1). Black TikTok creators are on strike to protest a lack of credit for their work. *NPR*. https://www.npr.org/2021/07/01/1011899328/black-tiktok-creators-are-on-strike-to-protest-a-lack-of-credit-for-their-work

Quashie, K. (2021). *Black aliveness, or a poetics of being*. Duke University Press.

Roberts, N. (2017). Theorizing freedom, radicalizing the Black radical tradition: On freedom as marronage between past and future. *Theory & Event, 20*(1), 212–230.

Schuschke, J., & Tynes, B. M. (in preparation). *Afro-Futuristic Selves: Black adolescents developing possible futures through engagements with popular media representations.* (in press).

Schuschke, J., Stewart, A. M., Tynes, B. M. (2023). Teacher expectations of Black students in a 1:1 technology program. *Journal of Negro Education* (in press).

Scott, K. A., Sheridan, K. M., & Clark, K. (2015). Culturally responsive computing: a theory revisited. *Learning, Media and Technology, 40*(4), 412–436.

Search Institute. (2018). *Developmental assets among U.S. youth (2018 update) [Data set]*. Search Institute. https://www.search-institute.org/wp-content/uploads/2018/01/DataSheet-Assets-x-Gender-2018-update.pdf

Spencer, M. B. (1995). Old issues and new theorizing about African-American youth:A phenomenological variant of ecological systems theory. In R. L. Taylor (Ed.), *African American youth: Their social and economic status in the United States* (pp. 37–69). Praeger.

Storm, S., & Jones, K. (2021). Queering critical literacies: Disidentifications and queer futurity in an afterschool storytelling and roleplaying game. *English Teaching: Practice & Critique, 20*(4), 534–548. https://doi.org/hp9j

Swanson, D. P., Spencer, M. B., Harpalani, V., Dupree, D., Noll, E., Ginzburg, S., & Seaton, G. (2003). Psychosocial development in racially and ethnically diverse youth: Conceptual and methodological challenges in the 21st century. *Development and Psychopathology, 15*(3), 743–771. https://doi.org/dj4rr2

Taffe, M. A., & Gilpin, N. W. (2021). Equity, diversity and inclusion: Racial inequity in grant funding from the US National Institutes of Health. *eLife, 10*(e65697), 1–11. https://doi.org/10.7554/eLife.65697

Taylor, K. Y. (Ed.). (2017). *How we get free: Black feminism and the Combahee River Collective*. Haymarket Books.

Tichavakunda, A. A. (2021). Black students and positive racialized emotions: Feeling Black joy at a historically white institution. *Humanity and Society*, 1–24. https://doi.org/d9fpmg

Toliver, S. R. (n.d.). *Defining afrofuturism*. Reading Black Futures. https://readingblackfutures.com/defining-afrofuturism/

Tyler, K. M., Boykin, A. W., Miller, O., & Hurley, E. (2006). Cultural values in the home and school experiences of low-income African-American students. *Social Psychology of Education, 9*(4), 363–380. https://doi.org/10.1007/s11218-006-9003-x

Tynes, B. M. (2022, in preparation). *Digital learning about race on Instagram*.

Tynes, B. M., & Schuschke, J. (2022, in preparation). *Black K-12 students' imagination and creativity in app design*.

Tynes, B. M., Schuschke, J., & Noble, S. U. (2016). Digital intersectionality theory and the #Blacklivesmatter Movement. In S. Noble & B. Tynes (Eds.), *The intersectional Internet* (pp. 21–40). Peter Lang.

Tynes, B. M., Willis, H. A., Stewart, A. M., & Hamilton, M. W. (2019). Race-related traumatic events online and mental health among adolescents of color. *Journal of Adolescent Health, 65*(3), 371–377. https://doi.org/10.1016/j.jadohealth.2019.03.006

Tynes, B. M., Stewart, A., Hamilton, M., & Willis, H. A. (2021). From Google searches to Russian disinformation: Adolescent critical race digital literacy needs and skills. *International Journal of Multicultural Education, 23*(1), 110–130. https://doi.org/10.18251/ijme.v23i1.2463

Van Steenis, E. (2020). Positive youth development through hip-hop music production. *Afterschool Matters, 31*, 51–59.

Walker, R. (2020). *The unapologetic guide to Black mental health: Navigate an unequal system, learn tools for emotional wellness, and get the help you deserve*. New Harbinger Publications.

Wildman, W., & Stockly, K. (2021). *Spirit tech: The brave new world of consciousness hacking and enlightenment engineering*. St. Martin's Press.

Williams, P. (1991). *The alchemy of race and rights: Diary of a law professor*. Harvard University Press.

Winchester, W. W., III. (2019). Engaging the black ethos: Afrofuturism as a design lens for inclusive technological innovation. *Journal of Futures Studies, 24*(2), 55–62. https://doi.org/10.6531/JFS.201912_24(2).0007

Chapter 12
Incorporating Diaspora into the Developmental Science of Immigrant Communities

Qurat-ul-ain Gulamhussein, Xiang Zhou, Adam Y. Kim, and Richard M. Lee

Immigrants throughout the world experience transnational diasporas that are both dynamic in nature and consequential in shaping human development over the lifespan. These individuals have dispersed from their homelands due to war, colonialism, poverty, oppression, and disaster, and demonstrate resilience as they navigate developmental and cultural processes and practices. To understand the nature and the lived experiences of diasporas, it is necessary to attend to the cultural and structural barriers, challenges, and oppression levied against immigrant youth and families (Bhatia & Ram, 2009). We introduce diaspora as a conceptual lens to examine individual differences within transnational, immigrant communities who experience racism, classism, and sexism in diverse ways. Incorporating diaspora into developmental science offers a unique perspective on some of the most frequently studied developmental domains for immigrant youth and families: acculturation and enculturation, school adjustment and peer friendships, parent-child relationships, intergenerational trauma, ethnic-racial identity development, and ethnic-racial socialization.

Qurat-ul-ain Gulamhussein (✉) · R. M. Lee
Department of Psychology, University of Minnesota–Twin Cities, Minneapolis, MN, USA
e-mail: gulam003@umn.edu

X. Zhou
Department of Psychology, Purdue University, West Lafayette, IN, USA

A. Y. Kim
Department of Psychology, Elon University, Elon, NC, USA

© The Author(s), under exclusive license to Springer Nature Switzerland AG 2023

D. P. Witherspoon, G. L. Stein (eds.), *Diversity and Developmental Science*,
https://doi.org/10.1007/978-3-031-23163-6_12

Diaspora Defined and Debated

Who and what constitutes a diaspora is contested across time and disciplines. In his foundational work, Safran (1991) defined a diaspora as a group of individuals who have dispersed from the homeland and created a collective narrative of the homeland. They hope to return to the homeland, invest emotionally or financially toward homeland preservation, and maintain ties to the homeland over the lifespan (Safran, 1991). This definition of diaspora also emphasizes a lack of acceptance, such as isolation, rejection, and discrimination, in the hostland[1] (Safran, 1991).

This group-level perspective, however, treats diasporas as static by-products of transnational migration (Tölölyan, 1996), without attending to heterogeneity across, and differential experiences within, diasporas. Anthias (1998) similarly critiqued earlier conceptualizations of diaspora for overlooking the diversity of lived experiences post-migration. Diasporic communities do not homogeneously experience the "troubled relationship" with their hostland (Berns-McGown, 2007, p. 6). In fact, individuals continuously negotiate their identities and adjust within shifting spaces, contexts, and time. A qualitative study of 15 Asian American adults who moved to the United States as children, for instance, found their acculturation experiences evolved in diverse ways based on the agency they held regarding moving to the United States (Tsong et al., 2021). Some described that over time, being part of transnational families facilitated their resilience and creativity in how they maintained ties with extended family in the homeland and developed relationships with partners and children in the United States.

We therefore define diaspora as any nature of dispersion from the homeland accompanied by an orientation to the real or imagined homeland and boundary maintenance of the diasporic community, without necessitating a strained relationship to the hostland (Brubaker, 2005; Cohen, 2008). We adopt a person-level perspective to center the diversity of individual experiences within diasporas and across time. While certainly not an exhaustive list, refugees, migrant workers, third culture children and adults (growing up in a culture and country other than their parents' culture or country of nationality), international students and scholars, and transnational adoptees all may understand their diaspora experiences through varying levels of connection with co-ethnics and attachment to their homelands and hostlands.

As developmental scientists, we specifically recognize the ongoing negotiation between homeland and hostland that shapes various developmental and cultural processes and practices for diasporic individuals and families, including the following:

- How does a refugee engage with acculturation over time when there are pressures to retain homeland and hostland practices and languages?

[1] Throughout this chapter, we use the term "hostland," borrowed from diaspora literature to refer to geographical or imagined lands that individuals engage with outside their homelands, without necessitating documented citizenship.

- How does a third culture child develop friendships with peers across physical borders and multiple homelands and sustain them as an adult?
- How does a transnational adoptee understand family while navigating the push and pull between the relationship with their adoptive family and desire to search for their birth family?
- How does a third-generation descendant of displaced refugees address intergenerational trauma and reflect on what homeland means?
- How does an immigrant parent approach ethnic-racial socialization with their child considering the family's diasporic journey over time?
- How does an international student develop their ethnic-racial identity within the context of the hostland where they may stay for short or long periods at a time?

We pose these questions as examples that highlight the need for a diaspora perspective on human development and family life.

In this chapter, we address how diasporic individuals and families navigate these developmental and cultural processes and practices in the context of structural policies and systems of racism, classism, and sexism in the homeland and hostland (Buchanan & Wiklund, 2020; Collins & Bilge, 2016; Crenshaw, 1991). We provide examples of how diaspora informs our understanding of key sociocultural and psychological constructs that are commonly studied in developmental science, that is, acculturation and enculturation, school adjustment and peer friendships, parent-child relationships, intergenerational trauma, ethnic-racial identity development, and ethnic-racial socialization. Later in this chapter, as a reflexive illustration, we share our diasporic positionality and briefly discuss how a person-level diaspora lens contributes to the understanding of developmental and cultural processes within our respective diasporas.

Developmental and Cultural Processes from a Diaspora Lens

For children, youth, and families, diaspora is a lifelong person-level social and psychological process and serves as a unique lens by which developmental and cultural processes and practices are experienced. It refers to more than a singular migration event or the short-term impact of immigration. This shift in understanding is similar to psychological perspectives on acculturation that assert acculturation is not simply contact and changes between two cultures but involves changes over time that occur within and between individuals who live in two or more cultural contexts (Titzmann & Lee, 2018).

We employ a strengths-based approach to studying diaspora for individuals and families embedded within structural systems (García Coll et al., 1996). We focus on how individuals within diasporas both experience challenges and demonstrate resilience in the hostland (Juang et al., 2018). Through visiting the homeland, engaging in local or international policy work to preserve the diaspora, sending remittances or other financial support, and teaching children about traditions and practices, individuals also can find homeland orientation to be protective for them and their families' well-being (Grossman, 2019). Therefore, it is vital to explore how the ebb and flow of homelands and hostlands in diasporic experiences shape developmental and cultural processes and practices.

Acculturation and Enculturation

A diaspora perspective conceptualizes acculturation and enculturation as a lifelong process of cultural adaptation that occurs not only after physical relocation (Berry et al., 2006), but also prior to and long after the migration of youth and families (Titzmann & Lee, 2018). Individuals can participate in remote acculturation and enculturation even while being physically distant from homelands and hostlands. For instance, research with adolescents and mothers in Jamaica indicates how they engage in remote acculturation to American culture through media consumption, eating habits, and feeling American (Ferguson et al., 2018). Similarly, transracially and transnationally adopted Koreans in the United States may practice remote enculturation and learn about Korean language and cultural traditions without exposure to Korean family or community, tapping into a diasporic yearning to experience homeland culture (Kim et al., 2021).

Applying diaspora as a conceptual lens highlights a lifespan perspective for how individuals may experience and process acculturation across different contexts and time. Traditional approaches to acculturation, by contrast, assume that migration and the acculturative process is experienced universally within and across diasporic communities. For example, Berry (1997) conceptualized acculturation to involve four strategies of how individuals adapt in new cultural contexts—assimilation, separation, marginalization, and integration. Individuals are often encouraged to pursue integration as their acculturation strategy (i.e., individuals adopt the hostland culture while also retaining the homeland culture; Berry, 1997). However, youth and families do not experience acculturation in one universal process and acculturation strategies are often adopted and discarded depending on the context.

A diaspora lens helps explain how communities transnationally co-construct relationships with others. A study of first-generation Indians in the United States found that they felt well integrated and accepted in the hostland based on their economic success, but the traumatic 9/11 events compelled them to re-evaluate their perceived integration (Bhatia & Ram, 2009). While they had previously assumed that their class privilege afforded them entry into hostland acceptance, they were now forced to prove their loyalty to the hostland because of their racial minority status in the United States. The Indian diaspora's dynamic experience of acculturation, which is further complicated by differences rooted in class, gender, and generational status, directly challenges the universalist assumption that acculturation is a linear process and that all individuals regardless of their positionality can achieve stable and permanent integration (Bhatia & Ram, 2001).

First-generation immigrants from African, Afro-Latinx, and Afro-Caribbean diasporas also have had varied experiences of stress while learning to navigate systems of immigration, healthcare, education, housing, and employment (Ndumu, 2019). Indeed, the methods through which individuals engage with systems in the hostland is negotiated by information that was accessible pre-migration and the pressures of life and success placed on individuals by homelands and hostlands post-migration. Certain individual differences influence how individuals within a diaspora navigate communication and decision-making. For instance, age and literacy levels shape accessibility to information and preferences for communication style. In Ndumu's (2019) focus groups, one Haitian immigrant participant described that his grandmother who could not read and write preferred engaging in oral tradition with co-ethnics to learn how to navigate various systems in the United States. A diaspora understanding of acculturation better accounts for non-linear processes, attends to the role of sociopolitical events and individual differences, and highlights how people creatively build alliances with other diasporic individuals.

School Adjustment and Peer Friendships

A diaspora perspective provides added insight into how diasporic youth navigate the school context and friendships over time. For instance, the ways in which a family actively participates in an ongoing relationship to the homeland and maintains boundaries around language and other cultural practices in the hostland can affect children's readiness for school, their language development and preferences, and their willingness and ability to form and sustain friendships in and out of school (Birman & Tran, 2017; Lijadi & Van Schalkwyk, 2018). While a younger immigrant child may learn the host language more quickly at school than an older immigrant child, two children of the same age can have vastly different experiences of school adjustment due to the varying importance placed on retaining one's heritage language, customs, and practices. This comparison is further complicated when peer relationships, ethnic and racial diversity, school climate, and structural policies are considered. Racial segregation and redlining create disparities in resources available to local schools and as a result, to the children who live in particular neighborhoods (Buchanan & Wiklund, 2020). Therefore, researchers need to carefully consider how academic and school competencies are defined and measured and how individual differences and diverse sociocultural contexts are taken into account (Lee et al., 2020). We provide two examples of contemporary diasporas below—Somali refugee youth and third culture children—to illustrate how children and youth's experiences of school adjustment and peer friendships are shaped by their constant negotiation of relationships to homeland and hostland.

In the case of Somali refugee youth, while many students quickly adapt to the new school environment, some may hold ambivalent feelings toward formal education in the hostland. For youth who have had limited previous formal education, what appears to be delayed competency achievement and behavioral issues in the classroom may actually reflect overlooked English language barriers and biases (Roxas, 2008). Birman and Tran (2017) also describe a lack of institutionalized diversity training of teachers who perceive Somali refugee students as "primitive," and an unrealistic expectation to assimilate into American school systems and schedules. For Somali refugee adolescents, who are Black and mostly Muslim, ethnic-racial and religious discrimination can also contribute to greater feelings of isolation and estrangement within the hostland (Ellis et al., 2010). It is important then to identify ways to enhance their sense of belonging at school in order to lower depression and raise self-efficacy. Kia-Keating and Ellis (2007) found this association even after controlling for previous traumatic experiences. Clearly, teachers and peers play a critical role in helping refugees with school adjustment.

The Somali refugee diaspora is not homogenous (Abdi, 2015). There is tremendous within-group diversity, which may better explain how Somali refugee youth succeed at school and build strong friendships. Notably, the Somali Bantu experience marginalization even within the larger Somali community because they are perpetually labeled and oppressed as descendants of slaves. Such experiences may influence how they interact with other Somali and non-Somali peers at school (Birman & Tran, 2017; Roxas, 2008). Similarly, it is important to understand gender-based differences within the larger diaspora of Somali refugee youth. Rooted in gendered societal norms, identifying more closely with Somali culture may be more protective for girls' mental health while identifying more closely with American culture may be more protective for boys' mental health (Ellis et al., 2010). Therefore, how youth choose to engage with peers and perform at school is intertwined with their positionality in terms of ethnicity and gender within the larger Somali diaspora.

Third culture children and young adults represent another contemporary diaspora whose school adjustment and peer friendships are often more nuanced than traditional immigrant peers of the same ethnicity. Growing up, third culture youth spend an extended period of time in hostlands that are not reflective of their parents' cultures or nationalities. While they build connections to both homeland and hostland, they are often perceived as outsiders in both cultures (Pollock & Van Reken, 2001). Many third culture adults reflect on the unique, often hidden, challenges they faced as children in learning new educational systems, languages, and cultural knowledge (Lijadi & Van Schalkwyk, 2018). They recall being perceived as perpetual foreigners and having difficulty building relationships in the different places in which they live, including their passport country (Kwon, 2019). Rather than associating their homeland with a concrete location, third culture children and young adults understand "home" as a dynamic space of belonging that is weaved into the transnational relationships they retain and cultivate virtually across geographical borders (Hannaford, 2016). Future research is needed to explore the diversity of school experiences and peer friendships for third culture children and adults based on the policies established within the hostland and the passport country, in addition to access to resources afforded to parents by their work organizations.

Parent-Child Relationships

Children and adults, including those who have family members in the United States, often experience physical and psychological distance from relatives in the homeland. Refugee children and adults may have migrated forcefully and permanently with some connection to relatives left behind, third culture children and adults may have moved voluntarily and temporarily with more realistic hopes for return, and transnationally adopted children and adults may live in adoptive families with little—or factually incorrect—information about their families and cultures of origin. As developmental scientists explore family in the diasporic context, it is necessary to understand how transnational family connections are imagined and sustained across time (Falicov, 2007).

Parent-child relationships can facilitate both support and conflict within diaspora families. For children and youth, maintaining secure attachment with caregivers can be protective, but it is equally necessary to consider the effects of stress experienced by caregivers before, during, and after migration on these relationships (Juang et al., 2018). Incorporating a diaspora lens complicates our understanding of how, for instance, Mexican immigrant families experience the socialization of cultural values such as familismo. Familismo involves acting with a strong sense of connectedness and commitment to family (Fuligni & Pedersen, 2002). Smith-Morris and colleagues (2013) conducted qualitative work to expand on the limited quantitative study of familismo and found that familismo is often experienced as a yearning for connection with the homeland and those living there. This cultural value can have protective effects on well-being throughout the lifespan, including greater self-esteem and stronger sense of belonging and cohesion within families and communities (Stein et al., 2014). In contrast, other studies have found that familismo can also be related to increased family conflicts and internalizing behaviors (East & Weisner, 2009; Kuhlberg et al., 2010). Thus, a person-level diaspora lens contributes to understanding how parents and children approach familismo in diverse ways based on their evolving attachments to the homeland.

Parent-child relationships for Mexican diaspora families are also complicated by how people relate to the homeland and hostland as individuals and as family units (Smith-Morris et al., 2013). A parent may engage in familismo in ways that were common in Mexico pre-migration. With passing time in the hostland, they struggle to imagine the homeland differently and to reconcile with shifting cultural practices in the homeland. To compensate for their loss of physical connection and to construct a diasporic connection, they invest efforts to retain pre-migration homeland norms in the United States. Simultaneously, their child experiences the push and pull between learning more about the homeland and hostland and identifying their own positionality and that of their parent's. How diasporic individuals engage in parent-child relationships is intertwined with complex person-level experiences of acculturation, cultural values, and imagined sense of diasporic community.

Filipina migrant workers, a unique racialized, gendered, and classed diaspora, represent another illustration of the way in which diasporic contexts shape parent-child relationships. The Philippines' government has established policies and programs to facilitate the employment of Filipinas abroad (Garabiles et al., 2017). This includes the socialization of Filipina women as mothers and wives to financially prioritize family and national interests (Tanyag, 2017). Filipina women who are more educated or skilled at domestic work are often hired to work overseas as substitute mothers for an employer's children. These women may then hire domestic help themselves—typically less educated and skilled Filipina women locally—to take care of their own children in the Philippines (Lan, 2003). In their gendered transformation as simultaneously an employer and employee, Filipina women build unique attachments with an employer's children overseas while negotiating relationships with their own children. Thus, these women experience parenting in the diaspora in nuanced ways based on governmental limitations on worker rights, the period of separation between themselves and their families, the support (or lack thereof) from families, and their own education and skill levels.

While noting how Filipina migrant workers navigate structural economic struggles and parent-child relationships, it is equally important to emphasize how they demonstrate resilience. For example, Filipina migrant workers along with family members transnationally published their strengths and challenges in a magazine that was widely shared within their "imagined (global) community" (Parreñas, 2001, p. 1145). These efforts represent how diasporic individuals and families actively engage in dialogue and coalition building (Crenshaw, 1991) that extends beyond the borders of homeland and hostland.

Intergenerational Trauma

As individuals navigate their relationships with peers and family in the homeland and hostland, experiences with intergenerational trauma are similarly shaped by these transnational contexts. Intergenerational trauma has been conceptualized as the negative health effects of trauma transmitted across generations (Sangalang & Vang, 2017). Awad and colleagues (2019) developed a model of cumulative racial-ethnic trauma for Americans of Middle Eastern and North African descent to account for historical trauma—experiences of violence, displacement, and genocide in the homeland—and ongoing institutional discrimination, fear, alienation, and physical and mental health conditions at the individual level.

Offering a person-level diaspora perspective allows for a more nuanced understanding of trauma across generations and time. Two third-generation Palestinian American adolescents may have different experiences of intergenerational trauma based on how their parents and grandparents discuss topics of ongoing Israeli colonial occupation, experience continued violence from media coverage of their homeland, and visit or stay in touch with relatives in the homeland over time. Similarly, members of diasporic families hold multiple identities that individually position them across societal axes of power and privilege (Buchanan & Wiklund, 2020; Collins & Bilge, 2016; Crenshaw, 1991) and shape their experiences of intergenerational trauma. For example, while Asian Americans in the United States have experienced intergenerational trauma from World War II, the Korean War, the Vietnam War, and the Cambodian genocide, Asian American women in particular have uniquely experienced sexism in their homelands and hostlands that compound the effects of intergenerational trauma (Kim-Prieto et al., 2018).

Most intergenerational trauma research focuses on effects of dispersion due to political turmoil, but trauma transmitted over generations due to oppressive systems within a single community has been largely overlooked. In the case of the Indian diaspora, caste-based trauma embedded in the homeland is perpetuated from one generation to the next and may be carried over to the United States. Second-generation Dalits are perceived as lower caste and often marginalized by other Hindu Indians. This communal and individual level marginalization limits their opportunities for education, employment, friendships, and marriage (Adur & Narayan, 2017). While caste-based trauma and its detrimental effects on health have been passed down across generations for centuries, this manifestation of intergenerational trauma has been neglected in psychology literature (Ahammed, 2019). Future research would be useful in empirically examining psychological distress experienced by Dalit families generationally in India and the United States, in addition to exploring individual experiences, based on whether and at what age one chooses to "come out" with their identity as Dalit within the larger Indian community in the United States, how they connect with the homeland, and engage in anti-caste activism and healing (Adur & Narayan, 2017; Ahammed, 2019).

Ethnic-Racial Identity and Ethnic-Racial Socialization

The operationalizations of ethnic-racial identity and ethnic-racial socialization do not take diaspora experiences explicitly into account. Yet, an individual's ethnic racial identity, or their sense of self related to their ethnic heritage and racial background (Umaña-Taylor et al., 2014), is often inherently rooted in a diasporic context. Individuals living in the diaspora are confronted with the task to understand oneself in relation to their homeland and home culture and their positionality in the hostland. For example, in Cross Jr.'s (1991) classic racial identity work on Nigrescence theory, Black individuals may turn to understand Black history and culture during the immersion-emersion stage. Similarly, in the operationalization of ethnic identity, scholars (Umaña-Taylor et al., 2014) have emphasized how ethnic-racial minority youth become connected to their group history, culture, and customs. In the active construction of their ethnic-racial identity, second-generation immigrant youth and emerging adults often experience a diasporic return (Tsuda, 2009) or return to their "homeland" (Suh, 2020). Yet, current ethnic-racial identity frameworks fail to address the transnational nature of ethnic-racial identity for these individuals.

Ethnic-racial identity labels (e.g., Chinese Americans, Mexican Americans) are usually wedded to their homeland and conflated with one's national identity (particularly for non-US-born individuals). An individual from China may have identified as Han within the Chinese context, but she may adopt a Chinese ethnic-racial identity upon immigrating to the United States due to hostland sociopolitics. Particularly, in cases where there may be a troubled relationship between diaspora groups and hostland (Cohen, 2008), diasporic individuals understand how their ethnic-racial group membership situates them within the hostland. This awakening to systemic racism and discrimination is often tied to ethnic-racial identity development (Meca et al., 2020).

We argue that the formation of a diasporic identity, in addition to one's ethnic-racial identity, can be a more developmentally appropriate task beyond childhood and adolescence. In this context, diasporic identity transcends beyond a self-identification to their ethnic-racial group and draws on the transnational nature of the diaspora itself as an identity (Kim et al., 2021). In contrast to ethnic-racial identity, diasporic identity emphasizes more on the roles that the homeland as well as co-ethnics across borders play in its formation process. Diasporic identity may have additional content domains such as co-ethnic solidarity or attachment to homeland not captured in traditional ethnic-racial measurements (Kim et al., 2021). Individuals with salient diasporic identity may especially engage in behaviors such as sending remittances to and preserving the homeland, seeking diasporic return during one's adulthood, or practicing ethnic-racial socialization which emphasizes a diasporic history. Black parents in France, for example, may connect their children to the #BlackLivesMatter movement in the United States to build solidarity and understand a shared racialized experience in post-colonial societies. Thus, applying a diaspora lens to ethnic-racial identity formation allows us to capture important behavioral processes that span across land borders.

Similarly, the conceptualization of family ethnic-racial socialization can be situated in a diaspora lens to further our understanding of families living in contemporary diasporas. Family ethnic socialization highlights the process in which beliefs, messages, and practices about their cultural heritage are transmitted (Hughes & Chen, 1999). Thus, ethnic socialization serves as an active means to make diasporic practices alive in a family context—for example, a Mexican immigrant parent may share not only Mexican history with their children, but also their diasporic history with regard to family migration or shifting nation-state borders (Portes et al., 2009). Family racial socialization emphasizes how parents help children understand intergroup relations and navigate a racist society.

Importantly, even though parents' ethnic-racial socialization practices take place within the family context to promote healthy ethnic-racial identity development, individuals are closely shaped by their diasporic experiences. In comparing African American and Black Caribbean youth in the United States, for example, researchers have found Black Caribbean parents are more likely to emphasize ethnic socialization and ethnic membership (as opposed to racial membership) and more likely to minimize racism (Lambert et al., 2020; Rong & Brown, 2002). This example illustrates the need to situate studies of ethnic-racial identity and ethnic-racial socialization in the unique diasporic histories of individuals (e.g., Falola & Oyebade, 2016). Again, the current conceptualization and measurement of ethnic-racial socialization fails to incorporate the transnational nature of contemporary parenting. From a diaspora lens, researchers may begin to address important questions such as how parents socialize youth to consider their co-ethnics in other countries and shared diaspora experiences.

Situating Ourselves Within Diasporic Contexts

In this section, the first three authors offer narratives on the different diasporas we identify with, in addition to offering person-level perspectives on relevant developmental and cultural processes.

Khojas of East Africa, Qurat-ul-ain Gulamhussein

To understand the experiences of children and families within the Khoja diaspora of East Africa, it is important to review the sociopolitical context and history of the Khoja community. Khojas traveled from what is now modern-day India to East Africa as merchants in the late eighteenth and early nineteenth centuries (Akhtar, 2015). The literature on Asians in East Africa, including Khojas, lacks attention to the diversity of lived experiences of "men and women, of traders and laborers, or of people from different religious, caste, linguistic, or regional backgrounds. The story of the 'Asian in East Africa' is the story of an Indian man who is a trader/shopkeeper/businessman" (Nagar, 1996, p. 63). This perceived homogeneity of Asians, specifically in Dar-es-Salaam, the capital of Tanzania with a large Khoja community, is conceptualized as the by-product of German and British colonialism (Nagar, 1997).

White colonizing powers established racially segregated structures for Europeans, Asians, and Africans and aftereffects persist in the post-colonial society (Smiley, 2013). Only a select group of wealthy Asians financially benefited from the existence of colonial powers in the region. Yet, mental maps drawn by Asian individuals, and particularly Khojas, in Dar-es-Salaam highlight their overall lack of socialization with children and adults of African families and the stark invisibility of predominantly African neighborhoods and communal spaces in their minds (Nagar, 1997; Smiley, 2013). These disparities may have contributed to why the majority of Asians were met with resentment by Africans after Tanzania won independence in 1964. Asians continued to be challenged regarding their loyalty and identification as indigenous Tanzanians (Nagar, 1996).

In response to increasing economic and political instability, Khoja communities transnationally developed The World Federation (Akhtar, 2014). While many Khoja groups survive today (Jaffer, 2012), I (Qurat-ul-ain Gulamhussein) focus on the Khoja Shia Ithna Asheri community of Tanzania, where I was born and raised. Over several decades, my community became part of "an imagined world that connected the worldwide Khoja jamat [community] based on an idealized Near Eastern Shiism through the cultural experience of Africa" (Akhtar, 2014, p. 33). It is this powerful organization that facilitated the escape and resettlement of Khojas when they were subjected to Idi Amin's 1972 expulsion order in Uganda and the 1991 political unrest targeting Asians in Somalia (Akhtar, 2014). When the Indian government refused to open borders at the time for Khojas after the 1972 Uganda expulsion order, "the Khoja of East Africa became diasporic and unable to permanently return to their homeland" (Akhtar, 2015, p. 13).

There remains a research gap in psychology and diaspora studies on the developmental science of Khoja youth and families who are now dispersed globally, with a large presence in the United States. Few Asians in East Africa acknowledge, "We have migrated here [to East Africa] because of the opportunities which were available…We were not given those opportunities [by Africans]. We took those opportunities and made ourselves at home" (Nagar, 1996, p. 75). Future research should examine how Khoja children and parents in the United States engage in ethnic-racial socialization and navigate the pushes and pulls of defining a homeland. Studies may also investigate the mental health effects of intergenerational trauma in both contributing to and experiencing historical oppression, racism, and Islamophobia across generations and physical land borders. These future directions will certainly be shaped by individual differences in class, gender, generational status, family dynamics, and opportunities to maintain ties with East Africa and the Indian subcontinent.

International Students and Scholars, Xiang Zhou

International education as a form of modern diaspora has not entered the main discourse of scholarly discussion. Seeking out education (from K-12 to post-doctoral training) across national borders is a *recent* phenomenon, dating back to the turn of the twentieth century, with a global exponential increase only occurring in the last few decades. Hallmark historical events such as the establishment of the Institute of International Education (IIE) in 1919 and a nonimmigrant visa in 1921 paved the way for the United States to be currently the largest receiving nation of international students in higher education (Institute of International Education, 2021).

On an individual level, international students or scholars do assume certain advantageous positions (i.e., educational attainment) and may have a homeland to which they can readily return. These features seemingly distinguish international students and scholars from the "forced" or "involuntary" migration experienced in other diasporic communities discussed in this chapter. In other words, they may be considered too "privileged" as a diasporic group.

However, on a societal level, the unidirectional flow of international students from the Global South to the Global North is oftentimes rooted in the educational and economic disparity that is a cascading effect of Euro-American colonization and imperialism. Contemporary indigenous scholars, especially in the humanities and social sciences, continue to grapple with finding their voices that are not translated for the White gaze (Shweder, 2000). Moreover, indigenous and international scholars are compelled to communicate their scholarship in English/their second-language and publish in academic outlets that are indexed by Euro-American publishers, which further perpetuates how educational resources are distributed globally. Yet, in other cases, resorting to international education for migration may be one of the few legitimate pathways in the United States, which has contributed to the "model minority stereotype." For example, the overall high educational attainment in Asian Americans has been partially attributed to the selective migration where only individuals with high cultural capital are allowed to migrate to the United States (Hsu, 2015).

As an international scholar and a former international student, I (Xiang Zhou) also wrestle with my own identity: When should I ever assume a hyphenated Chinese American identity? Is Americanness a cultural identification or an automatically granted identity with citizenship? Does an international student become "American" because an American-centered education etched their worldview? Does an international scholar *have to* identify as American after obtaining American citizenship? Or has this line of questioning rooted in individualism already made me American?

Recent studies have documented the impact of an international education diaspora on the racialization and ethnic-racial identity development of international students. For example, Afro-Caribbean international students noted that their experiences were different from those of African Americans, which in turn shaped their understanding about Blackness and developed new appreciation and commitment to home (Malcolm & Mendoza, 2014). The acculturation process among international students has also been well studied, which shares not only similarities with, but also notable contextual differences from, the acculturation process for immigrant and refugee populations (Smith & Khawaja, 2011). For example, higher education can be a unique source of acculturative stress (e.g., mismatch between expectations and reality, teaching styles) for international students (Smith & Khawaja, 2011). Lastly, many international scholars actively maintain ties to their homeland, sometimes through their work, for example, by building professional collaborations with scholars from their homeland (e.g., Zhai et al., 2014). Correspondingly, diaspora institutions have been increasing their efforts in recruiting diasporic international scholars for a reverse brain drain (e.g., Zweig & Wang, 2013).

Internationally Adopted Koreans, Adam Y. Kim

Like the broader Korean diaspora, the roots of the internationally adopted Korean diaspora are imperialism, war, and geopolitics.[2] First billed as war orphans and adopted as a form of humanitarian aid, the adoption of Korean children quickly

[2] For a comprehensive look at the history of Korean adoption, see: Choy (2013), Oh (2015), and Pate (2014).

became an economic engine that drove the global migration of Korean-born children (Condit-Shrestha, 2018; Pate, 2014). Starting in the early 1950s, Korea began sending children out of the country as a way to care for the children produced by foreign GIs and Korean women (Oh, 2015). This practice soon turned into a staple of Korea's local economies, bringing in foreign money through a cycle of "sex work economies tied to the US military, subsequent mixed-race children, [and] the practice of overseas adoption" (Condit-Shrestha, 2018, p. 369). However, this is not merely a story of foreign aid for a destabilized country. Unlike other displaced populations whose migrations are limited to a particular time or event, Korea is in its 68th continuous year of international adoption[3] (Condit-Shrestha, 2018; Kim, 2010; Oh, 2015; United States Department of State, n.d.). Further, the peak of international adoption from Korea occurred in 1985, roughly 30 years after the Korean War, with nearly 9000 children being internationally adopted (Kim, 2010). Roughly 200,000 Korean-born children have been internationally adopted since the 1950s (Nelson, 2019), with more being added to this list each year.

Transnationally adopted Koreans are a diaspora within a diaspora. While they are part of the greater Korean diaspora, their connection to the homeland and to non-adopted Koreans in the diaspora is tenuous due to their transnational adoptions into largely non-Korean families. Transnationally adopted Koreans report feeling a disconnect from non-adopted Koreans—both in the diaspora and in the homeland (Kim, 2010). This disconnect is tied to a lack of socialization in Korean cultural practices (e.g., language), experiencing discrimination from non-adopted Koreans, and having their own unique experiences of migration and homeland loss through their adoptions (Kim & Lee, 2020; Nelson, 2019). Therefore, transnationally adopted Koreans have formed extensive online and in-person communities with each other, including local and global events, social groups, and political organizations (Kim, 2010). In this way, they are both part of the larger "overseas Koreans" political project and also their own distinct diasporic community (McKee, 2016, p. 159). As a transnationally adopted Korean and an adoption scholar, this is my history and my field of study.

In my work, I (Adam Y. Kim) apply a diasporic lens to developmental research with transnationally adopted Koreans to acknowledge the impact that their migration has had on their relationship with non-adopted Koreans and Korea. For instance, the disconnect between transnationally adopted Koreans and non-adopted Koreans has the potential to cause psychometric problems for popular measures of ethnic-racial identity development, such as the Multigroup Ethnic Identity Measure (MEIM; Phinney, 1992; Roberts et al., 1999), which have not undergone rigorous psychometric evaluation with this population. This includes the affirmation/belonging dimension of the MEIM, which is meant to capture a cognitive understanding of group belonging and an affective sense of ingroup pride (Roberts et al., 1999). However, the lack of group cohesion that transnationally adopted Koreans express

[3] We arrived at this value by combining the data provided by Condit-Shrestha (2018), Kim (2010), Oh (2015), and the United States Department of State (n.d.).

with non-adopted Koreans means that items such as, "I have a strong sense of belonging to my own ethnic group" (Roberts et al., 1999, p. 319) may differentially load onto the affirmation/belonging factor for populations of adopted and non-adopted Koreans.

A diaspora lens not only highlights this measurement issue, but it points toward domains of research that we have not yet investigated in psychology. Rather than understanding themselves merely as Koreans, transnationally adopted Koreans incorporate their migration history into their identities. In order to capture this homeland orientation and within-group solidarity, the second, third, and fourth authors have developed a measure of diasporic identity for use with transnationally adopted Koreans (Kim et al., 2021).

A diasporic lens on transnationally adopted Koreans also complicates our understanding of family. Research shows that transnationally adopted Koreans think about their birth families (Kim et al., 2020). Importantly, thoughts about birth family are not related to psychological distress or maladjustment and are not related to a poor relationship between the adopted individual and their adoptive family (Kim et al., 2020). Further, thoughts about birth family are common across all ages, and therefore are not limited to a particular developmental period. This suggests that adopted individuals maintain an enduring psychological connection to their birth family that is not in conflict with their relationship with their adoptive family.

However, a diasporic lens also contests the dominant narratives that surround adoption and the adoptive parent. In stark contrast to the war orphan and savior narratives that are common in adoption, transnationally adopted Koreans are the product of an economic and political system in which babies were actively procured in order to create an influx of foreign capital for decades after the Korean War (Condit-Shrestha, 2018; Kim, 2010). It is in this context that transnationally adopted Koreans have forged their own understanding of family, not out of adherence to genetics or legal documents, but through their shared experience of transnational adoption (Kim, 2010).

Conclusion and Future Recommendations

In this chapter, we seek to advance diaspora as a conceptual lens to understand cultural and developmental processes and practices in diverse immigrant communities. Responding to Brubaker's (2005) call to move beyond conceptualizing diaspora in "substantialist terms as a bounded entity" (p. 12), we offer a person-level perspective on the diasporic individual. Rather than attempting to refashion diaspora, a person-level perspective complements extant group-level work on diasporas and necessarily attends to the structural forces of racism, sexism, and classism that inform human development. Using this person-level perspective of diaspora, we provided examples of its application to key development constructs of acculturation and enculturation, school adjustment and peer friendships, parent-child relationships, intergenerational trauma, ethnic-racial identity development, and ethnic-racial socialization. Lastly, through situating ourselves within our respective diasporic contexts, we offered additional contemporary examples of diaspora and emphasized the importance of research reflexivity in conducting anti-colonial research.

Taken together, we encourage developmental scientists to apply a diaspora lens that attends to both individual differences within diasporas and the structural mechanisms of systemic oppression (Buchanan & Wiklund, 2020; Collins & Bilge, 2016; Crenshaw, 1991). We offer three specific recommendations for developmental scientists to incorporate a diaspora lens in studying immigrant communities. First, we strongly encourage developmental scientists to adapt culturally sensitive research methodologies to gain an accurate understanding of diasporic individuals. Indeed, quantitative methods can provide valuable data, but it is necessary to ensure that measures have been previously validated and are relevant for use in diasporic community samples. Qualitative methods enable researchers to collect meaningful data that quantitative measures or existing theoretical models do not capture. In the case of third culture children and adults, it would be misleading to treat the entire group as financially privileged and quantitatively assess developmental milestones. Instead, qualitative studies can examine richer contexts of how pre-migration experiences, nature of parents' work, and alternative reasons for distance from the passport country contribute to developmental milestones (Kwon, 2019). For instance, third culture children in the United States who did not previously speak fluent English may experience heightened acculturative stress and pressure to learn English because of the prevalence and dominance of English across various American systems (Lijadi & Van Schalkwyk, 2018). This person-level research attends to the diversity of experiences as individuals and families navigate systemic forces. Thus, we urge developmental scientists to be intentional and culturally sensitive about selecting and implementing their research methodologies.

Second, we advocate for holistic strengths-based models to help pose culturally attuned research questions for various diasporas. When examining family relationships, for example, it is useful to incorporate the embedded contexts model to conceptualize intergenerational conflict as normative (Zhou et al., 2017). Further, researchers should question how structural barriers and historical oppression shape family relationships within and across diasporas. In addition, we recommend researchers to prioritize investigating how individuals draw on strengths and define success in their families and communities.

Finally, we urge researchers to reflect on their own positionality with the work they conduct. Which diasporas are selected to be researched, funded, and supported, and who makes those decisions? (Délano Alonso & Mylonas, 2019). Researchers are actors who are situated within the multiple axes of power and privilege. Therefore, we must reflect on how we conduct our own research, how we critique the research of others, and, ultimately, how we understand what is good and normative. For instance, researchers should consider why they include White "control" groups for research on the developmental and cultural processes experienced by racially marginalized children and families. We hope that careful consideration of contextual forces and individual differences within diasporas will further the field of developmental science.

Conflict of Interest We have no known conflict of interest to disclose.

References

Abdi, C. M. (2015). *Elusive jannah: The Somali diaspora and a borderless Muslim identity*. University of Minnesota Press.

Adur, S. M., & Narayan, A. (2017). Stories of Dalit diaspora: Migration, life narratives, and caste in the US. *Biography, 40*(1), 244–264.

Ahammed, S. (2019). Caste-based oppression, trauma and collective victimhood in erstwhile South India: The collective therapeutic potential of Theyyam. *Psychology and Developing Societies, 31*(1), 88–105.

Akhtar, I. S. (2014). Religious citizenship: The case of the globalised Khoja. *African Sociological Review/Revue Africaine de Sociologie, 18*(1), 27–48.

Akhtar, I. (2015). *The Khōjā of Tanzania: Discontinuities of a postcolonial religious identity*. Brill.

Anthias, F. (1998). Evaluating 'diaspora': Beyond ethnicity? *Sociology, 32*(3), 557–580.

Awad, G. H., Kia-Keating, M., & Amer, M. M. (2019). A model of cumulative racial–ethnic trauma among Americans of Middle Eastern and North African (MENA) descent. *American Psychologist, 74*(1), 76–87.

Berns-McGown, R. (2007). Redefining diaspora: The challenge of connection and inclusion. *International Journal, 63*(1), 3–20.

Berry, J. W. (1997). Immigration, acculturation, and adaptation. *Applied Psychology, 46*(1), 5–34.

Berry, J. W., Phinney, J. S., Sam, D. L., & Vedder, P. (2006). Immigrant youth: Acculturation, identity, and adaptation. *Applied Psychology, 55*(3), 303–332.

Bhatia, S., & Ram, A. (2001). Rethinking 'acculturation' in relation to diasporic cultures and postcolonial identities. *Human Development, 44*(1), 1–18.

Bhatia, S., & Ram, A. (2009). Theorizing identity in transnational and diaspora cultures: A critical approach to acculturation. *International Journal of Intercultural Relations, 33*(2), 140–149.

Birman, D., & Tran, N. (2017). When worlds collide: Academic adjustment of Somali Bantu students with limited formal education in a US elementary school. *International Journal of Intercultural Relations, 60*, 132–144.

Brubaker, R. (2005). The 'diaspora' diaspora. *Ethnic and Racial Studies, 28*(1), 1–19.

Buchanan, N. T., & Wiklund, L. O. (2020). Intersectionality research in psychological science: Resisting the tendency to disconnect, dilute, and depoliticize. *Research on Child and Adolescent Psychopathology, 49*, 25–31.

Choy, C. C. (2013). *Global families: A history of Asian international adoption in America* (Vol. 8). NYU Press.

Cohen, R. (2008). *Global diasporas: An introduction*. Routledge.

Collins, P. H., & Bilge, S. (2016). *Intersectionality*. Polity Press.

Condit-Shrestha, K. (2018). South Korea and adoption's ends: Reexamining the numbers and historicizing market economies. *Adoption & Culture, 6*(2), 364–400.

Crenshaw, K. (1991). Mapping the margins: Intersectionality, identity politics, and violence against women of color. *Stanford Law Review, 43*(6), 1241–1299.

Cross, W. E., Jr. (1991). *Shades of black: Diversity in African-American identity*. Temple University Press.

Délano Alonso, A., & Mylonas, H. (2019). The microfoundations of diaspora politics: Unpacking the state and disaggregating the diaspora. *Journal of Ethnic and Migration Studies, 45*(4), 473–491.

East, P. L., & Weisner, T. S. (2009). Mexican American adolescents' family caregiving: Selection effects and longitudinal associations with adjustment. *Family Relations, 58*(5), 562–577.

Ellis, B. H., MacDonald, H. Z., Klunk-Gillis, J., Lincoln, A., Strunin, L., & Cabral, H. J. (2010). Discrimination and mental health among Somali refugee adolescents: The role of acculturation and gender. *American Journal of Orthopsychiatry, 80*(4), 564–575.

Falicov, C. J. (2007). Working with transnational immigrants: Expanding meanings of family, community, and culture. *Family Process, 46*(2), 157–171.

Falola, T., & Oyebade, A. (2016). *The new African diaspora in the United States.* Routledge.

Ferguson, G. M., Muzaffar, H., Iturbide, M. I., Chu, H., & Meeks Gardner, J. (2018). Feel American, watch American, eat American? Remote acculturation, TV, and nutrition among adolescent–mother dyads in Jamaica. *Child Development, 89*(4), 1360–1377.

Fuligni, A. J., & Pedersen, S. (2002). Family obligation and the transition to young adulthood. *Developmental Psychology, 38*(5), 856–868.

Garabiles, M. R., Ofreneo, M. A. P., & Hall, B. J. (2017). Towards a model of resilience for transnational families of Filipina domestic workers. *PLoS One, 12*(8), e0183703.

García Coll, C., Crnic, K., Lamberty, G., Wasik, B. H., Jenkins, R., Garcia, H. V., & McAdoo, H. P. (1996). An integrative model for the study of developmental competencies in minority children. *Child Development, 67*(5), 1891–1914. https://doi.org/10.2307/1131600

Grossman, J. (2019). Toward a definition of diaspora. *Ethnic and Racial Studies, 42*(8), 1263–1282.

Hannaford, J. (2016). Digital worlds as sites of belonging for third culture kids: A new literacies perspective. *Journal of Research in International Education, 15*(3), 253–265.

Hsu, M. Y. (2015). *Good immigrants: How the yellow peril because the model minority.* Princeton University Press.

Hughes, D., & Chen, L. (1999). The nature of parents' race-related communications to children: A developmental perspective. In L. Balter & C. S. Tamis-LeMonda (Eds.), *Child psychology: A handbook of contemporary issues* (pp. 467–490). Psychology Press.

Institute of International Education. (2021). *Open Doors.* https://www.iie.org/en/Research-and-Insights/Open-Doorshttps, https://www.iie.org:443/Research-and-Insights/Open-Doors/Data/International-Students

Jaffer, H. A. M. (2012). *Endangered species.* Mulla Asghar Memorial Library & Islamic Resource Centre.

Juang, L. P., Simpson, J. A., Lee, R. M., Rothman, A. J., Titzmann, P. F., Schachner, M. K., et al. (2018). Using attachment and relational perspectives to understand adaptation and resilience among immigrant and refugee youth. *American Psychologist, 73*(6), 797–811.

Kia-Keating, M., & Ellis, B. H. (2007). Belonging and connection to school in resettlement: Young refugees, school belonging, and psychosocial adjustment. *Clinical Child Psychology and Psychiatry, 12*(1), 29–43.

Kim, E. J. (2010). *Adopted territory: Transnational Korean adoptees and the politics of belonging.* Duke University Press.

Kim, A. Y., & Lee, R. M. (2020). A critical adoption studies and Asian Americanist integrative perspective on the psychology of Korean adoption. In G. M. Wrobel, E. Helder, & E. Marr (Eds.), *The Routledge handbook of adoption* (pp. 120–134). Routledge.

Kim, A. Y., Kim, O. M., Hu, A. W., Oh, J. S., & Lee, R. M. (2020). Conceptualization and measurement of birth family thoughts for adolescents and adults adopted transnationally. *Journal of Family Psychology, 34*(5), 555–565.

Kim, A. Y., Zhou, X., & Lee, R. M. (2021). Theory and development of the diasporic identity scale with adopted Korean Americans. *Adoption Quarterly*, 1–25.

Kim-Prieto, C., Kim, G. S., Salvo Crane, L., Ming Lowe, S., Loan Le, P., & Dinh, K. T. (2018). Legacies of war: Asian American women and war trauma. *Women & Therapy, 41*(3–4), 203–218.

Kuhlberg, J. A., Peña, J. B., & Zayas, L. H. (2010). Familism, parent-adolescent conflict, self-esteem, internalizing behaviors and suicide attempts among adolescent Latinas. *Child Psychiatry & Human Development, 41*(4), 425–440.

Kwon, J. (2019). Third culture kids: Growing up with mobility and cross-cultural transitions. *Diaspora, Indigenous, and Minority Education, 13*(2), 113–122.

Lambert, S. F., Rose, T., Saleem, F. T., & Caldwell, C. H. (2020). Ethnic-racial socialization, perceived neighborhood quality, and psychosocial adjustment among African American and Caribbean Black adolescents. *Journal of Research on Adolescence, 31*(1), 1–19.

Lan, P. C. (2003). Maid or madam? Filipina migrant workers and the continuity of domestic labor. *Gender & Society, 17*(2), 187–208.

Lee, R. M., Titzmann, P. F., & Jugert, P. (2020). Towards a more dynamic perspective on acculturation research. In *Youth in superdiverse societies: Growing up with globalization, diversity, and acculturation* (pp. 74–91). Routledge.

Lijadi, A. A., & Van Schalkwyk, G. J. (2018). "The international schools are not so international after all": The educational experiences of Third Culture Kids. *International Journal of School & Educational Psychology, 6*(1), 50–61.

Malcolm, Z. T., & Mendoza, P. (2014). Afro-Caribbean international students' ethnic identity development: Fluidity, intersectionality, agency, and performativity. *Journal of College Student Development, 55*(6), 595–614.

McKee, K. D. (2016). Monetary flows and the movements of children: The transnational adoption industrial complex. *Journal of Korean Studies, 21*(1), 137–178.

Meca, A., Gonzales-Backen, M. A., Davis, R. J., Hassell, T., & Rodil, J. C. (2020). Development of the United States Identity Scale: Unpacking exploration, resolution, and affirmation. *Journal of Latinx Psychology, 8*(2), 127.

Nagar, R. (1996). The South Asian diaspora in Tanzania: A history retold. *Comparative Studies of South Asia, Africa and the Middle East, 16*(2), 62–80.

Nagar, R. (1997). Communal places and the politics of multiple identities: The case of Tanzanian Asians. *Ecumene, 4*(1), 3–26.

Ndumu, A. (2019). Linkages between information overload and acculturative stress: The case of Black diasporic immigrants in the US. *Journal of Librarianship and Information Science*, 1–11.

Nelson, K. P. (2019). Uri Nara, our country: Korean American adoptees in the global age. In T. Tsuda & C. Song (Eds.), *Diasporic returns to the ethnic homeland* (pp. 143–160). Palgrave Macmillan.

Oh, A. H. (2015). *To save the children of Korea: The Cold War origins of international adoption*. Stanford University Press.

Parreñas, R. S. (2001). Transgressing the nation-state: The partial citizenship and "imagined (global) community" of migrant Filipina domestic workers. *Signs: Journal of Women in Culture and Society, 26*(4), 1129–1154.

Pate, S. (2014). *From orphan to adoptee: US empire and genealogies of Korean adoption*. University of Minnesota Press.

Phinney, J. S. (1992). The multigroup ethnic identity measure: A new scale for use with diverse groups. *Journal of Adolescent Research, 7*(2), 156–176.

Pollock, D. C., & Van Reken, R. (2001). *Third culture kids: The experience of growing up among worlds*. Nicholas Brealey/Intercultural Press.

Portes, A., Fernández-Kelly, P., & Haller, W. (2009). The adaptation of the immigrant second generation in America: A theoretical overview and recent evidence. *Journal of Ethnic and Migration Studies, 35*(7), 1077–1104.

Roberts, R. E., Phinney, J. S., Masse, L. C., Chen, Y. R., Roberts, C. R., & Romero, A. (1999). The structure of ethnic identity of young adolescents from diverse ethnocultural groups. *The Journal of Early Adolescence, 19*(3), 301–322.

Rong, X. L., & Brown, F. (2002). Socialization, culture, and identities of Black immigrant children: What educators need to know and do. *Education and Urban Society, 34*(2), 247–273.

Roxas, K. (2008). Who dares to dream the American dream? The success of Somali Bantu male students at an American High School. *Multicultural Education, 16*(2), 2–9.

Safran, W. (1991). Diasporas in modern societies: Myths of homeland and return. *Diaspora: A Journal of Transnational Studies, 1*(1), 83–99.

Sangalang, C. C., & Vang, C. (2017). Intergenerational trauma in refugee families: A systematic review. *Journal of Immigrant and Minority Health, 19*(3), 745–754.

Shweder, R. A. (2000). The psychology of practice and the practice of the three psychologies. *Asian Journal of Social Psychology, 3*(3), 207–222.

Smiley, S. L. (2013). Mental maps, segregation, and everyday life in Dar es Salaam, Tanzania. *Journal of Cultural Geography, 30*(2), 215–244.

Smith, R. A., & Khawaja, N. G. (2011). A review of the acculturation experiences of international students. *International Journal of Intercultural Relations, 35*(6), 699–713.

Smith-Morris, C., Morales-Campos, D., Alvarez, E. A. C., & Turner, M. (2013). An anthropology of familismo: On narratives and description of Mexican/immigrants. *Hispanic Journal of Behavioral Sciences, 35*(1), 35–60.

Stein, G. L., Cupito, A. M., Mendez, J. L., Prandoni, J., Huq, N., & Westerberg, D. (2014). Familism through a developmental lens. *Journal of Latina/o Psychology, 2*(4), 224–250.

Suh, S. C. (2020). Racing "return": The diasporic return of US-raised Korean Americans in racial and ethnic perspective. *Ethnic and Racial Studies, 43*(6), 1072–1090.

Tanyag, M. (2017). Invisible labor, invisible bodies: How the global political economy affects reproductive freedom in the Philippines. *International Feminist Journal of Politics, 19*(1), 39–54.

Titzmann, P. F., & Lee, R. M. (2018). Adaptation of young immigrants. *European Psychologist, 23*, 72.

Tölölyan, K. (1996). Rethinking diaspora (s): Stateless power in the transnational moment. *Diaspora: A Journal of Transnational Studies, 5*(1), 3–36.

Tsong, Y., Tai, A. L., & Chopra, S. B. (2021). The emotional, cultural, and relational impact of growing up as parachute/satellite kids in Asian American transnational families. *Asian American Journal of Psychology, 12*(2), 147–157.

Tsuda, T. (2009). Why does the diaspora return home. In *Diasporic homecomings: Ethnic return migration in comparative perspective* (pp. 21–43). Stanford University Press.

Umaña-Taylor, A. J., Quintana, S. M., Lee, R. M., Cross, W. E., Jr., Rivas-Drake, D., Schwartz, S. J., et al. (2014). Ethnic and racial identity during adolescence and into young adulthood: An integrated conceptualization. *Child Development, 85*(1), 21–39.

United States Department of State. (n.d.). *Adoption Statistics*. https://travel.state.gov/content/travel/en/Intercountry-Adoption/adopt_ref/adoption-statistics-esri.html

Zhai, L., Yan, X., Shibchurn, J., & Song, X. (2014). Evolutionary analysis of international collaboration network of Chinese scholars in management research. *Scientometrics, 98*(2), 1435–1454.

Zhou, X., Wu, C. S., Juan, M. J. D., & Lee, R. M. (2017). Understanding and addressing parent-adolescent conflict in Asian American families. In *Asian American parenting* (pp. 143–163). Springer.

Zweig, D., & Wang, H. (2013). Can China bring back the best? The communist party organizes China's search for talent. *The China Quarterly, 215*, 590–615.

Chapter 13
Building the Bridge to Anti-Racist, Equitable, and Inclusive Practices: Translational Developmental Science for a Diverse Society

Emilie Phillips Smith, Velma McBride Murry, Melissa M. Yzaguirre, Catherine M. Gonzalez, and Chioma Kas-Osoka

The United States has historically and continues to be a country comprised of people indigenous to the land along with immigrant populations from Europe, Latin America, Africa, Asia, and beyond. Whereas earlier the country was characterized as a melting pot in which immigrating groups shed their ethnic and cultural background; current characterizations draw upon the imagery of a quilt in which diversity contributes to the fabric of the country (Portes & Rumbaut, 2014). Demographic portraits tell us that the racial-ethnic diversity of the US population will be more evident in the decades to come than ever. Already among children under 18, one out of four are immigrants, and by 2044, White children will comprise less than half of the country's population (Hernandez Donald & Napierala Jeffrey, 2013). In many ways, the fear of the browning of America fuels recent efforts in gerrymandering, the political backlash, and rises in overt racist acts (Yancy, 2018). Whereas some view issues of increasing diversity as a problem, we operate from the premise that it is welcome, ranging from biodiversity in our ecological systems to celebrating the contributions of people of color to industry, science, technology, arts, politics, and activism (Henry-Campbell & Hadeed, 2017; Schulze & Mooney, 2012). Valuing diversity, equity, and inclusion has the potential to enhance research, policies, and practices, elevating the voices of diverse racial-ethnic families, schools, and communities to ensure that research, policies, and practices are culturally relevant and designed to disrupt systemic racism, oppression, and structural inequities.

E. P. Smith (✉) · M. M. Yzaguirre
Michigan State University, East Lansing, MI, USA
e-mail: emsmith@msu.edu

V. M. Murry · C. M. Gonzalez
Vanderbilt University, Nashville, TN, USA

C. Kas-Osoka
Meharry Medical College, Nashville, TN, USA

© The Author(s), under exclusive license to Springer Nature Switzerland AG 2023

D. P. Witherspoon, G. L. Stein (eds.), *Diversity and Developmental Science*,
https://doi.org/10.1007/978-3-031-23163-6_13

This chapter provides an overview of practices and policies that can best foster the adaptation and success of a majority-minority population in the coming decades, poising the country with a better educated and healthy citizenry. Informed by a developmental-ecological model, we explore salient research-informed translational work. We align this chapter with several discussions raised in the previous chapters focused on shared and unique issues of development faced by majoritized-minoritized children, including contexts that promote healthy racial-ethnic socialization, identities, and adaptive acculturation that also foster civic engagement, positive youth development, equity, and social justice.

Fostering Healthy Development Within Families: Reducing the Impact of Poverty and Structural Inequities

A wealth of research has explored the impact of poverty and related stress on parenting. Results from longitudinal randomized trials have found significant improvements in youth socio-emotional development, behavior, academic achievement, and reduced substance use initiation and early sexual debut (Godwin & The Conduct Problems Prevention Research Group, 2020; Henry & The MVPP, 2012; Murry et al., 2007; Sanders & Turner, 2019). Prevention science is advancing broader implementation and dissemination strategies incorporated into community mental health systems, and statewide maternal and child home visiting (Kitzman et al., 2010; Parra-Cardona et al., 2017). These are important innovations, but primarily targeting micro-level processes, intervening on family factors to avert youth risk behavior. There remains a need to directly address systems level change to ameliorate the negative effects of poverty, structural racism, and discrimination on parental stress and, thereby, parent and youth outcomes. In their review of 38 randomized trials of parenting interventions for White, Black, Latino,[1] Asian, and Native American families, Garcia-Huidobro et al. (2018) reported that socio-economic status is often conflated with race-ethnicity, presenting barriers for families of color due to the lack of resources and increasing family stress. A lack of close attention to the pathways through which discrimination is linked to poverty and its consequences for families can result in continuing to "blame" families for situations that are consequences of systematic oppression. Studies that examine the cascading effects of upstream policies, programs, and practices are needed to more fully demonstrate how these systemic structural factors filter into the lives of families and ways to intervene to reduce their negative effects on functioning, adjustment, and development.

[1] Latino is an appropriate term to refer to populations that include Latina females and Latino males. To facilitate clarity of presentation and reading fluidity, we use the term Latino (or Latinos) to refer to both men and women in this chapter. The term Latinx, also used as a gender-neutral or nonbinary alternative, will be used to match language reported in previous literature.

In developmental science, there are a few notable large, multi-site studies employing both quantitative and qualitative methods that examine ways to address the root causes of poverty and, thereby, parental stress. The New Hope project sought to provide parents with temporary vouchers that increased family income and a host of other resources designed to reduce poverty and its associated stressors on parents (Huston et al., 2011). This project demonstrated the most impressive effects for boys, including greater future orientation, employment, and less cynicism about work and career preparation; positive results that unfortunately were not exhibited for girls. The temporary vouchers did increase maternal employment, as mothers were the focus, allowing families to broker important community program settings for their children that provided safe and supportive childcare and after-school opportunities. These quality and supportive settings fostered children's prospects, increasing behavioral and academic outcomes for young male participants, findings that were robust after 2 years (Huston et al., 2011).

Another impactful large-scale applied developmental research example is the Moving to Opportunity project (MTO). Through MTO, families received vouchers and support to move from an impoverished to a less impoverished neighborhood, with mixed but positive benefits, especially for boys. In contrast, for girls, initially the move led to experiences of harsher parenting from mothers who were presumably anxious about their child performing well in a context in which they were minoritized, both in terms of race and social class. Effects were also less beneficial for adolescents for whom the move was more disruptive (Leventhal et al., 2005). However, over time, living in less impoverished neighborhoods was related to higher college attendance and earnings (Chetty et al., 2016). Long-term effects on adult economic self-sufficiency, 5–7 years later, were not detected (Ludwig et al., 2008). For some scholars, this idea that relocation can improve child outcomes for pre-adolescent male children holds some promise. Others debate whether the effects were more demonstrative for families willing to make a move, alluding to the potential impacts of selection. The researchers have disputed this critique. Changes in neighborhood environments, where children receive less exposure to poverty and violence, represent one approach to addressing the issues facing families. However, for so many families who remain in these neighborhoods, for whom a move is infeasible, a level of change that more adequately addresses structural racism and inequalities more systemically holds promise for greater impact. Poverty-reduction strategies have some modest effects on income but mixed on other psycho-social elements (Duncan et al., 2019).

Strategies to Empower Families to Navigate Race-Related Experiences

Many studies document the effects of discrimination across developmental stages from childhood to adulthood (Seaton et al., 2011; Smith-Bynum, 2023). Racism, including discrimination experienced by immigrant Latino/a during the acculturation process, creates challenges that contribute to everyday life stress for minoritized families and youth. These experiences are systemic and pervasive, often manifested through lack of access to employment, housing, and quality schooling, as well as increased likelihood of being exposed to police profiling and system involvement (Goff & Rau, 2020; Massey, 2015; Noguera, 2008). Race-related experiences take a toll on parents and children (Murry et al., 2018). They may evoke negative emotions that comprise psychological (i.e., anxiety, depression, psychiatric disorders) and physical health functioning (i.e., elevated cortisol, vulnerability to chronic diseases), amplifying reliance on maladaptive coping responses. Ultimately, these challenges have been associated with ineffective parenting and parental substance use disorders (Berger & Sarnyai, 2015). In addition, Murry et al. (2018, 2021) illustrated ways in which chronic racial discrimination induces negative emotionality and heightened reactivity that spillover to adversely affects parent-child relationship quality.

Several stress-management practices, some of which have been characterized as cultural, strength-based, coping assets (Murry et al., 2018), have been shown to mediate the potential negative effects of stressors on families. For example, Bryant et al. (2010) found that psychosocial resources (e.g., religiosity, proactive problem-solving, sense of mastery, and control) may operate as protective mechanisms in the lives of African American families. In fact, optimism—that is, being hopeful for the future—may be especially relevant to the well-being of African Americans, especially during times of seemingly bleak situations (Mattis et al., 2004). Further, cognitive reframing, especially when life circumstances are challenging, may be an effective harm-reducing strategy that African American parents, as well as Latinx parents, employ to remain psychologically well-adjusted (James et al., 2018) as they attempt to exert control over adversity, such as racial discrimination. For families of color, African Americans families in particular, a strong sense of belongingness to their ethnic group of origin also is protective; these family processes are thought to buffer both parents and children from the potential harms associated with racial and socio-eco-political stressors (Anderson & Stevenson, 2019; Murry et al., 2018). Further, positive parent-child relationship quality also truncated the negative spillover effects of racism on youth, which fostered positive intrapersonal protective processes in their children. Racial socialization during middle childhood fostered heightened racial identity that buffered youth from internalized negative effects of racism, with sustaining effects from middle childhood through young adulthood (Murry et al., 2021). These findings have implications for preventive interventions that target supportive strategies to help minoritized parents navigate adversity as they prepare their children with the skills and capacities to navigate race-related challenges.

Parenting Approaches to Foster Positive Racial-Ethnic Identity, Socialization, and Adaptation

African American and Latino parents have the role of rearing healthy children who are confident and affirmed in their identities. These parents have the added responsibility of not only preparing their children for racial biases that they will encounter but also establishing a sense of pride as a member of their racial-ethnic groups, especially in the face of racism, discrimination, and stereotypical portrayals. Comprehensive reviews of the research have centered upon the role of racial-ethnic pride and affirmation, which have been found to be associated with self-esteem academic achievement and other positive developmental outcomes among children and adolescents from multiple racial-ethnic backgrounds (Rivas-Drake et al., 2023; Witherspoon et al., 2016). However, there is a clear need for more conclusive work on the role of racial-ethnic identity for API, Native American, and multiracial youth.

Reviews of the explosion of research on cultural socialization, inculcating children with values, beliefs, practices salient to their ethnic background, report positive findings of this dimension across racial-ethnic groups on identity and several psychosocial outcomes for children. In their recent review of the research, Umaña-Taylor and Hill (2020) offer significant insights of racial socialization across racial-ethnic groups, noting the protective nature of cultural-racial-ethnic socialization for African American, Asian American, American Indian/Native American, Latino, and multiracial youth, reducing reliance on substance use to cope with race-related stress.

In particular, race-related stress for immigrant families is often associated with discrimination relevant to language, documentation status, and the physical processes of migration that can fuel a sense of being an "insider or outsider" (Lorenzo-Blanco et al., 2023). The United States considers anyone not a US citizen at birth as international, identifying them as "naturalized U.S. citizens, lawful permanent residents (immigrants), temporary migrants (such as foreign students), humanitarian migrants (such as refugees and asylees), [or] unauthorized migrants" (U.S. Census, 2021). Estimates suggest that as many as 44 million immigrant populations reside in the United States, with Latinos making up nearly half (19 million) of the immigrant population (Vespa et al., 2018). Over 5.5 million children reside in families where parents are undocumented or hold mixed statuses. Parents facing issues with documentation are more likely to report a weakened family environment, depicted through fractured parent-child relationships, parents' emotional instability, and economic insecurity (Chaudry et al., 2010). The Asian American immigrant population is another important one to consider, that often suffers from issues of invisibility in the United States, in addition to a history of interment, and enduring hate and discrimination. Asian American immigrants have experienced large amounts of growth, 72%, in the United States (Yip et al., 2021). Immigrant parents, across ethnic subgroups, desire to socialize their children with an affirming sense of their cultural heritage and racial-ethnic identity that has been shown to be a protective factor in children's behavioral and academic outcomes (Cross et al., 2020). The importance of parental racial-ethnic socialization also suggests the need for more culturally relevant preventive interventions (Cobb et al., 2021).

Culturally Responsive Parenting Programs

Parenting programs are emerging that are especially attuned to the important and distinctive work that families of color must do pertaining to issues of race, ethnicity, and culture. It is no longer the standard to ignore the critical issues these families face as they rear children with healthy identities and adaptive coping strategies for discrimination, a task undertaken in family prevention approaches with families of color (e.g., Cooper et al., 2015; Murry et al., 2007). This body of work integrates racial-ethnic socialization strategies into parenting programs, finding that they enhance the effects of these programs, increasing positive parenting practices and more proactive racial socialization messages. These results have been demonstrated not only in programs targeted to mothers but also with minoritized fathers as well (Caldwell et al., 2014; Cooper et al., 2015). Parenting programs have been developed specifically with Latinx families that address not only ethnicity, culture, and language, but also issues of discrimination and documentation status that when combined are found to exhibit powerful, multi-faceted effects (Parra-Cardona et al., 2017). Culturally relevant and responsive models for racial and ethnic minoritized populations have been explored in the mental health literature for several decades. For those who contend with not only the issues of child and family development, but also racism, discrimination, and acculturation, systemic modifications to treatment help to support the therapist in utilizing approaches more likely to successfully engage minoritized populations (e.g., Bernal & Sáez-Santiago, 2006; Griner & Smith, 2006). Reese and Vera (2007) state that *cultural relevance* refers to the extent to which interventions are consistent with the values, beliefs, and desired outcomes of a particular community. The terminology *culturally relevant* versus *culturally responsive* forwards the notion that these intentional efforts to work with diverse populations come as a forethought, not reactive but proactive in addressing need. Culturally relevant and adapted prevention approaches can improve existing models for delivery to marginalized communities without disrupting fidelity or diminishing model integrity (Castro et al., 2004). Culturally relevant approaches resonate more with families and address unique issues with which they struggle, thereby increasing the effectiveness of treatment (Parra-Cardona et al., 2017; Bernal & Sáez-Santiago, 2006). Moreover, Cobb et al. (2021) have called for culturally appropriate prevention attuned to helping individuals cope with the stress of discrimination, socializing young people with a positive sense of identity, but also anti-racist approaches to disrupt the influence of structural factors, particularly upon the impoverished and less educated who report more discrimination experience and mental health disorders (Cobb et al., 2021).

Anti-Racist, Equitable, and Culturally Relevant Practices in School

An eco-developmental model acknowledges that multiple contexts, including families and schools, are key in child development. Schools in the United States vary dramatically in geographic locale, racial-ethnic composition, resources, and funding, with children in poorer neighborhoods receiving the brunt of educational inequities (Kozol, 1992). Decades of oppression barring African American children from learning and subsequently from quality education, assimilationist practices placing Native American children in settlement houses away from their families, not to mention decades of de jure and de facto segregation have all contributed to deeply rooted disparities in educational achievement. The structures for funding education are rooted in "redlining," discriminatory housing and lending practices that undervalue homes in minoritized communities (Massey, 2015), essentially eroding the tax base and funding for schools. Much educational debate has been dedicated to the origins of the achievement gap, essentially attacking the intelligence of minoritized children (Herrnstein & Murray, 2010). However, the public and scholarly discourse rarely if ever acknowledges the centuries-old and pervasive structural inequalities in the laws, policies, and practices in the United States contributing to these age-old disparities (Noguera, 2008).

Although much work focuses on a deficit approach to Black and Brown children and families, a strengths-based approach acknowledges the substantial gains over the past 2–3 decades. Several advances have been made in decreasing high school dropout; from 2000 to 2016, the Hispanic status dropout rate among 16- to 24-year-olds decreased from 28% to 9%, the Black rate decreased from 13% to 6%, and the White rate decreased from 7% to 5% (de Brey et al., 2019). The number of bachelor's degrees awarded to Hispanic students more than tripled between 2000–2001 and 2015–2016. During the same period, the number of degrees awarded also increased for students who were Asian/Pacific Islander (by 75%), Black (by 75%), and White (by 29%) (de Brey et al., 2019). Advances have been made decreasing the achievement gap, and increasing opportunities in terms of high school and college completion across race-ethnicity for young people in the United States

Yet, schooling in the United States remains more segregated than ever. While 71.7% of cities have become more integrated, 61.4% of school districts in these areas have become more segregated (Coughlan, 2018). Coughlan (2018) reports that the proportion of school districts with a majority White population shrunk from 45.5% in 1990 to 8.3% in 2015; the proportion of school districts with a majority Hispanic population grew from 6.8% in 1990 to 30.2% in 2015; the proportion of majority Black school districts remained relatively stable. The following section provides an overview of school-related structures and processes that hinder and facilitate academic affirming experiences of minoritized youth.

Charter Schools as an Approach to Diversify Schooling

Born out of school choice policies, charter schools often function as mechanisms of resegregation, allowing families in gentrifying neighborhoods to avoid more diverse public schools (Wells et al., 2019). Charter schools are incentivized by national public policy to select children of wealthier, more highly educated parents who are more likely to score higher on standardized tests (Wells et al., 2019). School practices and policies, particularly early in education before children have formed crystallized stereotypes and perceptions, can go far in informing interracial interactions in schools. Racially balanced, academically competitive magnet schools with leadership respectful and supportive of intergroup interactions, can foster these important interracial friendships (Williams & Graham, 2019). Schools in which teachers and leadership are clear in their value for diversity can go far in fostering meaningful interracial interactions among young people (Williams & Graham, 2019).

Immigration, Language, and the School Setting

Language is a prominent component of adaptation that evolves across generational status for international families living in the United States. Unfortunately, language discrimination is a festering problem in the United States, impacting parents and families in the workforce, schools, and homes. For example, 38% of Latinos in the United States have self-reported being criticized for speaking Spanish in public, made derogatory remarks to (i.e., told to return to their home country), or called offensive names (Lopez et al., 2018). There is often a level of dependency from immigrant parents on their children to translate or communicate on their behalf. Parents' Spanish language use has been reviewed in the literature regarding engagement within the school system. Zarate (2007) notes that because of the structural gap between parents' primary language and services or resources offered by the school, many parents have felt uncomfortable engaging with teachers, school activities, or helping their children with homework. For example, Latino parents have noted negative experiences with teachers, including feeling rejected, hindering relationship building (Olivos, 2004; Poza et al., 2014). The deficit lens by which the Spanish language is viewed within schools has impacted teachers' perceptions of the benefits and contributions that parents can make in their child's education through their native language. Less publicized forms of parent engagement can occur through socialization using families' native languages, valuing families' customs supportive of education (Zarate, 2007).

Latino parents often view academic learning as something that should occur at school, through teachers; and their role is to teach their children more about moral and civic responsibility (Rodríguez-Brown, 2009). Teachers can help by explaining different strategies to parents that will enable them to be co-facilitators in the academic development of their children. This also holds high relevance for dual language learners' academic success, as parent engagement has been positively linked to behavior (Hill et al., 2004; Wang et al., 2014), emotional functioning (Wang et al., 2014), and student attendance (Sheldon, 2007). Beginning these practices early on can successfully connect dual language learners and families to educational settings, integrating diversity, and inclusion that are championed in valuing multiple cultures and languages.

Identity Affirming Practices in Schooling

Schools are settings that offer important socialization that can foster cultural or ethnic identification. Brown (2017) explains that environmental factors within schools determine the extent to which children's racial-ethnic identities are strengthened or weakened. The adaptation process that children go through is highly relevant for immigrant children who may compromise their identification with their ethnic group in order to feel like they belong with the majority group (Nishina et al., 2010). Similarly, when faced with discrimination by peers, children are more likely to negatively evaluate their ethnicity and less likely to want to identify with that group (Turner & Brown, 2007). This process is also explained by Suárez-Orozco and Suárez-Orozco (2009) who coined the term "social mirroring" to explain how a school's messaging and culture heavily influence immigrant children's self-perceptions.

Brown (2017) found that children's ethnic identity is largely established by the end of elementary school, such that they are more likely to report and identify with the same ethnic group every year. This has important implications for early childhood education, as teachers and schools can begin to create learning environments that foster this identity formation early. In the same study, the author found that Latino children were more likely to embrace their ethnic identity over the span of time, when the school's demographic composition was primarily Latinos, both in teachers and students (Brown, 2017). Similarly, children were more likely to espouse their ethnic identity when schools valued or demonstrated support for multiculturalism.

School is an important developmental context in which students begin to encounter intergroup contact that can be helpful in learning and interacting with others. Cross-ethnic friendships in school have been found to play an important buffering role in the experience of discrimination (Benner & Wang, 2017). In studies examining Latino/a and African American adolescents, peer-perpetrated discrimination was found to be related to loneliness and depression and the experience of teacher-perpetrated discrimination fostered a sense of reduced school belonging and engagement. Cross-ethnic friendships were found to buffer experiences of both peer- and educator-perpetrated discrimination and were related to reduced loneliness and disengagement (Benner & Wang, 2017). Same-ethnic friendships were associated with stronger private regard in terms of racial-ethnic identity, while more ethnic diversity and cross-ethnic friendships were associated with less perceived vulnerability

(Graham et al., 2014). For African American students, positive perceptions of interracial interactions in school foster both a sense of belonging and academic motivation (Byrd, 2015). While this does not absolve students nor the teachers from responsibility for reducing discriminatory behavior (Rowley et al., 2008), in the wake of these experiences, it does identify cross-ethnic friendships as a potential helpful source for affirming discriminatory experiences of Black and Brown students, helping them to cope in adaptive ways.

School climates supportive of diversity and educators attuned to dismantling their own implicit biases and fostering culturally relevant practices go far in promoting equity in school settings (Byrd, 2017; Ladson-Billings, 2014). Culturally relevant and inclusive teaching practices play a critical role in students' identity development by helping them maintain their cultural integrity and increase opportunity for academic success. Ladson-Billings (2014) describes key areas of pedagogical practice that comprise culturally relevant teaching: a focus on academic success, that is students experiencing growth in learning and cultural competence and celebrating students' cultural heritage while helping them to appreciate the cultural origins of others. Ladson-Billings (2014) also integrates identifying, analyzing, and solving real-world problems, and socio-political consciousness in her conceptualization of culturally relevant and inclusive pedagogy.

Counter to the advances in culturally responsive pedagogy, integrating accurate, anti-racist, historical portrayals of systemic racism are currently being attacked under the guise of critical race theory. In actuality, critical race theory describes racism as systemic in laws, policies, and practices that discriminate, oppress, and disenfranchise people based upon race and ethnicity (Crenshaw et al., 1995). Bonilla-Silva (2009) describes "racism without racists" in which discriminatory practices against communities of color are "baked in" so as to encourage, for example, realtors, banks, and businesses to avoid communities of color presumably for reasons of finances and schooling, resulting in less enterprise, lower tax bases, less-resourced schools, and more neighborhood decline that contribute to a myriad of other issues.

Recent attacks on critical race theory seek to turn back the clock on progress in anti-racist and culturally relevant pedagogy. These attacks say to minoritized groups that they can be oppressed, but do not tell the story of the oppression and that curriculum cannot be used as a vehicle to educate the young, majority and minority alike, about historical facts of systemic racism, discrimination, and oppression. In one of the most notable examples in 2021, The State Legislature of Texas passed laws prohibiting the teaching of "critical race theory," enacting Bill HB 3979 stating that any "state agency, school district, or open-enrollment charter school may not: be required to engage in training, orientation, or therapy that presents any form of race or sex stereotyping or *blame on the basis of race or sex [italics added]…*" (Texas State Legislature, 2021). This law seriously limits educators from discussing the role of White majority people in the oppression of people of color in the United States, confusing educators in terms of what curriculum and books would be allowed in their classroom and school libraries, and even forbidding civic engagement of students for credit that includes activism, advocacy, and engagement. This legislation is a sweeping prohibition to anti-racist, culturally relevant pedagogy and civic engagement among young people. It has also cost several minoritized educational principals and school leaders their jobs (Green, 2021, in the *New York Times*). It is a critical reminder of the needed persistence of liberation efforts in schools and communities echoed by the words of abolitionist Frederick Douglas in 1857:

> If there is no struggle there is no progress. Those who profess to favor freedom and yet deprecate agitation are men who want crops without plowing up the ground; they want rain without thunder and lightning. They want the ocean without the awful roar of its many waters. This struggle may be a moral one, or it may be a physical one, and it may be both moral and physical, but it must be a struggle. Power concedes nothing without a demand. It never did and it never will.

The Neighborhood and Community Contexts of Development for Children of Color: Fostering Positive Youth Development, Engagement, and Social Justice

Youth identity, engagement, and activism for social justice are fostered not only in families and schools, but neighborhoods and communities are also influential socializing contexts for youth. For a number of children in the United States, their interactions with neighborhood and/or community are centered with peers, in after-school programs, or extra-curricular or faith-based activities. While neighborhood refers to a geographic area in which one resides in proximity to others, community is a broader social construction defined by the people and places that represent affinitive choices and preferences (Smith et al., 2016). In research exploring the interactions of the family-peer-community systems, we find that families play an important role in peer selection, leading to decreased delinquent peer affiliation and activities and increased positive peer interactions. This impact is often in neighborhood or community settings that allow youth to develop important prosocial skills not entirely facilitated by family contexts (Smith et al., 2016). Witherspoon et al. (2016) examine the different conceptual approaches to neighborhood and community, either as places that foster risks to safety due to crime and poverty or assets-based approaches that center upon the important interpersonal, social-cultural connections provided by peers, caring neighbors, organizations, and those working in broader neighborhoods and communities (McKnight & Kretzmann, 1996; Sampson et al., 1997).

National policies, such as the twenty-first Century Community Learning Centers, have sought to build upon the value of community settings that support working families by providing care for their children in the hours afterschool, a time of increased involvement in substance use, criminal and sexual activities if youth are left unmonitored and unsupported (Snyder & Sickmund, 2006). African American and Latino families are even more likely to avail themselves of afterschool resources for caring for their children. According to the Afterschool Alliance (2021), 10.2 million children (18%) participate in an afterschool program, a 15% increase from

2009. Nearly 1 in 4 families (23%) has a child in an afterschool program: 29% of Hispanic children, 24% of African American children and 12% of White children attend regular programming (Afterschool Alliance, 2021). Many estimate more families would likely participate if they had the resources.

Research demonstrates that quality afterschool programs, in particular those integrating cultural elements, are related to and/or affect youth cultural orientation, racial-ethnic identity, socio-emotional learning, and academic achievement (Riggs et al., 2010; Smith et al., 2017; Vandell et al., 2020). Although in some cases after school care may be provided by families or schools, increasingly community youth development organizations like the YM/YWCA, the Boys and Girls Club (BGC), and local organizations are in the business of providing safe and supervised care for young people. Afterschool programming is thought to function in two ways for children and youth, as a potential prevention strategy for risky behavior and as a method of promoting positive youth development and academic achievement. Relational developmental systems, one of the theoretical notions underlying positive youth development (PYD), posit that youth are agentic, influenced by *and* influencing their contexts (Lerner et al., 2021).

Leading scholars readily acknowledge that more of the work addressing PYD needs to attend to youth of diverse racial-ethnic backgrounds, youth who must cope with issues of racism, discrimination, equity, and social justice in their development (Barbarin et al., 2020; Lerner et al., 2021). Scholars examining aspects of PYD with examinations of racial-ethnic identity have found African American and Latino youth with more affirming identities to be higher in positive youth development (e.g., caring, connectedness, competence, responsibility) and associated with reduced internalizing symptoms (Williams et al., 2014). African American and Latino youth who are high in *both* positive racial-ethnic affirmation and PYD exhibit fewer problem behaviors and higher levels of academic achievement, even when they are aware of racial barriers (Yu et al., 2021). This finding holds important implications for attending to issues of both racial-ethnic identity and PYD in the goals and mission driving after school contexts.

More recent work in community-based afterschool settings has been exploring programs designed specifically for youth of color that incorporate cultural elements that seek to promote a positive identity among African American, Latino/a, and API children. Brittian Loyd and Williams (2017) have conducted a review demonstrating that these more culturally-specific approaches go far in helping youth to navigate a positive identity in the face of discrimination and stereotypical portrayals, fostering not only identity but also less problematic behavior and higher levels of academic achievement.

A notable report by the National Research Council emphasized the need for "one to take the local cultural context into account as programs are designed and evaluated" (Eccles & Gootman, 2002, p. 68). This is still deserving of attention in community-based youth programs, especially in defining approaches that include cultural influences as an evaluative metric as evidence of program quality. In a randomized study of over 70 programs statewide, Smith et al. (2017) examine aspects of quality, such as supportive relationships with adults and peers appropriate structure, and importantly, student engagement. These aspects of quality have been found not only to be related to reduced problem behavior, and increase PYD but also to a positive cultural orientation, that is, more collectivist attitudes and respect for adults, particularly for African American students. Yet, these measures of quality need to be expanded to explicitly include attention to issues of race, culture, diversity, and representation in these observational protocols.

Simpkins et al. (2017) have begun to lay out multiple dimensions of quality and ways in which they can be attuned to culture. They build upon definitions of culture put forth by Suárez-Orozco (2015) attuned to "…. The environment or social context, as a toolkit of symbols, beliefs, values, and practices (p. 13)". Simpkins et al. (2017) explore ways to integrate diversity, equity, and inclusion, for example in representing diversity visually and in program content, having shared behavioral norms that are culturally responsive to the communities being served, communicating linguistically in ways that are accessible to participating children and families, and being equitable, inclusive and welcoming by the staff and environment of all children in ways that help to embrace their racial-ethnic, and cultural backgrounds.

Supporting Youth Agency, Civic Engagement, and Voice

So much of the research on young people explores the ways in which adults have an impact upon their development, so seldom acknowledging that young people are developing their perspectives, opinions, and social justice stances that are expressed in many ways in their lives (Bañales et al., 2023). Civic engagement is defined as the degree to which young people are involved in activities related to their communities, government, and social justice. Hope and Spencer (2017; Mathews et al., 2023) consider civic engagement and commitment as adaptive responses to a world in which minoritized youth face discrimination and system racism in so many areas of their lives. Several authors have recognized the ways in which direct civic engagement might be inaccessible to youth of color given the gerrymandering that seeks to limit the influence of their voting, time, and resources that may limit action. For this group, Hope and Spencer (2017) argue that engagement may be expressed in commitment to family, on social media, in music and the arts (Watts & Flanagan, 2007; Wray-Lake & Abrams, 2020). Much less attention is given to the ways in which family, school, and community settings foster youth voice and engagement. What are the anti-racist and equitable practices that help them to act in intentional and impactful ways? These concepts point to the value of physically and psychologically safe spaces, that are diverse, welcoming of youth voice, inquiry, and youth participatory and community-engaged research.

Community-Engaged Research

Community engagement is defined as "the process of working collaboratively with groups of people affiliated by geographic proximity, special interests, or similar situations with respect to issues affecting their well-being" (CDC, 1997, p. 9). The rationale for community engagement stems from the ecological perspective, understanding that health and development is influenced by social, cultural, and environmental contexts (CDC, 2011). Community engagement is built on equitable relationships between community members and researchers that encompass the needs and priorities of the community through a culturally relevant lens (Ahmed & Palermo, 2010). With multiple models of community engagement, ranging from community-based participatory action research to youth participatory action research, community-engaged research focuses on involving the community throughout all stages of the research process (see Caughy et al., 2023). Training current and future educators, scientists, and practitioners to address systemic inequalities draws upon anti-racist, inclusive community-engaged research for developing sustainable partnerships among researchers and community members, building trust, and increasing participation in research among historically excluded populations. Community input is identified as necessary for translating existing research to implement and sustain programs, improve population health, and eliminate health disparities (CDC, 2011). Examples of community-engaged research include developing community advisory boards, collaborating with and training community health workers, and conducting focus groups with community members. Additionally, emphasis should be placed on policies requiring research institutions to actively train scientists and practitioners on the principles of community engagement and its importance in addressing the needs of marginalized communities.

Summary and Conclusion

With increasingly diversed society it is imperative that we consider preventive intervention approaches that bolster families of color, not in a "one-size fit all" approach that ignores issues of race, ethnicity, immigrant, and documentation status and culture, but instead are informed by their experiences and parenting practices. The next generation of prevention programs is and should be integrating important considerations of reducing poverty, and integrating cultural socialization, preparation for discrimination, culture and immigration into their programming in order to address pressing issues and amplify their important effects on parents and children, not only in terms of socio-emotional outcomes but also in terms of critical impacts upon their health and mental health. Similarly, school settings, even with threats upon their liberty to embrace diverse, equitable, inclusive, and welcoming practices, demonstrate more positive impact for children of color when they are attuned to dismantling discrimination from students and teachers thereby fostering an environment of intergroup contact and cross-ethnic friendships that can potentially help all to grapple with and reduce discrimination. Culturally responsive curriculum and pedagogy offer an environment in which we can learn about historical practices and chart a new future that is more embracing of the diversity that is our country's strengths. Community organizations and civic engagement offer important settings that facilitate positive youth development and engagement, not only in their communities but in their local, state, and national governments, through helping their families, and advocating for social justice and equity in communities. Coupled with these practices are the development of community-engaged developmental scholars who can tackle issues of inequality in partnership with communities for sustainable transformational change.

References

Afterschool Alliance. (2021). America after 3pm for black families and communities. http://afterschoolalliance.org/documents/AA3PM-2020/AA3PM-Black-Communities-2020-Fact-Sheet.pdf

Anderson, R. E., & Stevenson, H. C. (2019). RECASTing racial stress and trauma: Theorizing the healing potential of racial socialization in families. *American Psychologist, 74*(1), 63–75. https://doi.org/10.1037/amp0000392

Ahmed, S. M., & Palermo, A. G. S. (2010). Community engagement in research: Frameworks for education and peer review. *American Journal of Public Health, 100*(8), 1380–1387. https://doi.org/10.2105/AJPH.2009.178137

Bañales, J., Pinetta, B. J., Singh, S., Rodriguez, A. J., Aldanda, A., & Gutuirrez, F. J. (2023). Centering youth voice in developmental science: A research road map for partnerships with Latinx youth. In D. P. Witherspoon & G. L. Steins (Eds.), *Diversity and developmental science: Bridging the gaps between research, practice, and policy*. Springer.

Barbarin, O. A., Tolan, P. H., Gaylord-Harden, N., & Murry, V. M. (2020). Promoting social justice for African-American boys and young men through research and intervention: A challenge for developmental science. *Applied Developmental Science, 24*(3), 196–207. https://doi.org/10.1080/10888691.2019.1702880

Benner, A. D., & Wang, Y. (2017). Racial/ethnic discrimination and adolescents' well-being: The role of cross-ethnic friendships and friends' experiences of discrimination. *Child Development, 88*(2), 493–504. https://doi.org/10.1111/cdev.12606

Berger, M., & Sarnyai, Z. (2015). "More than skin deep": Stress neurobiology and mental health consequences of racial discrimination. *Stress (Amsterdam, Netherlands), 18*(1), 1–10. https://doi.org/10.3109/10253890.2014.989204

Bernal, G., & Sáez-Santiago, E. (2006). Culturally centered psychosocial interventions. *Journal of Community Psychology, 34*(2), 121–132. https://doi.org/10.1002/jcop.20096

Bonilla-Silva, E. (2009). *Racism without racists: Color-blind racism and the persistence of racial inequality in the United States* (2nd ed.). Rowman & Littlefield.

Brittian Loyd, A., & Williams, B. V. (2017). The potential for youth programs to promote African American youth's development of ethnic and racial identity. *Child Development Perspectives, 11*(1), 29–38. https://doi.org/10.1111/cdep.12204

Brown, C. S. (2017). School context influences the ethnic identity development of immigrant children in middle childhood. *Social Development, 26*(4), 797–812. https://doi.org/10.1111/sode.12240

Bryant, C. M., Wickrama, K. A. S., Bolland, J., Bryant, B. M., Cutrona, C. E., & Stanik, C. E. (2010). Race matters, even in marriage: Identifying factors linked to marital outcomes for African Americans. *Journal of Family Theory & Review, 2*(3), 157–174. https://doi.org/10.1111/j.1756-2589.2010.00051.x

Smith-Bynum, M. (2023). The theory of racial socialization in action for black adolescents and their families. *In D. P. Witherspoon & G. L. Steins (Eds.), Diversity and developmental science: Bridging the gaps between research, practice, and policy. Springer.*

Byrd, C. M. (2015). The associations of intergroup interactions and school racial socialization with academic motivation. *The Journal of Educational Research, 108*(1), 10–21. https://doi.org/10.1080/00220671.2013.831803

Byrd, C. M. (2017). The complexity of school racial climate: Reliability and validity of a new measure for secondary students. *British Journal of Educational Psychology, 87*(4), 700–721. https://doi.org/10.1111/bjep.12179

Caldwell, C. H., Antonakos, C. L., Assari, S., Kruger, D., Hill De Loney, E., & Njai, R. (2014). Pathways to prevention: Improving nonresident African American fathers' parenting skills and behaviors to reduce sons' aggression. *Child Development, 85*(1), 308–325. https://doi.org/10.1111/cdev.12127

Castro, F. G., Barrera, M., Jr., & Martinez, C. R., Jr. (2004). The cultural adaptation of prevention interventions: Resolving tensions between fidelity and fit. *Prevention Science, 5*(1), 41–45. https://doi.org/10.1023/B:PREV.0000013980.12412.cd

Caughy, M., Randolph Cunningham, S. M., & Calzada, E. (2023). Building collaborative teams and conducting ethical research in the Spirit of 2044: The complexity of conducting research in communities of color. In D. P. Witherspoon & G. L. Steins (Eds.), *Diversity and developmental science: Bridging the gaps between research, practice, and policy*. Springer.

Centers for Disease Control and Prevention. (1997). *Principles of community engagement*. CDC/ATSDR Committee on Community Engagement.

Centers for Disease Control and Prevention. (2011). *Principles of community engagement* (2nd ed.). NIH Publication.

Chaudry, A., Capps, R., Pedroza, J. M., Castaneda, R. M., Santos, R., & Scott, M. M. (2010). *Facing our future: Children in the aftermath of immigration enforcement*. Urban Institute (NJ1).

Chetty, R., Hendren, N., & Katz, L. F. (2016). The effects of exposure to better neighborhoods on children: New evidence from the moving to opportunity experiment. *American Economic Review, 106*(4), 855–902. https://doi.org/10.1257/aer.20150572

Cobb, C. L., Salas-Wright, C. P., John, R., Schwartz, S. J., Vaughn, M., Martínez, J., Charles, R., Awad, G., Pinedo, M., & Cano, M. Á. (2021). Discrimination trends and mental health among native- and foreign-born Latinos: Results from national surveys in 2004 and 2013. *Prevention Science, 22*(3), 397–407. https://doi.org/10.1007/s11121-020-01186-4

Cooper, S. M., Smalls-Glover, C., Metzger, I., & Griffin, C. (2015). African American fathers' racial socialization patterns: Associations with racial identity beliefs and discrimination experiences. *Family Relations, 64*(2), 278–290. https://doi.org/10.1111/fare.12115

Coughlan, R. W. (2018). Divergent trends in neighborhood and school segregation in the age of school choice. *Peabody Journal of Education, 93*(4), 349–366. https://doi.org/10.1080/0161956X.2018.1488385

Crenshaw, K., Gotanda, N., Pellar, G., & Thomas, K. (Eds.). (1995). *Critical race theory*. New Press.

Cross, F. L., Agi, A., Montoro, J. P., Medina, M. A., Miller-Tejada, S., Pinetta, B. J., et al. (2020). Illuminating ethnic-racial socialization among undocumented Latinx parents and its implications for adolescent psychosocial functioning. *Developmental Psychology, 56*(8), 1458–1474. https://doi.org/10.1037/dev0000826

de Brey, C., Musu, L., McFarland, J., Wilkinson-Flicker, S., Diliberti, M., Zhang, A., Branstetter, C., & Wang, X. (2019). *Status and trends in the education of racial and ethnic groups 2018* (NCES 2019-038). U.S. Department of Education. Washington, DC: National Center for Education Statistics. https://nces.ed.gov/pubsearch/

Douglass, F. (1857, May). *Two speeches, by Frederick Douglass; one on West India emancipation, delivered at Canandaigua, Aug. 4th, and the other on the Dred Scott decision, delivered in New York, on the occasion of the anniversary of the American Abolition Society*. CP Dewey, printer.

Duncan, G., & Le Menestrel, S. (2019). A roadmap to reducing child poverty. Consensus Study Report. National Academies Press. 500 Fifth Street NW, Washington, DC 20001.

Eccles, J., & Gootman, J. (2002). *Community programs to promote youth development*. National Academy Press.

Garcia-Huidobro, D., Doty, J. L., Davis, L., Borowsky, I. W., & Allen, M. L. (2018). For whom do parenting interventions to prevent adolescent substance use work? *Prevention Science, 19*(4), 570–578. https://doi.org/10.1007/s11121-017-0853-6

Godwin, J. W., & The Conduct Problems Research Group. (2020). The Fast Track intervention's impact on behaviors of despair in adolescence and young adulthood. *Proceedings of the National Academy of Sciences, 117*(50), 31748–31753. https://doi.org/10.1073/pnas.2016234117

Goff, P. A., & Rau, H. (2020). Predicting bad policing: Theorizing burdensome and racially disparate policing through the lenses of social psychology and routine activities. *The Annals of the American Academy of Political and Social Science, 687*(1), 67–88. https://doi.org/10.1177/0002716220901349

Graham, S., Munniksma, A., & Juvonen, J. (2014). Psychosocial benefits of cross-ethnic friendships in urban middle schools. *Child Development, 85*(2), 469–483. https://doi.org/10.1111/cdev.12159

Green, E. L. (2021). Black lives matter, she wrote. Then 'everything just imploded.' *New York Times*. https://www.nytimes.com/2021/10/10/us/politics/maryland-superintendent-racism-black-lives-matter.html

Griner, D., & Smith, T. B. (2006). Culturally adapted mental health intervention: A meta-analytic review. *Psychotherapy: Theory, Research, Practice, Training, 43*(4), 531–548. https://doi.org/10.1037/0033-3204.43.4.531

Henry, D. B., & The Multi-Site Violence Prevention Project. (2012). Mediators of effects of a selective family-focused violence prevention approach for middle school students. *Prevention Science, 13*(1), 1–14. https://doi.org/10.1007/s11121-011-0245-2

Henry-Campbell, S., & Hadeed, S. (2017). Managing a diverse workforce. In *Encyclopedia of strategic leadership and management* (pp. 953–964). IGI Global.

Hernandez Donald, J., & Napierala Jeffrey, S. (2013). *Diverse children: Race, ethnicity, and immigration in America's new non-majority generation*. Foundation for Child Development.

Herrnstein, R. J., & Murray, C. (2010). *The bell curve: Intelligence and class structure in American life*. Simon and Schuster.

Hill, N. E., Castellino, D. R., Lansford, J. E., Nowlin, P., Dodge, K. A., Bates, J. E., & Pettit, G. S. (2004). Parent academic involvement as related to school behavior, achievement, and aspirations: Demographic variations across adolescence. *Child Development, 75*(5), 1491–1509. https://doi.org/10.1111/j.1467-8624.2004.00753.x

Hope, E. C., & Spencer, M. B. (2017). Civic engagement as an adaptive coping response to conditions of inequality: An application of Phenomenological Variant of Ecological Systems Theory (PVEST). In N. J. Cabrera & B. Leyendecker (Eds.), *Handbook on positive development of minority children and youth* (pp. 421–435). Springer Science + Business Media. https://doi.org/10.1007/978-3-319-43645-6_25

Huston, A. C., Gupta, A. E., Walker, J. T., Dowsett, C. J., Epps, S. R., Imes, A. E., & McLoyd, V. C. (2011). The long-term effects on children and adolescents of a policy providing work supports for low-income parents. *Journal of Policy Analysis and Management, 30*(4), 729–754. https://doi.org/10.1002/pam.20613

James, A. G., Coard, S. I., Fine, M. A., & Rudy, D. (2018). The central roles of race and racism in reframing family systems theory: A consideration of choice and time. *Journal of Family Theory & Review, 10*(2), 419–433. https://doi.org/10.1111/jftr.12262

Kitzman, H. J., Olds, D. L., Cole, R. E., Hanks, C. A., Anson, E. A., Arcoleo, K. J., Luckey, D. W., Knudtson, M. D., Henderson, C. R., & Holmberg, J. R. (2010). Enduring effects of prenatal and infancy home visiting by nurses on children: Follow-up of a randomized trial among children at age 12 years. *Archives of Pediatrics & Adolescent Medicine, 164*(5), 412–418. https://doi.org/10.1001/archpediatrics.2010.76

Kozol, J. (1992). *Savage inequalities: Children in America's schools*. Perennial Harper Collins, Crown.

Ladson-Billings, G. (2014). Culturally relevant pedagogy 2.0: A. K. A. the remix. *Harvard Educational Review, 84*(1), 74–84. https://doi.org/10.17763/haer.84.1.p2rj131485484751

Lerner, R. M., Lerner, J. V., Murry, V. M., Smith, E. P., Bowers, E. P., Geldhof, G. J., & Buckingham, M. H. (2021). Positive youth development in 2020: Theory, research, programs, and the promotion of social justice. *Journal of Research on Adolescence, 31*(4), 1114–1134. https://doi.org/10.1111/jora.12609

Leventhal, T., Fauth, R. C., & Brooks-Gunn, J. (2005). Neighborhood poverty and public policy: A 5-year follow-up of children's educational outcomes in the New York City moving to opportunity demonstration. *Developmental Psychology, 41*(6), 933–952. https://doi.org/10.1037/0012-1649.41.6.933

Lopez, M. H., Gonzalez-Barrera, A., & Krogstad, J. M. (2018). *More Latinos have serious concerns about their place in America under Trump*. Pew Research Center.

Lorenzo-Blanco, E., Stein, G. L., Lee, R. M., & Ferguson, G. M. (2023). Acculturation and enculturation: The intersection of representational ethics, measurement, and conceptualization. In D. P. Witherspoon & G. L. Steins (Eds.), *Diversity and developmental science: Bridging the gaps between research, practice, and policy*. Springer.

Ludwig, J., Liebman, J. B., Kling, J. R., Duncan, G. J., Katz, L. F., Kessler, R. C., & Sanbonmatsu, L. (2008). What can we learn about neighborhood effects from the moving to opportunity experiment? *American Journal of Sociology, 114*(1), 144–188.

Massey, D. S. (2015). The legacy of the 1968 fair housing act. *Sociological Forum (Randolph, N.J.), 30*(Suppl 1), 571–588. https://doi.org/10.1111/socf.12178

Mathews, C. J., Josefina Bañales, N., Christophe, K., Briggs, A. S., & Hope, E. C. (2023). Action, but make it critical: The measurement and developmental processes of critical action for black and Latinx youth. In D. P. Witherspoon & G. L. Steins (Eds.), *Diversity and developmental science: Bridging the gaps between research, practice, and policy*. Springer.

Mattis, J. S., Fontenot, D. L., Hatcher-Kay, C. A., Grayman, N. A., & Beale, R. L. (2004). Religiosity, optimism, and pessimism among African Americans. *Journal of Black Psychology, 30*(2), 187–207. https://doi.org/10.1177/0095798403260730

McKnight, J. L., & Kretzmann, J. L. (1996). *Mapping community capacity*. Institute for Policy Research, Northwestern University.

Murry, V. M., Berkel, C., Brody, G. H., Gibbons, M., & Gibbons, F. X. (2007). The Strong African American Families program: Longitudinal pathways to sexual risk reduction. *Journal of Adolescent Health, 41*(4), 333–342. https://doi.org/10.1016/j.jadohealth.2007.04.003

Murry, V. M., Butler-Barnes, S. T., Mayo-Gamble, T. L., & Inniss-Thompson, M. N. (2018). Excavating new constructs for family stress theories in the context of everyday life experiences of Black American families. *Journal of Family Theory & Review, 10*(2), 384–405. https://doi.org/10.1111/jftr.12256

Murry, V. M., Gonzalez, C. M., Hanebutt, R. A., Bulgin, D., Coates, E. E., Inniss-Thompson, M. N., et al. (2021). Longitudinal study of the cascading effects of racial discrimination on parenting and adjustment among African American youth. *Attachment & Human Development, 24*, 1–17. https://doi.org/10.1080/14616734.2021.1976926

Nishina, A., Bellmore, A., Witkow, M. R., & Nylund-Gibson, K. (2010). Longitudinal consistency of adolescent ethnic identification across varying school ethnic contexts. *Developmental Psychology, 46*(6), 1389–1401. https://doi.org/10.1037/a0020728

Noguera, P. (2008). Creating schools where race does not predict achievement: The role and significance of race in the racial achievement gap. *The Journal of Negro Education, 77*(2), 90–103. http://www.jstor.org/stable/25608673

Olivos, E. M. (2004). Tensions, contradictions, and resistance: An activist's reflection of the struggles of Latino parents in the public school system. *The High School Journal, 87*, 25–35. https://doi.org/10.1353/hsj.2004.0012

Parra-Cardona, J. R., Bybee, D., Sullivan, C. M., Domenech Rodríguez, M. M., Dates, B., Tams, L., & Bernal, G. (2017). Examining the impact of differential cultural adaptation with Latina/o immigrants exposed to adapted parent training interventions. *Journal of Consulting and Clinical Psychology, 85*(1), 58–71. https://doi.org/10.1037/ccp0000160

Portes, A., & Rumbaut, R. G. (2014). *Immigrant America*. University of California Press.

Poza, L., Brooks, M. D., & Valdés, G. (2014). "Entre familia": Immigrant parents' strategies for involvement in children's schooling. *School Community Journal, 24*(1), 119–148.

Reese, L. R. E., & Vera, E. M. (2007). Culturally relevant prevention: The scientific and practical considerations of community-based programs. *The Counseling Psychologist, 35*(6), 763–778. https://doi.org/10.1177/0011000007304588

Riggs, N. R., Bohnert, A. M., Guzman, M. D., & Davidson, D. (2010). Examining the potential of community-based afterschool programs for Latino youth. *American Journal of Community Psychology, 45*(3–4), 417–429. https://doi.org/10.1007/s10464-010-9313-1

Rivas-Drake, D., Montoro, J., & Agi, A. (2023). Canaries and bellwethers: What can we learn about racial justice from studying ethnic-racial identity within and across groups? In D. P. Witherspoon & G. L. Steins (Eds.), *Diversity and developmental science: Bridging the gaps between research, practice, and policy*. Springer.

Rodríguez-Brown, F. V. (2009). Latino families: Culture and schooling. In E. G. Murillo, S. A. Villenas, R. T. Galvan, J. S. Munoz, C. Martinez, & M. Machado-Casas (Eds.), *Handbook of Latinos and education: Theory, research, and practice* (pp. 350–360).

Rowley, S. J., Burchinal, M. R., Roberts, J. E., & Zeisel, S. A. (2008). Racial identity, social context, and race-related social cognition in African Americans during middle childhood. *Developmental Psychology, 44*(6), 1537. https://doi.org/10.1037/a0013349

Sampson, R. J., Raudenbush, S. W., & Earls, F. (1997). Neighborhoods and violent crime: A multilevel study of collective efficacy. *Science, 277*(5328), 918–924. https://doi.org/10.1126/science.277.5328.918

Sanders, M. R., & Turner, K. M. T. (2019). The Triple P System: Parenting support for every family. In B. H. Fiese, M. Celano, K. Deater-Deckard, E. N. Jouriles, & M. A. Whisman (Eds.), *APA handbook of contemporary family psychology: Family therapy and training* (pp. 409–424). American Psychological Association. https://doi.org/hp9q

Schulze, E. D., & Mooney, H. A. (Eds.). (2012). *Biodiversity and ecosystem function*. Springer Science & Business Media.

Seaton, E. K., Neblett, E. W., Upton, R. D., Hammond, W. P., & Sellers, R. M. (2011). The moderating capacity of racial identity between perceived discrimination and psychological well-being over time among African American youth. *Child Development, 82*(6), 1850–1867. https://doi.org/10.1111/j.1467-8624.2011.01651.x

Sheldon, S. B. (2007). Improving student attendance with school, family, and community partnerships. *The Journal of Educational Research, 100*(5), 267–275. https://doi.org/10.3200/JOER.100.5.267-275

Simpkins, S. D., Riggs, N. R., Ngo, B., Vest Ettekal, A., & Okamoto, D. (2017). Designing culturally responsive organized afterschool activities. *Journal of Adolescent Research, 32*(1), 11–36. https://doi.org/10.1177/0743558416666169

Smith, E. P., Faulk, M., & Sizer, M. A. (2016). Exploring the meso-system: The roles of community, family, and peers in adolescent delinquency and positive youth development. *Youth & Society, 48*(3), 318–343. https://doi.org/10.1177/0044118X13491581

Smith, E. P., Witherspoon, D. P., & Wayne Osgood, D. (2017). Positive youth development among diverse racial-ethnic children: Quality afterschool contexts as developmental assets. *Child Development, 88*(4), 1063–1078. https://doi.org/10.1111/cdev.12870

Snyder, H. N., & Sickmund, M. (2006). *Juvenile offenders and victims: 2006 national report*. U.S. Department of Justice.

Suárez-Orozco, C. (2015). Migration between and within countries: Implications for families and acculturation. In L. A. Jensen (Ed.), *The Oxford handbook of human development and culture: An interdisciplinary perspective* (pp. 43–60). Oxford University Press.

Suárez-Orozco, C., & Suárez-Orozco, M. M. (2009). *Children of immigration*. Harvard University Press.

Texas State Legislature. (2021). H.B.A No. A3979 Section 28.002, Education Code, amended by Subsections (h-2), (h-3), (h-4), and (h-5). https://capitol.texas.gov/tlodocs/87R/billtext/pdf/HB03979F.pdf#navpanes=0

Turner, K. L., & Brown, C. S. (2007). The centrality of gender and ethnic identities across individuals and contexts. *Social Development, 16*(4), 700–719. https://doi.org/10.1111/j.1467-9507.2007.00403.x

U.S. Census Bureau. (2021). *Improved race and ethnicity measures reveal U.S. population is much more multiracial*. https://www.census.gov/library/stories/2021/08/improved-race-ethnicity-measures-reveal-united-states-population-much-more-multiracial.html

Umaña-Taylor, A. J., & Hill, N. E. (2020). Ethnic–racial socialization in the family: A Decade's advance on precursors and outcomes. *Journal of Marriage and Family, 82*(1), 244–271. https://doi.org/10.1111/jomf.12622

Vandell, D. L., Lee, K. T., Whitaker, A. A., & Pierce, K. M. (2020). Cumulative and differential effects of early child care and middle childhood out-of-school time on adolescent functioning. *Child Development, 91*(1), 129–144. https://doi.org/10.1111/cdev.13136

Vespa, J., Armstrong, D. M., & Medina, L. (2018). *Demographic turning points for the United States: Population projections for 2020 to 2060*. U.S. Department of Commerce, Economics and Statistics Administration, U.S. Census Bureau.

Wang, M. T., Hill, N. E., & Hofkens, T. (2014). Parental involvement and African American and European American adolescents' academic, behavioral, and emotional development in secondary school. *Child Development, 85*(6), 2151–2168. https://doi.org/ggs6dn

Watts, R. J., & Flanagan, C. A. (2007). Pushing the envelope on youth civic engagement: A developmental and liberation psychology perspective. *Journal of Community Psychology, 35*(6), 779–792. https://doi.org/10.1002/jcop.20178

Wells, A. S., Keener, A., Cabral, L., & Cordova-Cobo, D. (2019). The more things change, the more they stay the same: The resegregation of public schools via charter school reform. *Peabody Journal of Education, 94*(5), 471–492. https://doi.org/hp9s

Williams, S. M., & Graham, J. (2019). Cross-racial interactions in schools 65 years after Brown. *Peabody Journal of Education, 94*(5), 545–554. https://doi.org/hp9t

Williams, J. L., Anderson, R. E., Francois, A. G., Hussain, S., & Tolan, P. H. (2014). Ethnic identity and positive youth development in adolescent males: A culturally integrated approach. *Applied Developmental Science, 18*(2), 110–122. https://doi.org/ggk7hv

Witherspoon, D. P., Daniels, L. L., Mason, A. E., & Smith, E. P. (2016). Racial-ethnic identity in context: Examining mediation of neighborhood factors on children's academic adjustment. *American Journal of Community Psychology, 57*(1–2), 87–101. https://doi.org/f8n8wr

Wray Lake, L., & Abrams, L. S. (2020). Pathways to civic engagement among urban youth of color. *Monographs of the Society for Research in Child Development, 85*(2), 7–154. https://doi.org/10.1111/mono.12415

Yancy, G. (2018). *Backlash: What happens when we talk honestly about racism in America*. Rowman & Littlefield.

Yip, T., Cheah, C. S., Kiang, L., & Hall, G. C. N. (2021). Rendered invisible: Are Asian Americans a model or a marginalized minority? *American Psychologist, 76*(4), 575. https://doi.org/10.1037/amp0000857

Yu, D., Smith, E. P., & Oshri, A. (2021). Exploring racial–ethnic pride and perceived barriers in positive youth development: A latent profile analysis. *Applied Developmental Science*, 1–19. https://doi.org/10.1080/10888691.2019.1640607

Zarate, M. E. (2007). *Understanding Latino parental involvement in education: Perceptions, expectations, and recommendations*. Tomas Rivera Policy Institute.

Chapter 14
Building Collaborative Teams and Conducting Ethical Research in the Spirit of 2044: The Complexity of Conducting Research in Communities of Color

Margaret O'Brien Caughy, Suzanne M. Randolph Cunningham, and Esther Calzada

Changing Demographics and Health Disparities in the United States

Over the last 30 years, we have witnessed dramatic changes in the population of the United States, with the country becoming significantly more racially and ethnically diverse and less White. Based on 2014 data, the U.S. Census projects the proportion of the United States classified as White, non-Latinx would fall below 50% by the year 2044, making the United States officially a majority minority country (Colby & Ortman, 2015). Furthermore, because the child population in the United States is more diverse than the adult population, it is projected children of color would comprise more than half of the US child population at an earlier date than the US population as a whole. Using census estimates showing that the non-Latinx White, under-18 child population would drop from 53.7% in 2010 to 49.6% in 2020, a recent report suggests that the 2020 census data indeed show this demographic shift (Frey, 2021). The US population is also diversifying across the lifespan, with population gains for all groups of color and population loss for Whites. The undeniable diversification of the US population necessitates that all developmental scientists

M. O. Caughy (✉)
Department of Human Development and Family Science, University of Georgia, Athens, GA, USA
e-mail: margaret.caughy@uga.edu

S. M. Randolph Cunningham
Research and Evaluation Division, The MayaTech Corporation, Silver Spring, MD, USA

E. Calzada
Steve Hicks School of Social Work, University of Texas at Austin, Austin, TX, USA

© The Author(s), under exclusive license to Springer Nature Switzerland AG 2023

D. P. Witherspoon, G. L. Stein (eds.), *Diversity and Developmental Science*,
https://doi.org/10.1007/978-3-031-23163-6_14

attend to issues of diversity within the context of their research and do so in an ethical manner.

Furthermore, the persistence of health disparities faced by persons of color in this country represents an enduring challenge with which scholars, policy-makers, and practitioners have grappled for decades. These disparities are evident across all domains of health and well-being in both children and adults. For example, rates of diabetes are higher among non-Latinx Blacks and Latinx compared to non-Latinx Whites in this country, and rates of hypertension and preterm birth are highest among non-Latinx Blacks (CDC, 2013). Children of color in the United States are disproportionately impacted by asthma, obesity, and challenges to academic achievement, primarily due to higher rates of childhood poverty and systemic inequities in access to quality education, health care, built environments, and other resources (Federal Interagency Forum on Child and Family Statistics, 2019).

The global pandemic of SARS-CoV-2 laid bare these disparities in terms of the health as well as the economic impact of COVID-19. Analyses of data from a health care system serving Alaska, Washington, Oregon, Montana, and California indicate that Black, non-Latinx individuals are 1.5 times and Latinx individuals are twice as likely to contract COVID-19 compared to non-Latinx White individuals (Rozenfeld et al., 2020). Not only are people of color more likely to contract COVID-19, those who do are more likely to suffer severe disease and more likely to die (Townsend et al., 2020). There are striking disparities in the economic impact of the pandemic as well, with Latinx and Blacks more likely to experience job loss due to the pandemic, likely due to an over-representation of these populations in occupations negatively affected by social distancing measures and jobs that cannot be performed remotely (Montenovo et al., 2020). Moreover, COVID-19 mitigation guidelines imposed multiple structural barriers that further unveiled educational inequities for children and adolescents of color (Iruka et al., 2021; Oster et al., 2021). Although there remains much to be uncovered regarding the health, educational, and economic disparities of the pandemic, it is reasonable to surmise they reflect structural disparities in economic well-being, mental health, and physical well-being as well as the pernicious impact of systemic racism and discrimination in this country (Chowkwanyun & Reed, 2020).

Scholars in public health as well as in the developmental and family sciences have documented the role that systemic racism plays in creating and maintaining health disparities (Carter et al., 2017; Priest et al., 2013), and the weight of this evidence has resulted in calls to address the root causes of racism as a public health imperative in the United States (Malawa et al., 2021). Recent events such as the on-camera murder by police of George Floyd (an African American man) coupled with advances in technology have made the reality of the ongoing threat presented by White supremacy and systemic racism manifest to a significantly broader segment of the population. Cell phone images have proliferated across social media including images of police brutality against Black men, women and children, of immigrant children in cages, and of White men marching in Charlottesville, Virginia, chanting anti-Semitic slogans. Likewise, images of unprovoked violence against Asian Americans have highlighted the consequences of the pandemic-related anti-Asian rhetoric in which politicians and members of the media engage. Just as the television coverage of Bloody Sunday in the 1960s made the racism of the Jim Crow South salient to millions of Americans, these viral social media images of today have forced many Americans to recognize the pernicious nature of systemic racism and White supremacy in this country.

Centering Issues of Race and Racism in Developmental Science

Within this context, the inexorable demographic changes that will result in a majority "minority" United States within the coming decades require developmental scientists study the diversity of lived experiences of children, and the complexity of those experiences argues for the importance of interdisciplinary and community-engaged collaboration to effectively address disparities. However, the exclusion of communities of color including indigenous/tribal communities[1] as participants in developmental science is well-documented (see, e.g., McLoyd & Randolph, 1984), and recent analyses of trends in the diversity of study participants indicate this underrepresentation has not been rectified (Syed et al., 2018). The invisibility of communities of color in developmental science reflects the historical and contemporaneous power structures within the research academy that dictate which research questions are deemed to be of value and worthy of pursuit.

Syed et al. (2018) delineate how the ideological setting of developmental science results in explanations for race-related phenomena that are divorced from their race-related root causes. Syed et al. (2018) define ideological settings as "broader societal-level beliefs and values that underlie a normative life in a given context" (p. 814). In the United States, the predominant ideological setting is captured by the so-called "American Dream" narrative that emphasizes personal effort and persistence as the key to success and minimizes the role of contextual determinants such as social stratification. Syed et al. (2018) label this ideological setting as racial colorblindness and describe how it has structured developmental science and resulted in the invisibility of minoritized populations in the field.

Developmental science has always had a significant focus on identifying effective solutions to address real-world problems and to increase health equity. The confluence of current events has renewed what Dr. Martin Luther King, Jr. referred to as a "fierce urgency of now" and underscores the critical role of our research as developmental scientists. Meeting these challenges successfully requires that developmental scientists employ collaborative approaches to research with ethnically and racially diverse teams as well as with the communities directly affected by systemic

[1] Heretofore, we use the term "communities of color" or "children of color" to refer inclusively to Black, indigenous, Latinx, Asian, and other people of color. We have chosen this terminology over "Black, indigenous, and people of color" (BIPOC) because the latter "otherizes" any communities not specifically named such as Latinx and Asian populations.

racism. Rooting out racism and White supremacy will require ethical research with communities of color and research that takes a strengths-based approach rather than the deficit-based one that has dominated much of research with minoritized populations in the past (Syed et al., 2018).

Chapter Overview

We call for interdisciplinary research on children of color that is led by scholars of color, utilizes community-engaged approaches, and recognizes systemic racism as a primary cause of disparities. Scholarship on the resistance, healing, and joy of children of color is equally important. The purpose of this chapter is to provide important guidance for both scholars who are currently engaged in research with minoritized communities (especially those who do not have a background in the study of race and racism) as well as for scholars interested in expanding their research into these communities. We argue that ethical and collaborative research with minoritized communities requires a community-engaged research approach by a racially and ethnically diverse interdisciplinary team. We also posit that authentic collaborative research should include members on the team who are from the minoritized communities on which the research is focused.

Collectively, we have spent a number of years working in settings that require interdisciplinary and community-engaged work. Author Caughy is a White researcher who completed her training in public health and has collaborated with pediatricians, nurses, epidemiologists, and economists and with community partners addressing reproductive health and early childhood development. Calzada, a Latina clinical child psychologist by training, has collaborated with social workers, educators, anthropologists, and public health scholars. Through intervention and applied developmental studies, her research examines the unique context of Latinx families and how factors within the home and classroom shape learning and mental health of Latinx children. Author Randolph Cunningham is an African American developmental psychologist who has spent more than four decades conceptualizing, implementing, and disseminating research on African American children, families, and communities as well as mixed-race/ethnicity families and communities. Her collaborative research experiences have been conducted with federal and foundation grantees in community and clinical settings with collaborators in developmental and family sciences, public health, medicine, economics, social work, nursing, education, and nutrition. Further, Randolph Cunningham's work spans the continuum of community-engaged research from community-informed projects to community co-led research. As such, we have experienced firsthand both the benefits and the challenges of working with racially and ethnically mixed interdisciplinary teams and engaging with diverse community members as research partners. In this chapter, we will provide an overview of the challenges of conducting interdisciplinary and community-engaged research with diverse communities as well as outline the ethical considerations that should be addressed. Finally, we provide recommendations for best practices in conducting this research that scholars can use to guide their work in this area.

The Rise of Collaborative, Community-Engaged Research

For more than two decades, social and behavioral scientists in various fields such as public health, program evaluation, medicine, and prevention science have utilized research approaches to address social, behavioral, and public health problems that require researcher-community collaborations. Federal, other governmental, and foundation funders have also begun to require collaborative responses to their calls for grant applications, even for the most rigorous of scientific investigations such as randomized clinical trials (RCTs) (e.g., Clinical Translational Science Awards, CTSAs). Ancillary fields of science and guiding principles for such team and collaborative approaches to research have also grown as reflected in published guidelines for community engagement principles (CDC, 2011), the science of team science (Hall et al., 2018), implementation science (e.g., the Consolidated Framework for Implementation Research, CFIR, CFIR Research Team, 2021), collaborations and partnerships (Butterfoss, 2007), community-based participatory research (Wallerstein et al., 2017), and participatory action research (McIntyre, 2007). With increasing attention on determining the root causes for issues affecting minoritized communities at individual and systems levels, research has also been more concerned with social determinants of health, factors that disproportionately affect some communities and individuals due to inequities in social and economic resources (CSDH, 2008).

The field of developmental science has responded to these trends with spinoff specialties such as applied developmental science which embraces collaborative research (see, e.g., Oden, 2000). Furthermore, there has been an increase in peer-reviewed journals that have featured special issue volumes or supplements focused on community-engaged research (see, e.g., Zeldin, 2000). Journals specifically dedicated to community-engaged research have also been established as reputable outlets for scholarship (e.g., *Clinical and Translational Science* and *Progress in Community Health Partnerships: Research, Education and Action*). In addition, professional associations have incorporated principles of respect and cultural integrity to guide interactions with community members into their association's ethical standards. For example, the ethical standards of the American Public Health Association (APHA) specify that public health professionals and researchers must be sensitive to the unique needs of individuals and communities based on race, ethnicity, sex, sexual identity, sexual orientation, or gender identity (American Public Health Association, n.d.). Similarly, the American Psychological Association adopted guidelines for cultural competence in practice, research, consultation, and education (https://www.apa.org/monitor/2018/01/multicultural-guidelines).

Thus, this intersection of multiple factors in the United States has contributed to a new sense of urgency among developmental scientists to embrace principles of community engagement and interdisciplinary collaboration in their research efforts. Specifically, there has been increased interest in scholarship on minority health and health disparities that has expanded outlets for disseminating knowledge about collaborative research and its impact. In addition, legislation was passed elevating a former resource center to institute level at the National Institutes of Health (NIH) to specifically address minority health and health disparities issues (i.e., the National Institute for Minority Health and Health Disparities, NIMHD). The intersection of the COVID-19 pandemic, which further exposed the impact of race-based health disparities and inequities, along with the racial injustice pandemics (e.g., incidents of police violence against Blacks and random racist attacks on Asians), has shone an even brighter spotlight on this urgent need for interdisciplinary, collaborative research with community partners to address these growing needs. We recognize and appreciate the need to act but caution against doing so without careful consideration for the complexities inherent in conducting research with minoritized communities. In the remainder of this chapter, we present resources and strategies to guide future collaborative research in developmental science that seeks to focus on minoritized communities.

Creating Diverse and Equitable Interdisciplinary Teams

Interdisciplinary research is defined as research "that integrates information, data, techniques, tools, perspectives, concepts, and/or theories from two or more disciplines or bodies of specialized knowledge to advance fundamental understanding or to solve problems whose solutions are beyond the scope of a single discipline or field of research practice" (National Academy of Sciences et al., 2005, p. 26). For the purpose of this chapter, we define interdisciplinarity along a continuum ranging from collaborations of developmental scientists of differing expertise (e.g., cognitive development vs. social development) to collaborations across different academic disciplines (e.g., psychology, public health, sociology) to collaborations with individuals outside of academia including with community stakeholders.

As noted in the definition above, the impetus for interdisciplinary research is the need to address social problems for which there are multiple causes that exceed the expertise of a single academic discipline. The complexity of social problems is particularly true for issues affecting children and youth in general, and children and youth of color in particular. Childhood obesity is a case in point. Data from the National Health and Nutrition Examination Survey (NHANES) indicate the rate of obesity among children age 2–19 years has more than tripled since the early 1970s, from 5.2% in 1971–1974 to 18.5% in 2015–2016 (Fryar et al., 2018). Furthermore, child obesity disproportionally affects children of color, affecting 28% of Latinx boys, 25% of Black girls, 24% of Latinx girls, and 19% of Black boys compared to 15% and 14% of White boys and White girls, respectively (Fryar et al., 2018). The correlates of child obesity include factors at the child level such as individual differences in self-regulation, unhealthy eating habits, and lack of physical activity as well as family-level factors such as parent monitoring and characteristics of the home environment. However, these individual- and family-level differences are embedded in a broader social context such as communities that provide insufficient access to healthy food options (so-called food deserts) and which lack safe outdoor play spaces and schools that have reduced or eliminated physical activity opportunities during the school day.

At an even more macro level, social policies can influence the availability of healthy food options through federal subsidies of specific crops that some argue have led to the overproduction of processed, energy-dense foods (Franck et al., 2013), and nutritional assistance programs that may inadvertently increase the likelihood of childhood obesity (Basu et al., 2014; Kimbro & Rigby, 2010). Effectively tackling childhood obesity is a prime example of the demand for interdisciplinary collaboration between nutritionists, developmental scientists, public health specialists, and health policy experts and in collaboration with members of the affected communities (Harrist et al., 2012). An example of such an effort is the Coordinated Approach to Child Health (CATCH) program (www.catch.org), which has demonstrated that an interdisciplinary, community collaborative approach can be effective in reducing childhood obesity (see, e.g., Hoelscher et al., 2010).

The rise of interdisciplinary team science has led to the development of a cross-disciplinary research field, the "Science of Team Science" (SciTS), focused on identifying the factors that support effective team science such as the value of cross-organizational collaboration, the optimal size and composition of effective collaborative teams, and individual and team processes that support productivity and sustainability of teams (Hall et al., 2018). Guidance on interdisciplinary teamwork is essential to bridging differences in terminology and preferred approaches to data collection, analysis, and interpretation. For example, developmental scientists who have been trained in psychology tend to emphasize rich measures of individual attitudes, beliefs, and behaviors, whereas public health scientists emphasize population-level sampling and generalizability, using large sample sizes that by necessity limit the use of rich, individual-level measures. Ethnographers on a collaborative team may emphasize in-depth observations to qualitatively document environmental and personal factors that may contribute to understanding the issue being researched.

However, little of the scholarship of the SciTS field speaks to the role that racial-ethnic diversity in team composition plays in the effectiveness of the collaboration (Hall et al., 2018). We could identify only two studies focused on the impact of diversity in team science. Freeman and Huang (2015) reported that ethnically diverse authorship was less common than ethnically homogeneous author teams, although publications authored by more ethnically diverse teams were published in higher impact outlets, and Martins et al. (2003) found that ethnically diverse science teams function better when located in ethnically diverse units compared to ethnically homogeneous ones. The extremely limited research on racial-ethnic diversity in scientific collaboration provides little, if any, guidance on how best to configure interdisciplinary collaborations when organizing diverse teams. We turn to the recommendations Miller et al. (2019) set forth for research on race and racism that scholars (1) adopt a racially diverse team science approach, and (2) remain mindful of historical patterns of oppression and inequality.

Across disciplines, only 22% of full-time faculty identify as Black, Indigenous, Latinx, Asian, or Pacific Islander, with even less representation at higher ranks and in positions of leadership (National Center for Education Statistics, 2019). Furthermore, scholars of color are underrepresented among principal investigators of research grants awarded by the National Institutes of Health (Nikaj et al., 2018). The experiences of minoritized scholars are shaped not only by their race/ethnicity but also by other identities such as their gender, sexual orientation, academic rank, and historical trauma associated with their racial/ethnic group. Likewise, majority scholars are shaped not only by the benefits of White privilege but also by their other identities such as being a woman, a sexual minority, and/or a first-generation college student. Intersectionality theory asserts that the lived experiences of individuals, especially as related to oppression, cannot be understood without interrogating the multiple identities they inhabit (McCormick-Huhn et al., 2019; see also Smith-Bynum, Chap. 4, this volume). It is critical to be aware that the power dynamics embedded within a collaborative team depend on each individual's intersecting identities and to be intentional in mitigating the impact of oppressive practices on team members who are minoritized. Further, the historical experiences of the groups which these team members represent may shape their interest in addressing the "colonialism" of science in that minoritized team members may contest typical developmental science approaches (Nobles, 1986, 2015). For example, this may be done by incorporating what are referred to as "decolonized methodologies" (Keane et al., 2017; Zavala, 2013). These methodologies include non-traditional approaches that challenge the rigor and assumptions of our traditional academic research methods (e.g., incorporating cultural adaptations of evidence-based interventions or standardized measurement tools into the research design).

Recommendations for Creating Diverse and Equitable Interdisciplinary Teams

It is our assertion that a necessary component of conducting ethical developmental science research with minoritized communities is to implement such research with racially/ethnically diverse and interdisciplinary research teams. In this section, we set forth a set of best practices for forming and implementing research with such teams.

White Researchers Must Engage in Self-Reflective Work Regarding White Privilege

One aspect of White privilege is the ability to ignore the impact of social inequality and choose (intentionally or unintentionally) to not deal with racism and systemic oppression. White scholars are, first and foremost, humans and, as such, they are not exempt from the benefits of White privilege. White developmental scientists must interrogate their own lived experiences and how these experiences may bias the research questions they choose, the research methods they utilize to address those questions, and how they interpret the findings. Furthermore, White researchers must not rely on scholars of color to provide guidance in this journey of self-reflection, since scholars of color must already navigate a system largely defined by White supremacy (i.e., one that prioritizes the needs of White academics). In addition, it is important for White researchers to assume the role of "accomplice" in collaboration with communities of color in dismantling the power structures associated with White supremacy. Here, we purposively use the term "accomplice" instead of the often-used term "ally" because the former indicates a commitment to actively disrupt the system, whereas the latter does not.

Recently, a number of resources to support White researchers in this work have been developed, such as the Anti-racism and Ally White Working Group (AAWWG) of the Society for Research in Child Development (SRCD) and the Academics for Black Survival and Wellness (https://www.academics4blacklives.com). AAWWG is a working group comprised of White developmental scientists associated with SRCD's Ethnic Racial Issues (ERI) committee that sponsors self-reflection opportunities among White researchers and other activities to support the ERI committee. Likewise, A4BL provides opportunities to learn (e.g., through webinars and readings) about race and racism from a historical and contemporary lens and to apply those learnings to research, teaching, and service in the academy. Accountability groups encourage long-term reflections, conversations, and commitments in the move toward anti-racism.

White Researchers Must Recognize Their Limitations

White developmental science researchers collaborating with researchers from minoritized groups must recognize the limitations of their own ability to understand the lived experiences of minoritized children, youth, and adults. Regardless of seniority or scholarly achievements, White researchers must have the humility to recognize their personal limitations that arise from lack of firsthand experience with the pernicious and systemic racism affecting communities of color. In addition, White researchers who themselves experience bias related to their intersectional identities (e.g., women researchers who experience sexism or sexual minority researchers who experience homophobia) should recognize that their marginalized status in one domain does not imbue them with special insight into the experiences of people of color.

Collaborative Teams Must Elevate the Voices of Minoritized Scholars

To be effective, diverse teams must work to shift the power dynamics (e.g., tokenization, gas lighting) that are embedded in academia. White researchers must earn the trust of scholars of color through intentional anti-racist practices, a process that takes time. Moreover, it is insufficient to simply create a diverse team, and instead, collaborations should intentionally work to elevate the voices of scholars of color. Minoritized scholars should be in positions of leadership within the team. Given the disparity in numbers of more senior scholars of color, it is likely that the White researchers on the team will be of more senior rank. As such, it is incumbent upon White researchers on the team to cede power in explicit ways. Likewise, it is imperative that the methods and products of the team be grounded in the scholarship of minoritized scholars. For example, the work of minoritized scholars should inform theory development, the measures selected for the study, and the interpretation and dissemination of findings.

Minoritized Scholars Must Seize Opportunities to Elevate Their Voices and Express Their Own Limitations

Feelings such as being "the token" on the team or experiencing "imposter syndrome" can affect a minoritized scholar's contributions to the team (Chrousos & Mentis, 2020; Syed et al., 2018). However, in trusting, authentic collaborations, minoritized scholars should feel empowered to offer suggestions consistent with their worldviews with respect to enhancing the research effort. Although often invited late to the team, minoritized scholars have perspectives and ideas that enhance the overall research project and should be given license to lead the dissemination of scholarship based on those ideas. For example, in the NICHD Study of Early Child Care and Youth Development, minoritized scholars who were brought to the table after the initial award of this large multi-site study assumed key roles by leading the production of a *Child Development* manuscript focused on setting a research agenda for early care for children of color (Johnson et al., 2003).

Minoritized researchers should also self-reflect on how one's minoritized status can serve as both a benefit and a challenge in the research process (Egharevba, 2001; Manohar, 2013). For example, while having a shared vision for the research with White colleagues, the minoritized researcher may also be put in the position of experiencing or reliving experiences under investigation because they are a member of the group under study. Minoritized researchers must also engage in self-assessments of their limitations when engaging in community-based collaborations. As members of the academy, minoritized scholars hold identities (researcher, middle class status) that may be viewed with some hesitation by community members regardless of race/ethnicity. Even if the minoritized researcher's lived experiences were once similar to those of the study community, currently they may not experience similar adverse impacts that result from those circumstances. Also, one or two minoritized researchers cannot address the heterogeneity in their identity group. Thus, minoritized researchers may be limited in what perspectives they can lend to advance research designs or procedures. Acknowledging this limitation and engaging community members as partners who do have the lived experience under study (i.e., using the methodological approaches described below) can address this dilemma.

Systems-Level Strategies Must be Implemented to Facilitate Adoption of These Tenets

It is incumbent on the academy to support best practices to ensure the success of diverse interdisciplinary collaborative teams. The targets for these system-level changes should include, but are not limited to, appointment, tenure, and promotion processes; curricula used in training programs; grant reviewer and application selection processes; and editorial and reviewer selection processes of journals. Beyond the inclusion of minoritized researchers on review panels and editorial boards or in faculty and administrative positions to meet diversity, equity, and inclusion goals, radical transformations that explicitly address racism in the social sciences are needed. Examples of such transformations include infusing developmental course and practical content with issues of implicit bias, historical trauma, and contemporary social determinants of health for communities of color as well as explicitly adopting anti-racist pedagogical approaches. Some progress has been made (see, e.g., Marrero et al., 2013), but incidences such as the recent denial of tenure for a minoritized journalism scholar who founded and led the "1619 Project," a Pulitzer Prize-winning publication based on collaborative research describing the legacy of slavery in America,[2] show the ongoing and entrenched nature of racism in the academy.

Recognition of merit and financial investment in interdisciplinary collaborative research on minoritized populations is also needed for this work to thrive. For example, SCRD recently identified a set of priorities to dismantle systemic racism in early childhood care (Meek et al., 2020), and NIH established the UNITE initiative to "identify and address structural racism within the NIH-supported and greater scientific community" (https://www.nih.gov/ending-structural-racism/unite). An interdisciplinary journal in a related field, the *Early Child Care Research Quarterly*, recently published its recommendations for ensuring racial equity in its publications by including more research from minoritized scholars and on children and families in communities of color (ECRQ Ad Hoc Committee on Racial Equity in Publishing, 2021).

Summary of Recommendations

Creating diverse and equitable interdisciplinary teams requires intentional action to effect change at both the individual and systemic levels. Much of that action is required on the part of majority White scholars in addressing their personal growth, in elevating the voices of minoritized scholars, and in spearheading the changes needed in the academy that function as barriers to the scholarship trajectories of minoritized scholars. Although minoritized researchers play a role as well by

[2] *Nikole Hannah-Jones Issues Statement on Decision to Decline Tenure Offer at University of North Carolina-Chapel Hill and to Accept Knight Chair Appointment at Howard University.* https://tinyurl.com/38db6rem. Retrieved 07/08/2021.

seizing opportunities to elevate their work, the burden of change is with White scholars to dismantle the systemic barriers that exist.

Community-Engaged Collaborative Work in Minoritized Communities

Although we assert that diverse interdisciplinary research teams are a necessary component of conducting ethical developmental science research with minoritized communities, diverse teams alone are not sufficient: ethical research with minoritized communities should utilize a community-engaged approach. In this section, we discuss several principles of community engagement and cultural competency to guide developmental scientists in their pursuit of collaborative research. Where appropriate, we also provide examples of collaborations among researchers, and between researchers and community members, to illustrate these principles. Later in this chapter, we address ethical principles of community-engaged collaborative research with specific recommendations for best practices.

The Continuum of Community Engagement

Several models and frameworks for community engagement or collaboration are in the published literature (McIntyre, 2007; Wallerstein et al., 2017). These frameworks were primarily developed based on public health promotion efforts (e.g., coalitions to address tobacco cessation, substance abuse prevention, diabetes, obesity, infant mortality reduction, and cardiovascular disease). A number of texts on how to implement community-based research with community partners are available (McIntyre, 2007; Wallerstein et al., 2017), including a seminal publication by Butterfoss that includes a definition of collaboration, principles for effective community-based collaborations and partnerships, and case examples illustrating these principles (Butterfoss, 2007). More recently, our field has evolved to consider community partnerships as central to the implementation of randomized control trials (RCTs), traditionally the highest standard of evidence for intervention effectiveness and upon which policy decisions are made. Community partnerships help to address conceptual and practical concerns in recruiting and retaining community participants in RCTs. This requirement has been instituted by the federal government, such as for NIMHD grantees receiving Research Centers at Minority Institutions awards.

Collaborative efforts take place along a continuum, defined by the extent to which community members have or have not been involved or engaged in research, trust, the levels of expected impact, and the nature of communication flow. The least collaborative approach is *Outreach* (communication from the researcher to inform the community of the research and recruit participants), followed by *Consult* (bi-directional communication, sharing of information to get feedback on the approaches), *Involve* (bi-directional communication, more participatory on the part of the community with visible partnering), and *Collaborate* (a formal partnership and trust building, communication channels are formalized, outcomes are at the partnership level not community-wide). The most optimal level is *Shared Leadership*, defined by shared vision and decision making and mutually beneficial outcomes for the university/researcher and community.

Anchored at the non-optimal end of the continuum of community engagement is no involvement, which we do not consider ethical. When research is conducted in a community and there is no formal community involvement (or only a letter of commitment or support), we recommend that funders, peer reviewers, and readers question its utility and applicability. Examples would be studies in which the community is involved only because it is an access point through which a convenience sample can be recruited and induced with incentives to participate and in which there are no gatekeepers or cultural ambassadors by which the research team works with the community. In the middle of the continuum of community engagement are studies that identify a community gatekeeper or organization through which the research is introduced to the community, but once connections are made, the research team autonomously collects data and exits the community with no commitment to share findings or assist with the application of the findings to solve community challenges. The ultimate endpoint of the continuum is to have university-community partnerships that are community-driven (Pavao, 2012). In a community-driven collaboration, the community contacts the researcher to work together to conceptualize, develop, seek and secure funding, implement, evaluate, and disseminate the findings of an initiative (Holliday et al., 2018; Koblinsky et al., 2014). The community partner may be the recipient of the grant and share decision-making power with the university partners in terms of budget, policy making, and resources. The researchers, in turn, facilitate capacity-building, provide evaluation tools, and co-author and co-present study findings.

Where communities of color are concerned, the nature of involvement has been increasing due to foci of research on social determinants of health, the increasing diversity of our national and international communities, and requirements from funders for the inclusion of community collaborators in funding applications. This institutionalization of community engagement within the NIH research application and award process has resulted in a proliferation of scholarship on this topic (e.g., the journal *Clinical and Translational Science*). Also, the models for collaboration and community engagement in research have evolved to include a focus on topics related to communities of color such as health equity and social equity (Wallerstein et al., 2017), racial equity (French et al., 2020), and cultural competence (Burlew et al., 2019). Most recently, NIH launched the Community-Engaged Alliance (CEAL) Against COVID Disparities (https://covid19community.nih.gov/). CEAL has three areas of focus: increasing COVID-19 awareness and education research, promoting and facilitating diversity and inclusion in clinical trials, and supporting diversity in COVID-19 science and provides further evidence of NIH's commitment to community-engaged research.

The following resources and tips are offered to researchers looking to initiate or enhance their community-engaged collaborations to facilitate their research enterprise and scholarship. To assist scientists struggling to balance rigor with community engagement, the Centers for Disease Control and Prevention (CDC) produced a guidance document, *Community Engagement Principles,* to define community engagement, lay out principles, and provide strategies to ensure success. *Community engagement* was defined as, "the process of working collaboratively with and through groups of people affiliated by geographic proximity, special interest, or similar situations to address issues affecting the well-being of those people. It often involves partnerships and coalitions that help mobilize resources and influence systems, change relationships among partners, and serve as catalysts for changing policies, programs, and practices" (CDC, 2011, p. 9). This document was later revised to incorporate lessons learned from implementing the CDC guidelines in conducting a set of RCTs and now includes nine principles (CDC, 2011, pp. 46–53):

1. Be clear about the populations/communities to be engaged and the goals of the effort.
2. Know the community, including its norms, history, and experience with engagement efforts.
3. Build trust and relationships and get commitments from formal and informal leadership.
4. Collective self-determination is the responsibility and right of all community members.
5. Partnering with the community is necessary to create change and improve health.
6. Recognize and respect community cultures and other factors affecting diversity in designing and implementing approaches.
7. Sustainability results from mobilizing community assets and developing capacities and resources.
8. Be prepared to release control to the community and be flexible enough to meet its changing needs.
9. Community collaboration requires long-term commitment.

The Basics of Community-Engaged Research

Collaborative community-based research requires that developmental scientists take a strategic, comprehensive, and strengths-based approach to first learning about and valuing the culture and context in which the research will take place. For example, a preliminary study might include an assessment of not only the complex needs but also the assets of a community to understand its cultural values and the worldviews of its members. This approach will also convey to your community partners that a strengths-based (vs. deficit-based) approach is being undertaken. It is imperative to work directly with collaborators and community members who (with compensation) can guide these preliminary efforts. As the partnership develops, researchers must make meaningful but realistic and sustainable commitments to provide incentives, resources and other benefits back to the community. The knowledge generated to inform future practice, research, and policy must be of significance to the community and returned to the community in timely and usable ways.

Cultural Competence in Community-Engaged Research

The importance of understanding the culture and history of a community is probably the most important step to ensure success. When, where, how and through whom a research team enters and exits a community is as important as the research aims. Many of the best practices outlined above, particularly with regard to self-reflection, apply to community-engaged work and set the foundation for cultural competence.

The U.S. Office of Minority Health (OMH) defines cultural and linguistic competence as, "a set of congruent behaviors, attitudes, and policies that come together in a system, agency, or among professionals that enables effective work in cross-cultural situations" (Cross et al., 1989). Culture shapes how community members define their community (by geographic boundaries, social identities, or personal attributes), perceive their community, relate to each other, build trust, assign meaning to various concepts and actions, ascribe power and status, and are willing to partner, communicate, and share information. For example, a guiding principle based on an African worldview is preference for subjective vs. objective orientation. That is, if one has a bad feeling about a researcher or research approach (a subjective criterion), regardless of its rigor or monetary incentive (objective criteria), members of an African American community may not feel comfortable because "it does not carry the feel."

There are five essential elements of cultural competence that researchers and their teams should consider in their approaches: (1) valuing diversity; (2) having the capacity for cultural self-assessment; (3) being conscious of the dynamics inherent when cultures interact; (4) institutionalizing cultural knowledge through policies and procedures manuals; and (5) developing adaptations to research approaches to reflect diversity between and within cultures.

Attending to these elements allows a researcher, research team or organization to move from enacting harm toward creating meaningful, effective and equitable collaborations. In the absence of cultural competence, researchers risk cultural destructiveness, incapacity or blindness. *Cultural destructiveness* is characterized by attitudes, practices and policies that harm a specific population or cultural group such as offering evidence-based intervention opportunities only if participants shed some honored cultural tradition (e.g., not wearing tribal clothing to session). *Cultural incapacity* is illustrated by a lack of capacity of the research team to respond effectively to the needs and preferences of a specific population or cultural group such as by neglecting institutional or systemic bias and/or having lower expectations for some racial/ethnic groups. *Cultural blindness* involves viewing and treating all people as the same, which leads to a lack of appreciation for diversity and how race, ethnicity, and culture influence individuals. Other examples of cultural blindness include ignoring cultural strengths, blaming individuals/families for their circumstances, and placing little value on training and resource development that could facilitate acquiring cultural knowledge.

As cultural competence grows, researchers move from cultural pre-competence to cultural competence and ultimately, to cultural proficiency. Proficiency indicates the institutionalization of the essential elements of cultural competence into the system in which the research project is embedded.

Recommendations for Collaborating with Minoritized Communities

The types of collaborative research partnerships described above are not second nature to most researchers, especially if one also has to navigate the tricky circumstances of respecting cultural integrity, showing humility, and showing deference to the community's traditions and cultural values. In this section, we provide some tools to assist researchers in enhancing their collaborative research efforts with communities of color. These may be useful even to researchers who self-identify as a member of the study community since the "researcher" role and attendant responsibilities change how we behave as well as how we may be perceived, even in/by our own community. With that goal, we offer a set of best practices for community-engaged research with minoritized communities.

Research Team Members and Organizations Must Conduct a Self-Assessment to Determine Their Cultural Competency

Periodic re-assessment is also important to continually increase awareness, reduce bias, and enhance competencies. A good source for information on cultural competency self-assessments is the National Center for Cultural Competence at Georgetown University (https://nccc.georgetown.edu/assessments/).

Teams Must Actively Seek Out Resources to Address Gaps in Understanding and Training as Identified in the Self-Assessment

This requires a commitment to professional development to enhance cultural competence, self-assess, and remediate biases. National professional associations or caucuses for specific minoritized groups (e.g., the SRCD Tri-Caucus) can provide valuable information. Minoritized researchers and researchers from other disciplines like ethnography can also serve vital roles in promoting understanding and competencies for working with communities of color. NIMHD's Research Centers at Minority Institutions, the National Research Mentoring Network for minoritized scholars, and The Alliance of National Psychological Associations for Racial and Ethnic Equity all support these goals.

Teams Must Critically Evaluate All of Their Research Processes Through an Anti-Racist Lens to Ensure Equity and Inclusion

For example, researchers must develop the capacity to select, collect, and analyze data using variables that have meaningful impact culturally. Researchers can review websites, journals, conference agendas, and proceedings to identify relevant emergent topics, as well as consult and collaborate with subject matter experts. Moreover, traditional approaches should be challenged, including the race-comparative method and RCTs as standards for quality, aggregated data analyses that obscure inequities, vague descriptions of the racial/ethnic composition of samples, and the exclusion of community member input as a legitimate source of knowledge.

Teams Must Commit to the Transfer of Knowledge and Skills Between All Stakeholders on the Research Team and in the Community

This requires humility and recognition of various types of expertise (e.g., non-academic) and includes honoring the strengths of the community. This is also reflected in advocacy work and a commitment to sustain best practices through the active pursuit of resources to support the community and expand its capacities.

Teams Must Support Systemic Change Within the Academy

Community-engaged work must be respected and rewarded in promotion and tenure processes and valued by professional associations and in top-tier journals. Time and financial support should be allotted to scholars who engage in community-based research, including resources for fair compensation of community partners.

Historically, academic ways of knowing and collaborating/partnering have been university researcher-driven (e.g., community-based without community participation or university-community partnerships that have their own research agenda). New models of collaboration include participatory action research (McIntyre, 2007), empowerment evaluation (Fetterman & Wandersman, 2005), utilization-focused evaluation (Patton, 2008), and community-based participatory research projects (Wallerstein et al., 2017). Although designed to engage communities in authentic ways, these approaches have often been researcher-initiated, directed, and sustaining. Authentic engagement requires that researchers put in time, energy, money, and other investments in communities, especially those at risk for every "next worse thing" (e.g., a public health crisis like the COVID-19 pandemic, direct or vicarious exposure to police violence, or a natural disaster such as a hurricane). Researchers must recognize that communities of color have legacies of historical traumas they also must navigate to remain mentally, socially, and economically viable. However, the central challenge in collaborative research with communities of color remains the tension between rigor and practicality, and this tension needs to be addressed successfully if there are to be mutual benefits. The question is, how does one transform developmental science approaches to respect the cultural integrity of communities while balancing the need for the scientific "rigor" required by funders and peer-reviewed outlets?

Summary of Recommendations

Similar to efforts to create equitable and diverse collaborative teams, creating authentic and equitable collaborations with minoritized communities requires intentional action at both the individual and systemic levels. Researchers must engage in self-assessment regarding their own cultural competency and then take actions to address any deficits revealed therein. Furthermore, researchers must invest the time necessary to build authentic collaborations that respect community strengths and prioritize the benefits that are returned to the community. However, systemic changes in the promotion and tenure process are needed in order to support faculty engaged in this work. Although some universities have taken such steps to reform promotion and tenure processes (see, e.g., Pelco & Howard, 2016), these reforms are still relatively rare.

The following are a few examples of enhanced models and frameworks for community participation that have been used at the optimal end of the engagement continuum to address the rigor/practicality dilemma and yield substantive research productivity.

Examples of Successful Collaborations with Community Partners

Our first example is of a community-engaged research project to conduct a needs assessment of the COVID-19 pandemic's impact on mental health for which the research team included community voices to plan for data collection and better understand the data in multiple Black communities across the nation. Communities were experiencing not only the COVID-19 pandemic but also had vicarious exposure to the racial injustice pandemic via media stories on police violence against Blacks. Therefore, in addition to measures of changes in health, mental health, family life, and child behavior, we also included measures of attitudes toward police violence, experiences with everyday discrimination, Black identity, and cultural mistrust to better understand these race- and culturally based influences on child, individual, and family mental health outcomes. Two minoritized developmental scientists and their collaborative team of minoritized researchers (Iruka et al., 2021) created specific measures in their study to examine racism, racial trauma, and discrimination among Black families. These race- and culture-specific variables allowed the researchers to capture more nuanced data about the experiences of Black parents with very young children during the intersection of the COVID-19 pandemic with racial injustice events.

In other projects, we have culturally adapted existing assessment protocols and published findings in the *Journal of Black Psychology* (Caughy et al., 2002), and culturally adapted an evidence-based parenting program for families at high risk for substance abuse by incorporating values from the Africentric worldview, Kwanzaa principles, African proverbs, and Adinkra symbols (Roberts et al., 1999). Some of these efforts are led by White researchers, such that the responsibility is not always left to minoritized scholars.

Conclusion

As the United States continues to diversify, it has become critical that the field of developmental science accelerates research focused on Black, indigenous, and other children of color. Since the field's inception, developmental science has operated from an ideological setting that centers the development of White children as normative and optimal. Although there is developmental research centered on the lived experiences of minoritized children and youth, such as ethnic-racial socialization research and research on ethnic-identity development, the majority of developmental science fails to adequately address issues of diversity.

In this chapter, we assert that the future of developmental science requires the adoption of an interdisciplinary, community-engaged approach focused on the study of the diverse lived experiences of children and youth. Centering developmental science on this diversity requires that scholars explicitly interrogate the ideological settings underlying their assumptions regarding developmental processes and its determinants. One critical means for successfully interrogating, and hopefully changing, these ideological settings is through ethnically diverse research teams. However, engaging ethnically diverse scholars in collaborative research in an equitable and ethical manner requires intentionality on the part of White scholars on the team.

Likewise, ethical research with minoritized communities requires a community-engaged approach. In this chapter, we have provided an overview of the continuum of forms of community-engaged collaboration as well as recommendations for engaging in collaboration with minoritized communities in an ethical manner. However, it is important to be aware that these recommendations, as well as those for implementing ethnically diverse collaborative teams, are not "one and done" recommendations. The goal of engaging in ethical collaborative research with minoritized communities is a goal that must be continually strived for. The journey of self-reflection necessary for ethical research of this type is ongoing, regardless of one's level of experience.

Children and youth today face many challenges to their well-being not experienced by previous generations. Emerging technologies such as digital social media and rapidly changing society norms around gender identity are but a few examples of new areas of inquiry for developmental science. At the same time, disparities in health and well-being that were identified decades ago persist, illustrating the shortcomings of developmental science to promote equity in child development. These circumstances demand that developmental scholarship be centered on issues of diversity, equity, and inclusion. The continued relevance of the field to inform our understanding of how to support the healthy development of children and youth depends upon it.

References

American Public Health Association. (n.d.). *Public health code of ethics*. Retrieved 20 Aug 2021, from https://www.apha.org/-/media/files/pdf/membergroups/ethics/code_of_ethics.ashx

Basu, S., Seligman, H. K., Gardner, C., & Bhattacharya, J. (2014). Ending SNAP subsidies for sugar-sweetened beverages could reduce obesity and type 2 diabetes. *Health Affairs, 33*(6), 1032–1039. https://doi.org/10.1377/hlthaff.2013.1246

Burlew, A. K., Peteet, B. J., McCuistian, C., & Miller-Roenigk, B. D. (2019). Best practices for researching diverse groups. *American Journal of Orthopsychiatry, 89*(3), 354–368. https://doi.org/10.1037/ort0000350

Butterfoss, F. D. (2007). *Coalitions and partnerships in community health*. John Wiley & Sons.

Carter, R. T., Lau, M. Y., Johnson, V., & Kirkinis, K. (2017). Racial discrimination and health outcomes among racial/ethnic minorities: A meta-analytic review. *Journal of Multicultural Counseling and Development, 45*(4), 232–259. https://doi.org/10.1002/jmcd.12076

Caughy, M. O., Randolph, S. M., & O'Campo, P. J. (2002). The Africentric Home Environment Inventory: An observational measure of the racial socialization environment of the home for preschool children. *Journal of Black Psychology, 28*, 37–52. https://doi.org/10.1177/0095798402028001003

CDC. (2011). *Principles of community engagement* (2nd ed., NIH Publication No. 11-7782). Retrieved 10 June 2021, from https://www.atsdr.cdc.gov/communityengagement/pdf/PCE_Report_508_FINAL.pdf

CDC. (2013). CDC health disparities and inequalities report – United States, 2013. *Morbidity and Mortality Weekly Report, 62*(Suppl. 3), 1–186. http://www.cdc.gov/mmwr/pdf/other/su6203.pdf

CFIR Research Team. (2021). *Consolidated framework for implementation research (CFIR)*. Retrieved 7 July 2021, from https://cfirguide.org/

Chowkwanyun, M., & Reed, A. L. (2020). Racial health disparities and Covid-19: Caution and context. *New England Journal of Medicine, 383*(3), 201–203. https://doi.org/10.1056/NEJMp2012910

Chrousos, G. P., & Mentis, A.-F. A. (2020). Imposter syndrome threatens diversity. *Science, 367*(6479), 749–750. https://doi.org/10.1126/science.aba8039

Colby, S. L., & Ortman, J. M. (2015). *Projections of the size and composition of the U.S. population: 2014 to 2060*. Retrieved 22 Jan 2021, from https://www.census.gov/content/dam/Census/library/publications/2015/demo/p25-1143.pdf

Cross, T., Bazron, B., Dennis, K. W., & Isaacs, M. (1989). *Toward a culturally competent system of care: A monograph on effective services for minority children who are seriously emotionally disturbed*. Retrieved 4 Apr 2021, from https://spu.edu/~/media/academics/school-of-education/Cultural%20Diversity/Towards%20a%20Culturally%20Competent%20System%20of%20Care%20Abridged.ashx

CSDH. (2008). *Closing the gap in a generation: Health equity through action on social determinants of health. Final report of the Commission on sociall Determinants of Health*. Retrieved 6 July 2021, from file:///C:/Users/mcaughy/Downloads/9789241563703_eng.pdf

ECRQ Ad Hoc Committee on Racial Equity in Publishing. (2021). Towards racial equity in publishing at Early Childhood Research Quarterly (ECRQ). *Early Childhood Research Quarterly, 55*, A1–A3. https://doi.org/10.1016/j.ecresq.2021.01.003

Egharevba, I. (2001). Researching an-'other' minority ethnic community: Reflections of a Black female researcher on the intersections of race, gender and other power positions on the research process. *International Journal of Social Research Methodology, 4*(3), 225–241. https://doi.org/10.1080/13645570010023760

Federal Interagency Forum on Child and Family Statistics. (2019). *America's children: Key national indicators of well-being, 2019*. Retrieved 31 Jan 2020, from https://www.childstats.gov/pdf/ac2019/ac_19.pdf

Fetterman, D. M., & Wandersman, A. (2005). *Empowerment evaluation principles in practice*. Guilford Press.

Franck, C., Grandi, S. M., & Eisenberg, M. J. (2013). Agricultural subsidies and the American obesity epidemic. *American Journal of Preventive Medicine, 45*(3), 327–333. https://doi.org/10.1016/j.amepre.2013.04.010

Freeman, R. B., & Huang, W. (2015). Collaborating with people like me: Ethnic coauthorship within the United States. *Journal of Labor Economics, 33*(S1), S289–S318. https://doi.org/10.1086/678973

French, B. H., Lewis, J. A., Mosley, D. V., Adames, H. Y., Chavez-Dueñas, N. Y., Chen, G. A., & Neville, H. A. (2020). Toward a psychological framework of radical healing in communities of color. *The Counseling Psychologist, 48*(1), 14–46. https://doi.org/10.1177/0011000019843506

Frey, W. H. (2021). *All recent U.S. population growthe comes from people of color, new census estimates how*. Retrieved 6 July 2021, from https://www.brookings.edu/research/all-recent-us-population-growth-comes-from-people-of-color-new-census-estimates-show/

Fryar, C. D., Carroll, M. D., & Ogden, C. L. (2018). *Prevalence of overweight and obesity among children and adolescents: United States, 1963–1965 through 2015–2016*. Retrieved 1 June 2021, from https://www.cdc.gov/nchs/data/hestat/obesity_child_15_16/obesity_child_15_16.pdf

Hall, K. L., Vogel, A. L., Huang, G. C., Serrano, K. J., Rice, E. L., Tsakraklides, S. P., & Fiore, S. M. (2018). The science of team science: A review of the empirical evidence and research gaps on collaboration in science. *American Psychologist, 73*(4), 532–548. https://doi.org/10.1037/amp0000319

Harrist, A. W., Topham, G. L., Hubbs-Tait, L., Page, M. C., Kennedy, T. S., & Shriver, L. H. (2012). What developmental science can contribute to a transdisciplinary understanding of childhood obesity: An interpersonal and intrapersonal risk model. *Child Development Perspectives, 6*(4), 445–455. https://doi.org/10.1111/cdep.12004

Hoelscher, D., Springer, A., Ranjit, N., Kelder, S., Evans, A., Stigler, M., & Perry, C. (2010). Reduction in child obesity among disadvantaged school children with community involvement: The Travis County CATCH trial. *Obesity, 18*(1), 202–205. https://doi.org/10.1038/oby.2009.430

Holliday, C. E., Wynne, M., Katz, J., Ford, C., & Barbosa-Leiker, C. (2018). A CBPR approach to finding community strengths and challenges to prevent youth suicide and substance abuse. *Journal of Transcultural Nursing, 29*(1), 64–73. https://doi.org/10.1177/1043659616679234

Iruka, I. U., Curenton, S. M., Sims, J., Escayg, K.-A., Ibekwe-Okafor, N., & RAPID-EC. (2021). *Black parent voices: Resilience in the face of the two pandemics—COVID-19 and racism*. Retrieved 24 Feb 2021, from https://fpg.unc.edu/sites/fpg.unc.edu/files/resource-files/RISER%20Black%20Parent%20Voices%202Pandemics_FINAL.pdf

Johnson, D. J., Jaeger, E., Randolph, S. M., Cauce, A. M., Ward, J., & NICHD Early Child Care Research Network. (2003). Studying the effects of early child care experiences on the development of children of color in the United States: Toward a more inclusive research agenda. *Child Development, 74*(5), 1227–1244. https://doi.org/10.1111/1467-8624.00604

Keane, M., Khupe, C., & Seehawer, M. (2017). Decolonising methodology: Who benefits from indigenous knowledge research? *Educational Research for Social Change, 6*(1), 12–24. https://doi.org/10.17159/2221-4070/2017/v6i1a2

Kimbro, R. T., & Rigby, E. (2010). Federal food policy and childhood obesity: A solution or part of the problem? *Health Affairs, 29*(3), 411–418. https://doi.org/10.1377/hlthaff.2009.0731

Koblinsky, S. A., Leslie, L. A., & Cook, E. T. (2014). Treating behavioral health conditions of OEF/OIF veterans and their families: A state needs assessment of civilian providers. *Military Behavioral Health, 2*(2), 162–172. https://doi.org/10.1080/21635781.2014.890884

Malawa, Z., Gaarde, J., & Spellen, S. (2021). Racism as a root cause approach: A new framework. *Pediatrics, 147*(1), 1–6. https://doi.org/10.1542/peds.2020-015602

Manohar, N. N. (2013). 'Yes you're Tamil! but are you Tamil enough?' An Indian researcher interrogates 'shared social location' in feminist immigration research. *International Journal of Multiple Research Approaches, 7*(2), 189–203. https://doi.org/10.5172/mra.2013.7.2.189

Marrero, D. G., Hardwick, E. J., Staten, L. K., Savaiano, D. A., Odell, J. D., Comer, K. F., & Saha, C. (2013). Promotion and tenure for community-engaged research: An examination of promotion and tenure support for community-engaged research at three universities collaborating through a clinical and translational science award. *Clinical and Translational Science, 6*(3), 204–208. https://doi.org/10.1111/cts.12061

Martins, L. L., Milliken, F. J., Wiesenfeld, B. M., & Salgado, S. R. (2003). Racioethnic diversity and group members' experiences: The role of the racioethnic diversity of the organizational context. *Group & Organization Management, 28*(1), 75–106. https://doi.org/10.1177/1059601102250020

McCormick-Huhn, K., Warner, L. R., Settles, I. H., & Shields, S. A. (2019). What if psychology took intersectionality seriously? Changing how psychologists think about participants. *Psychology of Women Quarterly, 43*(4), 445–456. https://doi.org/10.1177/0361684319866430

McIntyre, A. (2007). *Participatory action research*. Sage Publications.

McLoyd, V. C., & Randolph, S. M. (1984). The conduct and publication of research on Afro-American children: A content analysis. *Human Development, 27*(2), 65–75. https://doi.org/10.1159/000272904

Meek, S., Iruka, I. U., Allen, R., Yazzie, D. A., Fernandez, V., Catherine, E., McIntosh, K., Gordon, L., Gilliam, W., Hemmeter, M. L., Blevins, D., & Powell, T. (2020). *Fourteen priorities to dismantle systemic racism in early care and education*. Children's Equity Project. https://childandfamilysuccess.asu.edu/cep

Miller, A. L., Stern, C., & Neville, H. (2019). Forging diversity-science-informed guidelines for research on race and racism in psychological science. *Journal of Social Issues, 75*(4), 1240–1261. https://doi.org/10.1111/josi.12356

Montenovo, L., Jiang, X., Rojas, F. L., Schmutte, I. M., Simon, K. I., Weinberg, B. A., & Wing, C. (2020). *Determinants of disparities in COVID-19 job losses 0898-2937*. Retrieved 22 Jan 2021 from http://www.nber.org/papers/w27132

National Academy of Sciences, National Academy of Engineering, & Institute of Medicine. (2005). *Facilitating interdisciplinary research*. National Academies Press. https://doi.org/10.17226/11153

National Center for Education Statistics. (2019). *Integrated Postsecondary Education Data System (IPEDS), Spring 2016 through Spring 2019 Human Resources component, Fall Staff section*. Retrieved 10 June 2021, from https://nces.ed.gov/programs/digest/d19/tables/dt19_315.20.asp

Nikaj, S., Roychowdhury, D., Lund, P. K., Matthews, M., & Pearson, K. (2018). Examining trends in the diversity of the U.S. National Institutes of Health participating and funded workforce. *The FASEB Journal, 32*(12), 6410–6422. https://doi.org/10.1096/fj.201800639

Nobles, W. W. (1986). *African psychology: Toward its reclamation, reascension & revitalization*. Institute for the Advanced Study of Black Family Life and Culture.

Nobles, W. W. (2015). From Black psychology to Sakhu Djaer: Implications for the further development of a Pan African Black psychology. *Journal of Black Psychology, 41*(5), 399–414. https://doi.org/10.1177/0095798415598038

Oden, S. (2000). How researchers can support community efforts for change: Illustrations from two case studies. *Applied Developmental Science, 4*(supl), 28–37. https://doi.org/10.1207/S1532480XADS04Suppl_4

Oster, E., Jack, R., Halloran, C., Schoof, J., McLeod, D., Yang, H., Roche, J., & Roche, D. (2021). Disparities in learning mode access among k–12 students during the COVID-19 pandemic, by race/ethnicity, geography, and grade level—United States, September 2020–April 2021. *MMWR. Morbidity and Mortality Weekly Report, 70*(26), 953–958. https://doi.org/10.15585/mmwr.mm7026e2

Patton, M. Q. (2008). *Utilization-focused evaluation*. Sage Publications.

Pavao, C. A. O. (2012). Community engagement efforts at NIH: Examining best practices to bridge community and research agendas. In *NIEHS/OHRP SANC Conference*, Raleigh, NC, United States. https://www.niehs.nih.gov/news/assets/docs_a_e/community_engagement_efforts_at_nih_examining_best_practices_to_bridge_community_and_research_agendas_508.pdf

Pelco, L. E., & Howard, C. (2016). Incorporating community engagement language into promotion and tenure policies: One university's journey. *Metropolitan Universities, 27*(2), 87–98. https://doi.org/10.18060/21129

Priest, N., Paradies, Y., Trenerry, B., Truong, M., Karlsen, S., & Kelly, Y. (2013). A systematic review of studies examining the relationship between reported racism and health and well-being for children and young people. *Social Science & Medicine, 95*, 115–127. https://doi.org/10.1016/j.socscimed.2012.11.031

Roberts, D. D., Koblinsky, S. A., & Randolph, S. M. (1999). A University-Head Start partnership to promote positive outcomes for children, parents, and educators in high-violence neighborhoods. *NHSA Dialog, 2*(1), 104–105. https://doi.org/10.1207/s19309325nhsa0201_17

Rozenfeld, Y., Beam, J., Maier, H., Haggerson, W., Boudreau, K., Carlson, J., & Medows, R. (2020). A model of disparities: Risk factors associated with COVID-19 infection. *International Journal for Equity in Health, 19*(1), 126. https://doi.org/10.1186/s12939-020-01242-z

Syed, M., Santos, C., Yoo, H. C., & Juang, L. P. (2018). Invisibility of racial/ethnic minorities in developmental science: Implications for research and institutional practices. *American Psychologist, 73*(6), 812–826. https://doi.org/10.1037/amp0000294

Townsend, M. J., Kyle, T. K., & Stanford, F. C. (2020). Outcomes of COVID-19: Disparities in obesity and by ethnicity/race. *International Journal of Obesity, 44*(9), 1807–1809. https://doi.org/10.1038/s41366-020-0635-2

Wallerstein, N., Duran, B., Oetzel, J., & Winkler, M. (Eds.). (2017). *Community-based participatory research for health* (3rd ed.). Jossey-Bass.

Zavala, M. (2013). What do we mean by decolonizing research strategies? Lessons from decolonizing, Indigenous research projects in New Zealand and Latin America. *Decolonization: Indigeneity, Education & Society, 2*(1), 55–71. http://digitalcommons.chapman.edu/education_articles/106

Zeldin, S. (2000). Integrating research and practice to understand and strengthen communities for adolescent development: An introduction to the special issue and current issues. *Applied Developmental Science, 4*, 2–10. https://doi.org/10.1207/S1532480XADS04Suppl_1

Chapter 15
Developmental Science in the Twenty-First Century: Moving Forward to Integrate Cultural and Racial Processes into Research

Dawn P. Witherspoon and Gabriela Livas Stein

Developmental science can no longer be "segregated science"; it must be inclusive and incorporate cultural and racial processes at its center. Though the integrative model (García Coll et al., 1996), the phenomenological variant of ecological systems theory (PVEST, Spencer 2006), and other cultural-developmental theories that focus on specific cultural processes (e.g., ethnic-racial identity—Umaña-Taylor et al., 2014; immigration—Suárez-Orozco et al., 2018) have called for more centralized examination of racial and ethnic processes and their derivatives in developmental research, there is more to be desired. The purpose of this book was to provide a comprehensive, though not exhaustive, road map and blueprint for scholars interested in actively engaging in a diverse developmental science. With each of the preceeding chapters, the authors have focused on specific developmental competencies and processes (e.g., ethnic-racial identity, ethnic-racial socialization, acculturation), specific understudied populations (e.g., multiracial, transnational immigrants), oppressive forces (i.e., discrimination), contexts (e.g., schools and neighborhoods, online environments), methods (e.g., qualitative designs and youth voice centered approaches), and translation efforts (e.g., practice and policy, collaboration) to elucidate how a meaningful diverse developmental science can be done. As each of the chapters provide foundational information regarding the concepts, the authors also provide practical recommendations so that scholars can *do the work* to be an inclusive science to improve the lives of racial-ethnic diverse children, youth, and families. In this final chapter, we will highlight some overarching themes as a practical guide to make doing the work a bit easier. Sharing these recommendations, we do

D. P. Witherspoon (✉)
The Pennsylvania State University, State College, PA, USA
e-mail: dpw14@psu.edu

G. L. Stein
University of North Carolina at Greensboro, Greensboro, NC, USA

© The Author(s), under exclusive license to Springer Nature Switzerland AG 2023
D. P. Witherspoon, G. L. Stein (eds.), *Diversity and Developmental Science*,
https://doi.org/10.1007/978-3-031-23163-6_15

not intend to be proscriptive or exhaustive, but rather we want to provide concluding thoughts on the few things we can attempt to do to get it more right than wrong.

Acknowledge Power, Privilege, and Whiteness

As clearly evident in the chapters in this volume, researchers need to start by acknowledging their own positionality both in terms of the populations that are included in their samples, but also in terms of their research teams. In that examination of positionality, it is important that researchers grapple with their own racial-ethnic identity and how that shapes their worldview and life experiences—both inside and outside of academia. Societies are inherently stratified by social class, ethnicity and race, religion, immigrant status, and many other characteristics. In the United States, this stratification is evident in the fact that 18.8% of the Black population and 15.7% of the Latinx population live below the poverty line relative to 7.3% of White and Asian populations (Creamer, 2020). As so elegantly described in Garcia Coll et al. (1996), this stratification resonates across developmental contexts in terms of schools, neighborhoods, health care settings, and workplaces impacting the developmental outcomes of youth. Yet, researchers need to further consider not just how this influences the developmental trajectorics of their samples– but also how stratification has made an impact in their own lives as scientists and academics.

This stratification is apparent in academia. Data from the National Center for Educational Statistics (2022) documents that only 4% of full-time faculty were Black females with 3% each for Black males, Latinx males, and Latinx females (data not reported for non-binary faculty of color). This stratification is even more stark at the Full Professor rank where representation falls to 2% respectively for each group and, in 2017, 88% of University Provosts/Academic Leaders were White (Bichsel et al., 2018). Meanwhile, in 2017, 36% of the college student population was Black or Latinx (US Census, 2018). This means that White faculty tend to hold positions of power in academic institutions, may not have many faculty of color as colleagues, and likely are in positions of power relative to non-White others (e.g., students, faculty, staff). Due to economic and racial/ethnic segregation, many White faculty likely also do not live in diverse neighborhoods. There also may be further segregation in research teams that approach research using a colorblind perspective seeking "universal truths" without acknowledging race, ethnicity, and immigration status as central to their investigations. Together, this suggests that developmental scientists need to be purposeful in considering their approaches both to the science and their research teams—aware of how power and privilege may be impacting the production and dissemination of knowledge.

To do so, chapters in this book highlight the need for researchers to reflect on their own biases. Bañales et al. (Chap. 9) provide questions that research teams working with Latinx populations should consider prior to engaging in their work. These questions can be modified to consider multiple marginalized populations. In the same vein, Lorenzo Blanco et al. (Chap. 5) consider positionality and representational ethics when it comes to work with immigrant populations. Together, these chapters invite us to be introspective at all points in the scientific endeavor when it comes to how we approach science, form hypotheses, engage our samples, and disseminate knowledge. We also need to consider our epistemological positions in terms of how we create knowledge and the reification of certain methods that may be limiting our understanding of others. Coppens and Coppinger (Chap. 7) argue for the need to consider ecological validity of our methods for all populations. Similarly, Caughy et al. (Chap. 14) highlight the intentionality that is necessary in building diverse, collaborative research teams that are well-positioned to pursue a developmental science that is cognizant of how stratification, race and ethnicity, and segregation influence the developmental outcomes of all youth.

For too long the science of families of color has focused on deficits—whether clearly intentional as in the 1960s or through a focus on risk mechanisms and disparities in the 1990s. These chapters aimed to also highlight the need to focus on assets—not just at the individual level as in the science of resilience but at the community level as well. Tynes et al. (Chap. 11) invite us to envision a developmental science built upon assets focused on liberation, freedom, and possibility. We need to follow their lead by listening to communities who have dreamed and thrived to inform our theory, methods, and questions.

Describe Samples

A relatively low cost and simple strategy to move toward a diverse developmental science is to appropriately describe research samples. Professional societies in the social sciences, and developmental science in particular, have provided guidance to and implemented policies for authors to follow when describing samples, making it a point that authors must provide sociocultural information that is theoretically relevant for the developmental phenomenon under consideration. For example, in 2020, the Society for Research in Child Development (SRCD) implemented a journal-wide sociocultural policy that requires disclosure of data collection dates, selection and recruitment procedures, and demographic characteristics of samples, including socioeconomic information as well as the places from which the sample comes. Requiring demographic information of many types about study participants moves toward an intersectional understanding of the social identity and location of the sample and its participants. Providing this type of information increases consumers' ability to critically evaluate the external validity and generalizability of the findings to minoritized and marginalized populations as well as the global world. Further, The American Psychological Association (APA, 2021) developed an equity, diversity, and inclusion (EDI) framework to guide their systematic efforts that lead to "transformative and sustainable change" by focusing on the organization, the field of psychology, and broader society. Within the APA (2022), they focus on inclusive policies and practices as well as access and equity that include a toolkit for journal editors (https://www.apa.org/pubs/authors/equity-diversity-inclusin-toolkit-journal-editors.pdf) and the dissemination of science. Exemplars and tools such as these are immensely helpful for scholars engaged in a diverse developmental science.

Additionally, the description of samples is critically necessary for populations that may be hidden (Witherspoon, Bámaca-Colbert, Stein, & Rivas-Drake, 2020)—minoritized and marginalized populations that are underrepresented in developmental science. One of those groups, multiracial populations, only make up about 4% of the United States' population but is projected to be one of the fastest growing populations in the next decades (Vespa et al., 2020). Yoo and colleagues (Chap. 6) provide a critical multiracial theory, the model of multiracial racialization, for scholars to utilize to better understand the experiences of multiracial individuals. Providing this theory, the authors provide a blueprint or suggestions about racial and cultural processes that should be explored (i.e., racial identity, racial identification, and racial category) when describing and characterizing developmental phenomena across multiple ecological levels—individual to time. Critical to this discussion is the importance of using multiple and diverse methods to meaningfully capture the unique developmental trajectories of multiracial people. This heuristic underscores the need for culturally anchored methodologies in a diverse developmental science.

Engage in a Culturally Anchored Methodology

A central thesis in this collection of work is that race, ethnicity, and culture must be carefully considered in developmental research. Several of the chapters explicitly provide recommendations on how to include cultural processes into the investigation of developmental phenomena and children and youth's developmental competencies. Some of the specific cultural processes and behaviors explored with detail are ethnic-racial identity (Chap. 3), ethnic-racial socialization (Chap. 4), and accult uration/enculturation (Chap. 5) as they are generally theorized to be developmental assets, promotive of youth outcomes (Neblett Jr et al., 2012). Conversely, given stratification and white supremacy, discrimination (Chap. 2) is a risk factor for development; critical action (Chap. 10) by youth of color is an attempt to disrupt systems of oppression such as discrimination. To capture these meaningful developmental processes, assets, and vulnerabilities, the methods developmental scientists use to understand diverse children and youth's development must be ecologically valid and move away from deficit-focused frameworks (Chap. 7) and incorporate youth voice (Chap. 9).

Hughes and colleagues (Hughes et al., 1993; Hughes & Seidman, 2002) assert that culture is embedded in and intersects with all phases of the research process. It impacts how scholars define the problem, create research questions, choose research methods and designs, and interpret data. Further, Yasui and Dishion (2007) describe how the importance of culture must be infused and considered in intervention practices. As carefully considered in this book, it is imperative that developmental scientists consider culture and its associated processes when doing their research. Hughes et al. (1993) assert that all individuals develop within a cultural context (Yasui & Dishion, 2007); that cultural values, norms, behaviors, and processes are intergenerational transmitted through socialization practices that are adaptive to the environment (e.g., García Coll et al., 1996); that some cultural aspects are abstract; and that cultural patterns exist that are shaped by the larger sociohistorical context. Given the pervasive impact of culture, it is critical that diversity-minded developmental scientists explore cultural variability (Yasui & Dishion, 2007) and consider various elements of culture, especially those cultural processes and practices tied to race and ethnicity, throughout the research process—even at the stage of problem definition and formulation, what we observe and how we measure it, as well as how we interpret what we observe from data. These principles extend to the creation of adaptive interventions that are culturally sensitive and responsive (Yasui & Dishion, 2007; Smith et al., this volume). An important guideline that resonates profoundly in the pursuit toward a diverse developmental science that integrates cultural and racial processes is to be proactive and include multiple stakeholders at the table when formulating the research question, defining the population(s), identifying constructs, and determining research designs. With diverse voices contributing to the creation of the research project, the exploration of alternative pathways and the centering of the experiences of the underrepresented may become more normative rather than an exceptional practice. To achieve this goal, multiple ecologically valid, strengths-based (e.g., Black "thrival"—see Tynes et al., Chap. 11), community-engaged research epistemologies and approaches must be utilized.

As we suggest a diverse developmental science, a great example of moving towards the integration of cultural and racial processes is through considering developmental and cultural processes from a diaspora lens (Chap. 12; Gulamhussein et al.). This perspective attends to power, stratification, and systemic forms of oppression—racism, classism, etc.—and provides interesting insights into how to investigate the constructs/cultural processes of acculturation, enculturation, ethnic-racial identity, and ethnic-racial socialization as well as youth outcomes such as school adjustment. Further, Witherspoon et al. (Chap. 8) push scholars to reconsider how school and neighborhood ethnic-racial compositions are measured, allowing for greater specificity in setting-level variability in racial, ethnic, or cultural composition and how it relates to youth behavioral, academic, and psychosocial outcomes. Importantly, the scholars assert that it is important to consider the demographic composition of populations present within the setting (e.g., race, ethnicity, immigrant status) as well as the group norms and signs and symbols present in the setting. For example, for an adolescent who is becoming more cognitively mature and has increasing autonomy and independence, the ethnic-racial composition of the school (e.g., level of diversity or concentration) may be (dis)aligned with the prevailing group norms (e.g., stereotypes, prejudices) reinforced by those in power or the symbols related to race or ethnicity in the school (e.g., posters, flyers about school events, flags hung). The experience of the ethnic-racial compositions of the school may have differential impacts on school adjustment (e.g., belonging, achievement, engagement). Together these examples of providing more nuanced consideration of race, ethnicity, and culture offer a potential blueprint for other scholars to engage similarly. With a culturally anchored methodology, we recognize that all cultural processes may not be measured within in study; however, we urge scholars to consider—based on theory or empirical evidence—what specific cultural processes and stratification variables might be at play for the developmental phenomenon under investigation. State those considered; measure some of them, and provide the rationale for inclusion or exclusion. Such an approach can incrementally advance our understanding of the developmental experiences of diverse children and youth.

Expand Research Teams to Include Expertise of Diverse Populations of Interest

A continuing narrative, represented across multiple chapters in this book, focuses on collaboration among interdisciplinary teams that *include* and are led by scholars of scholar to realize a diverse developmental science. Caughy and colleagues (Chap. 15) provide a perfectly delineated call to action for scholars and raised the importance of community-engaged approaches, recognizing systemic racism as a central contributor to disparities and elevating ethical considerations when doing diverse developmental science. Engaging in steps provided within and across chapters in this book, practice and policy innovations can be actualized and representative collaborative research teams can be formed.

Central to this recommendation is the notion of equity and access. Interestingly and importantly, Caughy and colleagues (Chap. 15) defined interdisciplinary teams as those with collaborations among developmental scientists with varying expertise, collaborations across different disciplines, and collaborations with individuals outside of academia. Forming interdisciplinary teams in this way can provide access to community stakeholders who might not otherwise be involved in the research process. Expanding research teams to include diverse community stakeholders requires an understanding of the continuum of community-engaged research approaches and a pursuit of cultural competency in these interactions. It is also imperative that scholars engage in authentic relationships where all parties are transparent about possibilities and limits as an avenue to build and maintain trust. In addition, from an equity and access lens, Caughy and colleagues assert that it is necessary for resource procurement and allocation to be carefully examined identifying problematic patterns and strengths-based solutions and that the research process must be approached from an anti-racist perspective. Implementing these recommendations have the potential to redistribute power and begin to level the research enterprise.

Furthermore, equitable interdisciplinary research teams should include scholars of color who are members of the populations of interest while also considering, from an intersectionality perspective, the other identities of those individuals. Creating an inclusive interdisciplinary research team recognizes the limitations of privileged scholars and elevates the voices of minoritized scholars while simultaneously allowing minoritized scholars to identify their limitations. In addition to advancing science in meaningful ways, the inclusion of underrepresented minorities as members of research teams may chip away at the enduring disparities in external federal funding from the National Institutes of Health (NIH; Ginther et al., 2011; Hoppe et al., 2019) and the National Science Foundation (NSF; Chen et al., 2022). These problems are structural and require a multipronged approach to elevate, sustain, and maintain diversity-based developmental science.

Conclusion

Developmental science must be the science of all populations—a science that fully integrates culture, race, ethnicity, and marginalization into all of our work. We appreciate that this call can present obstacles for research teams as many of us were not trained with these methodologies, approaches, or questions. We also know that change is difficult, but we are accustomed to change and expanding our skill sets. As a field, we must value growing in these methods and approaches as much as we value a new statistical analysis or neuroscientific method. We cannot grow as science until we fully commit to accurately and holistically representing the lived experiences of all populations. We need to build our science in partnership with communities with principles of equity and inclusion as our guiding force. Only then will we have a developmental science that is truly universal.

References

American Psychological Association (APA). (2021). *Equity, diversity and inclusion framework.* Retrieved from https://www.apa.org/about/apa/equity-diversity-inclusion/framework.pdf

American Psychological Association (APA). (2022, October 7). *APA Publishing's equity, diversity, and inclusion toolkit for journal editors.* https://doi.org/10.31219/osf.io/wst4q

Bichsel, J., Pritchard, A., Li, J., & McChesney, J. (2018). *Administrators in higher education annual report: Key findings, trends, and comprehensive tables for 2017–18 academic year. Research report.* Knoxville, TN: CUPA-HR.

Chen, C. Y., Kahanamoku, S. S., Tripati, A., Alegado, R. A., Morris, V. R., Andrade, K., & Hosbey, J. (2022). *Decades of systemic racial disparities in funding rates at the National Science Foundation.* https://doi.org/10.31219/osf.io/xb57u

Creamer, J. (2020). *Inequalities persist despite decline in poverty for all major race and Hispanic origin groups.* Retrieved from https://www.census.gov/library/stories/2020/09/poverty-rates-for-blacks-and-hispanics-reached-historic-lows-in-2019.html

García Coll, C., Crnic, K., Lamberty, G., Wasik, B.H., Jenkins, R., García, H.V. and McAdoo, H.P. (1996), An Integrative Model for the Study of Developmental Competencies in Minority Children. *Child Development,67*: 1891–1914. https://doi.org/10.1111/j.1467-8624.1996.tb01834.x

Ginther, D. K., Schaffer, W. T., Schnell, J., Masimore, B., Liu, F., Haak, L. L., & Kington, R. (2011). Race, ethnicity, and NIH research awards. *Science, 333*(6045), 1015–1019. https://doi.org/10.1126/science.1196783

Hoppe, T. A., Litovitz, A., Willis, K. A., Meseroll, R. A., Perkins, M. J., Hutchins, B. I., et al. (2019). Topic choice contributes to the lower rate of NIH awards to African-American/Black scientists. Science. *Advances, 5*(10), eaaw7238. https://doi.org/10.1126/sciadv.aaw7238

Hughes, D. L., & Seidman, E. (2002). In pursuit of a culturally anchored methodology. In T. A. Revenson et al. (Eds.), *Ecological research to promote social change* (pp. 257–290). Kluwer Academic/Plenum Publishers. https://doi.org/10.1007/978-1-4615-0565-5_10

Hughes, D., Seidman, E., & Williams, N. (1993). Cultural phenomena and the research enterprise: Toward a culturally anchored methodology. *American Journal of Community Psychology, 21*(6), 687–703.

National Center for Education Statistics. (2022). *Characteristics of postsecondary faculty. Condition of education.* U.S. Department of Education, Institute of Education Sciences. Retrieved May 31, 2022, from https://nces.ed.gov/programs/coe/indicator/csc

Neblett, E. W., Jr., Rivas-Drake, D., & Umaña-Taylor, A. J. (2012). The promise of racial and ethnic protective factors in promoting ethnic minority youth development. *Child Development Perspectives, 6*(3), 295–303. https://doi.org/10.1111/j.1750-8606.2012.00239.x

Society for Research in Child Development. (2020). *New sociocultural policy enacted across all SRCD journals.* Retrieved from https://www.srcd.org/news/new-sociocultural-policy-enacted-across-all-srcd-journals

Spencer, M.B. (2006). Phenomenology and ecological systems theory: Development of diverse groups. In W. Damon, & R. Lerner (Eds.), Handbook of child psychology, vol. 1 (R. Lerner, Ed.): Theoretical models of human development 6, 829–893. New York: Wiley

Suárez-Orozco, C., Motti-Stefanidi, F., Marks, A., & Katsiaficas, D. (2018). An integrative risk and resilience model for understanding the adaptation of immigrant-origin children and youth, *American Psychologist, 73*(6), 781–796. https://doi.org/10.1037/amp0000265.

Umaña-Taylor, A. J., Quintana, S.M., Lee, R.M., Cross, W.E. Jr, Rivas-Drake, D., Schwartz, S.J., Syed, M., Yip, T., Seaton, E., & Ethnic and Racial Identity in the 21st Century Study Group (2014). Ethnic and racial identity during adolescence and into young adulthood: An integrated conceptualization. *Child Development, 85*(1), 21–39. https://doi.org/10.1111/cdev.12196.

United States Census Bureau. (2018). *Current population survey, school enrollment supplement, October 2007 and 2017.* Retrieved from www.census.gov/programs-surveys/cps.html

Vespa, J., Medina, L., & Armstrong, D. M. (2020). *Demographic turning points for the United States: Populations projects for 2020 to 2060.* Retrieved from https://www.census.gov/content/dam/Census/library/publications/2020/demo/p25-1144.pdf

Witherspoon, D. P., Bámaca-Colbert, M. Y., Stein, G. L., & Rivas-Drake, D. (2020). Hidden populations: Uncovering the developmental experiences of minoritized populations across contexts. *Developmental Psychology, 56*(8), 1425–1430. https://doi.org/10.1037/dev0001055

Yasui, M., & Dishion, T. J. (2007). The ethnic context of child and adolescent problem behavior: Implications for child and family interventions. *Clinical Child and Family Psychology Review, 10*, 137–179. https://doi.org/10.1007/s10567-007-0021-9

Index

© The Author(s), under exclusive license to Springer Nature Switzerland AG 2023
D. P. Witherspoon, G. L. Stein (eds.), *Diversity and Developmental Science*, https://doi.org/10.1007/978-3-031-23163-6